# CLINICAL IMPLICATIONS OF AIR POLLUTION RESEARCH

**AMERICAN MEDICAL ASSOCIATION**
**Air Pollution Medical Research Conference**
**December 5-6, 1974**

**Asher J. Finkel and Ward C. Duel**
**Editors**

Publishing Sciences Group, Inc.
Acton, Massachusetts
a subsidiary of CHC Corporation

Distributed in

Canada
McAinsh & Company, Ltd.
1835 Yonge Street
Toronto, Ontario M4S 1L6
Canada

Australia and New Zealand
Butterworths Pty. Ltd.
586 Pacific Highway
Chatswood, NSW 2067
Australia

Asia
Toppan Company Ltd.
5, 1-Chome, Taito-Ku
Tokyo 110
Japan

Toppan Company Ltd.
No. 38 Liu Fang Road
Jurong Town
Singapore-22

India and Sri Lanka
Affiliated East-West Press Pvt. Ltd.
9 Nizamuddin East
New Delhi 110013
India

United Kingdom and Europe
South Africa
Medical and Technical
Publishing Co., Ltd.
P.O. Box 55
St. Leonardgate
Lancaster LA1 1PE
England



Printed in the United States of America.

International Standard Book Number: 0-88416-017-3

Library of Congress Catalog Card Number: 75-12024

# CONTRIBUTORS

**Wilbert S. Aronow, M.D.**
Chief, Cardiovascular Sections, Veterans Administration Hospital,
Long Beach, California
Professor of Medicine and Chief, Cardiovascular Research,
University of California,
Irvine, California

**Joseph F. Boyle, M.D.**
Member, AMA Council on Environmental, Occupational, and Public Health
Los Angeles, California

**Congressman George E. Brown, Jr.**
38th District of California
U.S. House of Representatives
Washington, D.C.

**Benjamin Burrows, M.D.**
Professor of Internal Medicine
Director, Division of Respiratory Sciences
College of Medicine
University of Arizona,
Tucson, Arizona

**Robert S. Chapman, M.D.**
Medical Officer, Epidemiology Branch
Human Studies Laboratory
National Environmental Research Center, USEPA
Research Triangle Park, North Carolina

**T. Timothy Crocker, M.D., Chairman**
Department of Community and Environmental Medicine
California College of Medicine
University of California
Irvine, California

**Domenic A. DeBias, Ph.D.**
Professor and Chairman
Department of Physiology and Pharmacology
Philadelphia College of Osteopathic Medicine
Philadelphia, Pennsylvania

**E. R. Dirksen, Ph.D.**
Associate Research Physiologist
Cancer Research Institute
University of California Medical Center
San Francisco, California

**Shigeru Funabashi, M.D.**
School of Medicine
Chiba University
Chiba, Japan

**John R. Goldsmith, M.D.**
Head, Environmental Epidemiology Unit
California Department of Public Health
Berkeley, California

**Gareth M. Green, M.D.**
Professor of Medicine
College of Medicine
University of Vermont
Burlington, Vermont

**Herschel E. Griffin, M.D.**
Dean and Professor of Epidemiology
Graduate School of Public Health
University of Pittsburgh
Pittsburgh, Pennsylvania

**E. Groll-Knapp, Ph.D.**
Institute of Environmental Hygiene
University of Vienna
Vienna, Austria

**Edward Haak, M.D.**
Human Studies Laboratory
U.S. Environmental Protection Agency
Research Triangle Park, North Carolina

**Jack D. Hackney, M.D.**
Chief, Environmental Health Laboratory

Rancho Los Amigos Hospital
Downey, California

**Manfred Haider, M.D.**
Professor of Environmental Hygiene
Lehrkanzel für Umwelthygiene
Der Universität Wien
Vienna, Austria

**Douglas I. Hammer, M.D.**
Epidemiologist, Human Studies Laboratory
Division of Health Effects Research
National Environmental Research Center, USEPA
Research Triangle Park, North Carolina

**Victor Hasselblad, Ph.D.**
Human Studies Laboratory
National Environmental Research Center
U.S. Environmental Protection Agency
Research Triangle Park, North Carolina

**Carl G. Hayes, Ph.D.**
National Environmental Research Center, USEPA
Research Triangle Park, North Carolina

**Robert I. Henkin, M.D.**
Chief, Section on Neuro-endocrinology
National Heart and Lung Institute,
National Institutes of Health
Bethesda, Maryland

**Takeshi Hirayama, M.D.**
Chief, Epidemiology Division
National Cancer Center Research Institute,
Tokyo, Japan

**H. Höller, Ph.D.**
Institute of Environmental Hygiene,
University of Vienna
Vienna, Austria

**Barry L. Johnson, Ph.D.**
Chief, Behavioral Studies Laboratory
National Institute for Occupational Safety and Health
Cincinnati, Ohio

**Jun Kagawa, M.D.**
Department of Environmental Medicine and Occupational Health
School of Medicine
Tokai University
Isehara-shi, Kanagawa-ken, Japan

**John H. Knelson, M.D.**
Chief, Biomedical Research Branch
Division of Health Effects Research, USEPA
Research Triangle Park, North Carolina

**Joel D. Koppel, M.D.**
Chief, Chronic Dialysis Unit VA
Wadsworth Hospital Center
Assistant Professor of Medicine and Public Health,
School of Medicine, UCLA,
Los Angeles, California

**T. R. Lewis, Ph.D.**
National Institute for Occupational Safety and Health,
Cincinnati, Ohio

**George W. Meyer, M.D.**
David Grant USAF Medical Center
Travis Air Force Base, California

**F. J. Miller, M.S.**
Environmental Protection Agency
Research Triangle Park, North Carolina

**Ken Motomiya, M.D.**
School of Medicine
Chiba University
Chiba, Japan

**M. Neuberger, M.D.**
Institute of Environmental Hygiene,
University of Vienna
Vienna, Austria

**Masahiro Nakaza**
Department of Environmental Medicine and Occupational Health

School of Medicine
Tokai University
Isehara-shi, Kanagawa-ken, Japan

**J. G. Orthoefer, D.V.M.**
National Institute for Occupational Safety and Health
Cincinnati, Ohio

**Michael Rabinowitz, Ph.D.**
Wadsworth Hospital Center,
Veterans Administration
Los Angeles, California

**Hiroyasu Saito, M.D.**
School of Medicine
Chiba University
Chiba, Japan

**Irving J. Selikoff, M.D.**
Professor, Department of Community Medicine
Mount Sinai School of Medicine
New York City, New York

**Philippe Shubik, M.D.**
Director, Eppley Institute for Research and Cancer
University of Nebraska Medical Center
Omaha, Nebraska

**Carl M. Shy, M.D., Dr. P.H.**
Director, Institute for Environmental Studies
University of North Carolina
Chapel Hill, North Carolina

**Robert E. Smith, M.D.**
McGovern Allergy Clinic and Texas Allergy Research Foundation
Houston, Texas

**Frank E. Speizer, M.D.**
Chief, Division of Clinical Epidemiology
Harvard Medical School
Boston, Massachusetts

**A. G. Stead, M.S.**
Environmental Protection Agency,
Research Triangle Park, North Carolina

**H. Stidl, Sc.D.**
Institute of Environmental Hygiene,
University of Vienna
Vienna, Austria

**Ira B. Tager, M.D.**
Channing Laboratory
Harvard Medical School
Boston, Massachusetts

**Malcolm C. Todd, M.D.**
President, American Medical Association
Long Beach, California

**Toshio Toyama, M.D.**
Department of Environmental Medicine and Occupational Health
School of Medicine
Tokai University
Isehara-shi, Kanagawa-ken, Japan

**Paul F. Wehrle, M.D.**
Hastings Professor of Pediatrics,
University of Southern California and
Director of Pediatrics, Pediatric Pavilion,
Los Angeles County — University of Southern California Medical Center
Los Angeles, California

**George Wetherill, Ph.D.**
Veterans Administration
Wadsworth Hospital Center
Los Angeles, California

**John V. R. Williams, B.S.**
National Environmental Research Center
U.S. Environmental Protection Agency
Research Triangle Park, North Carolina

**C. Xintaras, Sc.D.**
National Institute for Occupational Safety and Health
Cincinnati, Ohio

**Ryo Yoshida, M.D.**
Professor of Public Health
School of Medicine
Chiba University
Chiba, Japan

# AMERICAN MEDICAL ASSOCIATION

Council on Environmental, Occupational, and Public Health

Henri F. Wendel, M.D.
Chairman
New Castle, Delaware

Otto K. Engelke, M.D.
Vice Chairman
Ann Arbor, Michigan

Mario C. Battigelli, M.D.
Chapel Hill, North Carolina

Joseph F. Boyle, M.D.
Los Angeles, California

Melvin H. Evans, M.D.
St. Thomas, Virgin Islands

Sam A. Nixon, M.D.
Floresville, Texas

Edward Press, M.D.
Portland, Oregon

Frank W. Barton, LL.B.
Secretary
Chicago, Illinois

Division of Scientific Activities

Asher J. Finkel, M.D., Ph.D., Director
John C. Ballin, Ph.D., Assistant Director

Department of Environmental, Public, and Occupational Health

Asher J. Finkel, M.D., Ph.D.
Director

Frank W. Barton, LL.B.
Associate Director

Henry F. Howe, M.D.
Associate Director

Jack A. Bell, M.P.H.
Assistant Director

Ward C. Duel, M.P.H.
Assistant Director

Robert H. Wheater, M.S.
Assistant Director

Lucille E. Morrisey
Administrative Assistant

Christine C. Kelly, B.A.
Research Assistant

Proceedings of the 1974
AMA Air Pollution Medical Research Conference
San Francisco, California, December 5–6, 1974.

Sponsored by
Council on Environmental, Occupational, and Public Health and Department of Environmental, Public, and Occupational Health, Division of Scientific Activities, American Medical Association with the support of the U.S. Environmental Protection Agency.

# CONTENTS

**PART TWO**
**Cardiovascular Diseases**



**PART THREE**
**Immunology, Hypersensitivity, and Host Defense**

# PREFACE

The health of a significant portion of the population of the United States, the jobs of thousands of workers, and a great deal of money are all involved when clean air objectives are chosen and the options for meeting these standards are selected. With the improvement of the national air monitoring network, we are beginning to define the kinds and amounts of air pollutants that are present in ambient air. The contributors to the 1974 AMA Air Pollution Medical Research Conference, the proceedings of which constitute this volume, addressed the medical aspects of the air pollution problem. The 1974 Conference was unique in focusing on sets of disease systems in its search for the clinical significance of recent air pollution medical research. Special attention was given to the illnesses of children in a separate session. Some of the individual technical papers provide the results of recent research efforts in the field of air pollution medical investigation, while the introductory and summary papers for each session, respectively, provide a review and an evaluation of the clinical significance of the contributions. Selected portions of the discussion in each session have also been included here.

The Conference that produced the material for this book represents a portion of the effort of the American Medical Association to meet a public health and preventive medicine responsibility with respect to unnecessary illness from air pollution among the American people. These efforts are part of an ongoing program that is already looking forward to the organization of the next biennial Conference in 1976.

The American Medical Association is grateful to the authors for their participation in the Conference, for their very substantial contributions to our collective knowledge, and for their generous cooperation in matters relating to the Conference as well as to the preparation of this volume. The success of this meeting, as with the previous conferences in this series, was a tribute to the efforts of Mr. Frank W. Barton, Associate Director of the AMA Department of Environmental, Public, and Occupational Health, as well as the other members of the AMA staff. Special acknowledgement is made to Mrs. Lucille Morrisey for her careful editorial efforts and to Mss. Mary Boegel, Barbara Jansson, Nancy O'Connor, Mary Stelmach, and Leatha Tiggelaar for their diligent efforts in the preparation of the typescript.

**A.J.F.** and **W.C.D.**

# FOREWORD

For many years the American Medical Association has demonstrated continuing concern about the known and potential adverse health effects from air pollution. As early as 1923 an article in *The Journal of the American Medical Association* warned that "the contamination of the air in the more congested streets of American cities during hours of heavy traffic reaches the upper limit and for a shorter period even exceeds the upper limit of a well-founded health standard." With the adoption in June 1973 of a resolution by its House of Delegates, the Association endorsed the present levels and time schedules as promulgated by the Clean Air Act Amendments of 1970 (P.L. 91-604), and encouraged Congress to preserve those levels and time schedules as necessary public health measures. In March 1975 these views were made known through AMA testimony before the Subcommittee on Public Health and Environment, Committee on Interstate and Foreign Commerce, U.S. House of Representatives, in hearings that were conducted on the health effects of air pollution.

Beginning in 1966 Air Pollution Medical Research Conferences have been sponsored biennally by the American Medical Association. These Conferences have contributed to a better understanding of the impact of air pollution on human health. Physicians and others who have attended the Conferences have acquired increased knowledge from the research findings that were reported. Publication of the Conference papers in scientific journals has enhanced the ready availability of the material within the scientific community and has sustained interest in, and the need for, continuing research efforts.

The Conferences in 1972 and 1974 were partially funded by a grant from the Human Studies Laboratory, National Environmental Research Center, U.S. Environmental Protection Agency. The three earlier Conferences (1966, 1968, 1970) were funded, in part, by the U.S. Public Health Service. The contrast between the 1966 and 1974 Conferences reflects the greater store of information on the adverse health effects from air pollution that is now known and reported in the scientific literature.

The goal of the 1966 Conference was fairly simple: "We shall decide the important things that we must accomplish in the future, and we shall establish communications between disciplines so that the work to be accomplished can be interrelated, recognizing that no one is capable of all of the skills necessary for the solution of this complex problem."

The program outline of the 1974 Air Pollution Medical Research Conference characterizes the extent of scientific advances that have followed since 1966. The papers and discussions gathered together in this volume focus on the clinical implications of air pollution research for human disease.

Still, the relationship between human illness and air pollution exposures is neither simple nor fully understood. Death and disease represent only the extreme end of a spectrum of responses. When the effects of pollutant exposure are long delayed or dependent upon events distantly separated in time, the causal relationship between pollutant exposures and adverse human health effects is exceedingly difficult to document. Likewise, it is difficult to assess the effects of long-term, low-level pollutant exposure. Carefully designed, systematic, coordinated laboratory, clinical, and epidemiological investigations are required to quantitate the health effects of air pollution exposures.

Science is a dynamic process and research raises as many questions as it answers. Moreover, the acquisition of new information and understanding is a long process. If we are to meet the environmental challenges, including the adverse consequences of air pollution to public health and welfare, we must continue to sustain health effects research. Such research offers the best opportunity to evaluate the significance of the degree or character of air pollutant hazards.

Special recognition and appreciation are expressed to the members of the planning committee for their excellent contributions to the program of the 1974 Conference. These members are Joseph F. Boyle, M.D., a private practitioner in Los Angeles and member of the AMA Council on Environmental, Occupational, and Public Health; John R. Goldsmith, M.D., Medical Epidemiologist of the California Department of Health; Carl M. Shy, M.D., of the Human Studies Laboratory, National Environmental Research Center, Environmental Protection Agency; Frank C. Speizer, M.D., Assistant Professor of Medicine, Harvard Medical School; and Ward C. Duel, M.P.H., Assistant Director, Department of Environmental, Public, and Occupational Health, American Medical Association. Mr. Duel also served as the Conference Coordinator; his capable and effective contributions in coordinating the administrative aspects of the Conference are gratefully acknowledged.

**Frank W. Barton, Project Manager**
Air Pollution Medical Research Conference
American Medical Association
(Research Project Grant R803448-01-0)*

*Grant from Human Studies Laboratory, National Environmental Research Center, Environmental Protection Agency, Research Triangle Park, North Carolina, in support of 1974 Conference.

# KEYNOTE ADDRESS: Clinical Applications of Air Pollution Research: The Data Base

Herschel E. Griffin, M.D.

"The primary basis for urban air pollution control in the United States has been its adverse effects on human health."[1] This was the first sentence of the keynote address by Dr. Vaun Newill to the American Medical Association Air Pollution Medical Research Conference four years ago. It needs little modification today. I would make only one:

"The primary basis for urban air pollution control in the United States has been — *and still is* — its adverse effects on human health."

In the language of the Clean Air Act, as amended, national primary air quality standards "shall be ambient air quality standards the attainment and maintenance of which . . . are requisite to protect the public health."[2]

Implicit in Dr. Newill's statement and in the law is the assertion that there is something about the pollution of urban air that adversely affects the health of human beings. (The same could be said about the pollution of non-urban air.) This assertion — which is by now essentially non-controversial — makes it not only appropriate, but in fact makes it essential, that clinical practitioners of medicine be deeply concerned and involved in both the study and management of the problem. Hence the theme of this conference, "Clinical Applications of Air Pollution Research."

The first necessary step to be taken by such clinicians is to become familiar with the available technical and professional knowledge regarding the effects of atmospheric pollution on human health. The next step is the determination of how to apply this knowledge in clinical practice. This meeting in San Francisco stands at the interface of these two components.

It would seem rational, then, to begin by describing some of the recent and current efforts directed toward the first of these components. Each session will then consider research relative to specific clinical areas and its clinical significance. After that we should have a better idea of the degree to which currently available research findings are useful to the clinician. We should, conversely, become more specifically aware of the areas needing further studies. Perhaps we might even agree as to who should assume the responsibility for conducting them. Finally, I hope we will have a better grasp of the knotty problem: What is the clinician to do about the adverse health effects of air pollution?

Let me direct my contribution primarily to an account of the work being done under the aegis of the National Academy of Sciences (NAS) and, in particular, its committee on the Medical and Biological Effects of Environmental Pollutants. (It is only fair to note that much of this work is being done by some of the people on the program of this conference.)

The Academy and the National Research Council got intensively involved in this area of activity in 1970. The Clean Air Act of 1967[3] had required the development of air quality criteria documents that would collect and present available knowledge about selected pollutants. The data in these documents would allow the establishment of air quality standards and the development of control programs to meet these standards. Initially, this project was assigned to the National Air Pollution Control Administration. Soon it became apparent, however, that a broader scientific base was needed.

The National Academy of Sciences seemed to be the appropriate agency for the task. From its founding in 1863, its charter had provided that the Academy "shall, whenever called upon by any department of the Government, investigate, examine, experiment and report upon any subject of science and art . . . "[4] The Environmental Protection Agency (EPA) turned in this direction.

In the rather flamboyant language of government contracts, the Academy was asked to "prepare an open-ended series of comprehensive state-of-the-art reports which will reflect the latest scientific knowledge useful in indicating the kind and extent of all identifiable effects on health and welfare which may be expected from the presence of a variety of pollutants in the environment." The contract specified a great many factors to be considered. Of major relevance to the concerns of this meeting are matters such as:

1. Effects of pollutants on human health and well-being.

2. Dose-response relationships.
3. The margins of safety to be used in establishing environmental quality standards.
4. Identification of groups in the general population known or likely to be particularly sensitive to exposure.

To date, reports* have been published on asbestos, fluorides, airborne lead, manganese, particulate polycyclic organic matter, chromium, and vanadium. Reports on nickel, selenium, chlorine and hydrogen chloride, and vapor phase polycyclic organic material are nearly completed. Others under way include copper, arsenic, zinc, and the platinum group metals. Also in process are carbon monoxide, hydrocarbons, nitrogen oxides, and photochemical oxidants. A listing of several others has been submitted by EPA for consideration.

Closer to the needs of the clinician, and somewhat more congruent to his interests, is a second major effort of the Academy. This was a complex group of studies to find answers posed by the United States Senate Committee on Public Works, the committee with cognizance over the Clean Air Act in the Senate.

On April 30, 1971, the Administrator of the Environmental Protection Agency had promulgated air quality standards,[5] in accordance with the Clean Air Act of 1970, for six pollutants — sulfur oxides (sulfur dioxide), particulate matter, carbon monoxide, nitrogen dioxide, photochemical oxidants, and hydrocarbons. Except for the first, these pollutants were largely associated with automotive emissions.

By the summer of 1973, many questions had been raised regarding the validity of these standards and the need for their application. These questions particularly focused on the costs of implementation and energy implications. Some had suggested, though, that new health effects evidence suggested that the standards might be more stringent — or less stringent — than necessary to protect the public health.

Because of this concern, the Committee on Public Works turned once more to the National Academy of Sciences to carry out its 100-year-old mission.[6] The Academy was requested to undertake a thorough review of the health effects of those air pollutants for which national ambient air quality standards had been published. In particular, the Academy was asked to respond to a long list of questions (Appendix I). Most of them sought information concerning the data base regarding the health effects of the pollutants, the basis for the present standards, the margin of safety in the standards, and the identification of groups in the population who are particularly susceptible.

---

*Reports are made to EPA, but separately reviewed and published by the NAS using its own review and publication procedures.

A phased approach was envisioned. First, a colloquium would be held early in October 1973 on the health effects of the pollutants specified, to be documented in a report due not later than October 31. Second, the Academy would undertake an intensive investigation of these air pollution-related health effects and submit a final report within a year's time. Thus, the Academy was to enlist the expertise of the scholars and scientists most knowledgeable on the subject who would review, evaluate, and report upon the evidence needed to answer the Committee's questions. The results of this effort, then, should provide a compilation of the data base we need.

This was an immense undertaking, especially in view of the time limits. Suffice it to say that it was accomplished and the reports have been submitted.[7,8] I commend them to you for study and reference.

I would like to highlight just a few aspects of the study. First, I would briefly describe the nature of the available data base, with emphasis on some of its peculiarities and limitations. Second, I would summarize the evaluations of the air quality standards as protection against human health effects. Last, I would report to you on several significant issues that emerged during the course of the study.

Three complementary approaches have been used to study health effects: (1) laboratory toxicological studies on animals, (2) laboratory clinical investigations on humans, and (3) epidemiological studies under naturally occurring conditions. Each has certain unique advantages as well as its own limitations. *Toxicological studies* allow for rigid control of conditions, dosage over a wide range, restriction of variables, and post-mortem examination. Limitations of the utility of this approach include species variation, lack of appropriate animal models in some cases, and the artificiality associated with study of a single pollutant — a condition not usually found in nature. *Clinical investigations* can provide information regarding human responses and controlled exposures, and sophisticated measurement techniques may be utilized. Within limits, especially susceptible subjects can be studied. Unfortunately, long-term exposure studies are hard to conduct. Ethical considerations likewise impose constraints. *Epidemiological studies* have the major advantage of being real-world events. The effects of complex short- and long-term exposures to population segments of varying susceptibility may be investigated. As we know, though, rigid experimental design is usually not possible, correlations of exposure and effect may be difficult, measurement methods are relatively crude, and ranges of exposures are limited.

Evaluations can be made and conclusions drawn with substantially more assurance when all three approaches are utilized and integrated than when reliance is placed on one alone. Each of the panels that contributed to the health effects reports used this multiple approach. The data base thus contains input in all three forms.

What, then, can one say about the health effects associated with the air pollutants under study and the adequacy of the standards set "to protect the public health." The short answer[9] is that the evidence that has accumulated since 1971 supports the standards set at that time. None of the individual panels identified any substantial basis for changing them, in either direction. This is not to say that there is complete satisfaction with the data base for these conclusions. As one panel put it "The technical data base for the . . . standard was inadequate at the time the standard was set (1971) and remains inadequate . . . "[10] But the only major change suggested was that an hourly as well as an annual standard for $NO_2$ be considered.

Let me indicate the approaches and conclusions of each of the panels.

Knowledge of the pathological changes associated with exposure to carbon monoxide is more extensive and better documented than that for any other pollutant. Its effects stem from its propensity to interfere with oxygen transport in the body by displacing oxygen from hemoglobin and other proteins. This propensity is so constant that an equation can be developed to describe the relationship between the concentration of inspired carbon monoxide and the resultant blood carboxyhemoglobin at equilibrium. The most sensitive measures that detect the effects of carboxyhemoglobin concentrations fall in three categories:[11]

a. *Effects on vigilance.* Carboxyhemoglobin concentrations in the range of 3% to 5% may degrade the ability to detect small unpredictable environmental changes.

b. *Effects on exercise.* Concentrations as low as 5% will decrease maximal oxygen consumption during exercise in healthy young males.

c. *Exacerbation of symptoms in patients with cardiovascular disease.* Patients with exertional angina pectoris develop chest pain earlier in the course of exercise at COHb concentrations of 2.5% to 3.0% than at 1%. Further, COHb concentrations of 3% have hastened the onset of leg pain during exercise in patients with peripheral arteriosclerosis.
(The COHb concentration at equilibrium associated with exposures to CO at the primary one-hour standard is about 5%.)

Only a marginally lower level of confidence is associated with assessment of the health effects from exposure to sulfur oxides and suspended particulate matter. In this case, the most significant data have come from epidemiologic studies. Controlled laboratory exposure of animals to $SO_2$ has only limited relevance to the setting of standards. It is known that exposure to $SO_2$ in atmospheric concentrations many times that of the air quality standard is required to elicit any detectable adverse health effect. On the other hand, it is believed that the health effects observed in epidemiologic studies are caused by a complex set of pollutants, for which $SO_2$ and total suspended particulates serve as convenient indices. Unfortunately, this relationship is not constant. Certain observations seemed relatively reliable and reproducible, though. Excess mortality and hospital admissions were observed when 24-hour concentrations of particulates and/or $SO^2$ rose to 500 $\mu g/m^3$ or more. The condition of patients with chronic respiratory disease worsened when concentrations of particulates and $SO_2$ rose above 300 and 400 $\mu g/m^3$ respectively. The 24-hour primary standard for $SO_2$ is 365 $\mu g/m^3$ and for particulates is 260 $\mu g/m^3$. Likewise, respiratory symptoms became more prevalent and sickness absence due to bronchitis and respiratory illness in children became more frequent when average annual concentrations of either pollutant were 100 $\mu g/m^3$ or over. The annual standards for $SO_2$ and particulates are 80 $\mu g/m^3$ and 75 $\mu g/m^3$ respectively. Thus, only a modest "margin of safety" is provided.

There is a rich body of literature concerning photochemical oxidants and ozone.[12] Nevertheless, there are several major difficulties that complicate this subject. First, the substances that possess the chemical properties that lead to their being termed "oxidant" are many, complex, and, in the main, as yet poorly characterized. Second, they are almost always secondary pollutants created by chemical reactions in the atmosphere. Third, the air quality standard depends on measurement of ozone, which is not associated with some of the most noticeable effects of smog (e.g., eye irritation).

There are, however, adverse health effects that are regularly observed at measurable levels of these agents:

1. Two-hour laboratory exposures of healthy smokers and nonsmokers to ozone at 3 to 4 times the standard have led to reduced pulmonary function. Linear extrapolation of these data suggests that effects might occur at 0.17 ppm; the one-hour standard is 0.08 ppm.
2. Eye irritation occurs at ambient concentrations at or somewhat below the standard.
3. Decreased lung tissue elasticity has been seen in rats exposed for 30 days to ozone at concentrations 2.5 times the standard.
4. Some persons with asthma or advanced chronic obstructive pulmonary disease are particularly susceptible to photochemical oxidants. Occasional individuals may show marked sensitivity.

Two summary conclusions were drawn:

1. Adverse health effects from short-term exposure to photochemical oxidants and ozone at the standard concentrations have not been observed in man.

2. There are human and animal data that suggest that adverse health effects might be expected at concentrations near the standard, especially under conditions of long-term exposure or in the presence of co-pollutants.

There is less general agreement on the correlation of adverse health effects with specified ambient atmospheric concentration in the case of the oxides of nitrogen than with the other air pollutants. The existing primary $NO_2$ standard of 100 $\mu g/m^3$ was based on studies of school children in Chattanooga. Exposure to concentrations between 117 and 205 $\mu g/m^3$ for six months and longer was associated with measurable health effects. These included: (1) an increase in acute respiratory illness, (2) significantly decreased lung function in schoolgirls, and (3) increased morbidity from bronchitis in school children exposed for 2 to 3 years. These results were supported by finding changes in the structure of lung tissue in rabbits, rats, and mice, and increased mortality in the latter, under several sets of exposures where the concentrations were substantially higher, though, in the order of 940 $\mu g/m^3$.

The Chattanooga studies have been widely criticized. There were problems with reliability of monitoring measurements, the simultaneous presence of other pollutants, and somewhat less than a straight-line dose-response relationship. Certainly, there is a manifest need for extensive further epidemiologic studies.

Nevertheless, the panel concluded[13] "As regards the present health standard for nitrogen dioxide, the available data, including those most recently reported, indicate that this standard of 0.05 ppm per year does afford reasonable protection from the risk of air pollution in healthy, as well as inordinately susceptible, individuals." (0.05 ppm = 100 $\mu g/m^3$)

During the conduct of these studies, several significant issues emerged that warrant consideration by all concerned. Each is worthy of substantial discussion; at this time, however, I will merely introduce them.

Inherent in the entire standard-setting process is the "threshold-margin of safety" issue. Present standards imply the existence of threshold levels for ambient concentrations (i.e., levels below which essentially the entire population will remain free from adverse health effects). The standard would be set somewhat below that level, the difference being considered the margin of safety. Many have questioned whether this concept has any physiologic validity. The dose-response relationship is generally continuous, with no lower limit below which no effects at all occur. One might argue that effects occur at all levels, but become manifest only after an "incubation period" that may be measured in years, or after diagnostic techniques are developed to the point where they might be detected. Further, the Clean Air Act does not define the term "health effect." In some cases, the effect of exposure on human health may be trivial. To what extent and expense should controls be imposed to prevent this class of effects? Perhaps the judgment of the clinician could help guide public policy.

Somewhat related to this issue is the problem of the uniquely susceptible individual or segment of the population. For example, of the 3% to 5% of the population with bronchial asthma, only 5% can relate their attacks to photochemical oxidant pollution, yet occasional individuals are markedly sensitive.[14] Why? Are they responding to some unidentified pollutant? Do they have a different disease? Is there something unique about their immune mechanism? These questions need answers.

Easier to understand is the susceptibility caused by pre-existing conditions. It is clear that some patients would be especially affected by exposure to increased levels of CO (e.g., those with respiratory disease that interferes with oxygen uptake, those with reduced oxygen transport capability in the blood, those with impaired blood flow to the tissues). What is not so clear is whether the standard should be lowered to protect all these individuals, or perhaps whether they should rely on the advice of their physicians as to protective action they might take when ambient concentrations reach levels that are dangerous for them.

Another kind of variability — that of geographic difference — leads to a similar issue. A very large part of what is known about air pollution, perhaps a majority in some cases, has been gained from studies in southern California. Difficulties arise when findings from this location are applied elsewhere. For example, Chambers[7] has described the problems generated by attempting to develop control strategies in Houston based on Los Angeles data. Again, the CO panel noted[8] that "Ambient air standards set at sea level are not applicable for high altitude sites." They point out that a cubic meter contains 18% less air in Denver than at sea level. Therefore, any given allowable concentration expressed in weight/volume terms would actually be 22% higher at that altitude. The point is that blind application of nation-wide standards will lead to different results in different locations. Local interpretation is needed. If the purpose of air pollution control is the protection of public health, there must be a role for the clinician here.

We have already noted what might be called the "surrogate measure" problem. Sulfur dioxide is used as an index of other sulfur oxides, acid sulfate aerosols, and some particulate sulfates. The measure of total suspended particulates is an index of those particles that cause adverse health effects. Ozone is used as an indicator for the whole oxidant mixture, many of whose individual components have not been identified. The toxic properties of these indicators are not always the same as the properties of the agents for which they act as surrogates; they do not even always vary in concentration at the same rate, in the same direction, or to the same magnitude as the other members of the class they represent. Once again, the relationship of variation in the indicator to the incidence of health effect must be considered in each locality, in the interest of both effectiveness and economy.

Many other issues remain to be elucidated. Among them are the roles of synergism, transformation, and chemical interaction. The degree to which air pollutants are mutagenic or carcinogenic is inadequately understood. The Committee report[8] mentions many others.

I have tried to avoid the extremes of being overly simplistic in drawing conclusions about what is known, as well as being overly negativistic — emphasizing the areas in which our knowledge is limited. There is, in fact, a large body of information on the health effects of air pollutants. There is both the legal and societal requirement for action. But in taking further action, wisdom is needed. I would call upon the clinical community to be a part of the action, and to contribute the wisdom and understanding they possess about the health of the public to the management of this most important problem.

## Appendix I

Questions Posed by the U.S. Senate Committee on Public Works to the National Academy of Sciences

1. What data base exists for present health-related ambient air quality standards? Is it adequate? What areas would be receiving study and analysis? To what extent are other than human health data a reasonable basis for extrapolating human health effects? Are present health standards supported by existing data — do new data support the existing standards? Please evaluate the relationship of new data to those standards. To what extent do the "margins of safety" built into existing standards represent a reasonable measure of protection against scientific unknowns?

2. Are the existing standards based on the health of normal, healthy adults or on the health of susceptible groups within the population? Which is a reasonable basis? Can population segments that are considered susceptible to the referred-to pollutants be identified within the population of any given area? What proportion of the population in major urban areas is within susceptible group categories?

3. Are the standards based on "threshold levels" plus a margin of safety or are they established at levels that will protect public health with a margin of safety? Are there assumed to be "threshold" effects levels for $NO_x$, CO, and oxidants? What is the best information available as to any "threshold" levels of the referred-to pollutants? What is a reasonable margin of safety below those "threshold" levels?

4. What improvement in air quality has actually occurred as a result of automobile emission controls to date? Are there conclusive monitoring data to quantify any change in ambient concentrations of these pollutants? What improvement in ambient air quality will result from meeting present and anticipated emission standards? Can we quantify this improvement in advance of application of controls?

5. What groups within the population will still be subjected to health hazards in 1985–86 (assuming statutory standards are met? Are changed?) How are these groups characterized (socio-economic level, unusual sensitivity to pollutants, old, young, ill, geographical location, seasonal exposure, etc.)? What is the most appropriate means of expressing ambient air quality standards to reflect human health effects (number of exposures, frequency, duration, pollutant concentration)? Can you identify the uncertainties in setting health effect standards for nitrogen oxides, carbon monoxide, and oxidants (confidence limits, standard deviation, errors in analyses, margin of safety concept)?

6. What information is available about synergism or antagonism in the set of air pollutants found in cities?

7. What proportion of total health hazard to the city dweller comes from air pollution? From automobile emissions? What is the proportion of total health hazard from urban air pollution that is due to these pollutants in the amounts that they are emitted from automboiles?

8. What are the limitations (to predictive judgments on alternative pollution control strategies) that are dictated by the probabilistic nature of science, the complexity of environmental health relationships, the variability among persons, and the vagaries of human nature?

## REFERENCES

1. Newill VA: Air Pollution Control: The Physician's Role. Oct 5, 1970 (unpublished).

2. US Government. Clean Air Amendments of 1970. Public Law 91–604, Sec. 109 (b) (1). Dec 31, 1970.

3. US Government. The Clean Air Act of 1967. Public Law 90–148. 90th Congress, S 780, Nov 21, 1967. Government Printing Office, Washington 1969.

4. US Government. An Act to incorporate the National Academy of Sciences. Thirty-seventh Congress, March 3, 1863.

5. *Federal Register,* vol 36, No. 84, Part II, pp 8186–8190. April 30, 1971.

6. Authorizing Supplemental Expenditures by the Committee on Public Works for Inquiries and Investigations. Report No. 93–369, US Senate. 93rd Congress, 1st Session.

7. Proceedings of the Conference on Health Effects of Air Pollutants. National Academy of Sciences-National Research Council. Oct 3–5, 1973. Prepared for the Committee on Public Works, US Senate, November 1973. Serial No. 93–15, 93rd Congress, 1st Session, US Government Printing Office, Washington 1973, 709 p.

8. Air Quality and Automobile Emission Control. A Report by the Coordinating Committee on Air Quality Studies. National Academy of Sciences, National Academy of Engineering. Prepared for the Committee on Public Works, US Senate, Sept 1974, Serial No. 93–24, 93rd Congress, 2d Session. US Government Printing Office, Washington, 1974.

Vol. 1 Summary Report, 129 p.

Vol. 2 Health Effects of Air Pollutants, 511 p.

Vol. 3 The Relationship of Emissions to Ambient Air Quality, 137 p.

Vol. 4 The Costs and Benefits of Automobile Emission Control, 470 p.

9. Ibid. Vol. 1, pp 6–8.

10. Ibid. Vol. 1, p 49.

11. Ibid. Vol. 2, p 3.

12. Ibid. Vol. 2, pp 317–389.

13. Ibid. Vol. 1, p 39.

14. Schoettlin CE, Landau E: Air pollution and Asthmatic Attacks in the Los Angeles Area. *Public Health Reports* 76:545–548, 1961.

PART ONE

# RESPIRATORY DISEASES

# 1 Lung Cancer and the Urban Environment; A Review

Carl M. Shy, M.D., Dr. P.H.

## INTRODUCTION

The purpose of this paper is to review the evidence for a relationship between lung cancer and non-occupational exposure to air pollutants in the general environment. I deliberately exclude occupational lung cancers from this review because the relationship between lung cancer and asbestos, nickel carbonyl, chromates, and several other industrial chemicals is well established and accepted by the scientific community, even though many issues are unresolved concerning dose-response functions, mechanisms of action, and environmental standards. However, there is no similar general agreement that urban air pollutants contribute to development of lung cancer in human populations. Emphasis on the strong and well characterized role of cigarette smoking in pulmonary carcinogenesis has perhaps inhibited identification of considerably weaker carcinogens in the general environment. Also, the virtually overwhelming contribution of cigarette smoking to lung cancer has made comparisons of lung cancer rates in high and low pollution areas quite meaningless unless the smoking habits of residents could be quantified or at least assumed to be equally distributed.

In the first part of this review, the age, race, sex, and smoking aspects of lung cancer in the U.S. are briefly characterized. This summary is followed by a section on experimental induction of lung cancer in animals with chemicals found in ambient air; these data provide the evidence that atmospheric pollutants can cause lung cancer in appropriate animal models, albeit at much higher concentrations than normally occur. The sources and atmospheric concentrations of benzo(a)pyrene are discussed next to indicate the extent of human exposure to an identified chemical carcinogen that is frequently cited as an index of other atmospheric pulmonary carcinogens. The larger portion of the paper is given to a detailed review of the epidemiologic body of evidence that supports the concept that factors in the general environment, other than smoking and occupational exposures, contribute to population differences in lung cancer risk.

This evidence is based on studies of (a) foreign immigrants who move from an environment with a given lung cancer risk to a country with a different level of risk, (b) urban-rural differences in lung cancer risk, and (c) quantitative or qualitative differences in air pollution and associated lung cancer rates. In order to provide the reader with sufficient evidence to reach his own conclusions, I have presented much of the actual data from various studies in tabular form, and have minimized the amount of interpretation given to each study. A more exhaustive and interpretive review of the literature on atmospheric carcinogens can be found in the recent monograph *Particulate Polycyclic Organic Matter*[15] prepared by a working panel for the Committee on Biologic Effects of Atmospheric Pollutants of the National Academy of Sciences.

## DESCRIPTIVE EPIDEMIOLOGY: AGE, SEX, RACE, SMOKING

Between the years 1950 and 1967, lung cancer mortality rates (ICD No. 162, age-adjusted over all ages) for males in the United States (Figure 1)[1] increased among whites and nonwhites from 9.0 and 7.3 deaths per 100,000 respectively in 1950 to 19.7 and 22.1 per 100,000 respectively in 1967. The changes reveal, over the 17-year period, a 2.2-fold increase in mortality rates for white males and a 3-fold increase for nonwhite males. During the same period, annual lung cancer mortality rates for females showed no racial differences and increased gradually from 1.6 deaths per 100,000 in 1950 to 2.9 deaths per 100,000 in 1967, a 1.8-fold increase over the 17 years. Higgins[2] examined time trends in age-specific death rates for respiratory system mortality (ICD 160–164) in the United States from 1940 through 1967 and noted a marked contrast in white male and female rates. White male death rates increased at a declining rate; the rate for white females increased markedly among 35 to 54-year-olds beginning in 1955 and among 55- to 74-year-olds beginning in 1960. These rate changes correspond with data showing little change among men in the proportion of cigarette smokers between 1955 and 1965, while the proportion of regular cigarette smokers among women 17 years and older increased from 23.6% in 1955 to 33.2% in 1964–1965.[2]

Age specific lung cancer death rates (ICD 162) averaged over the period 1950–1967[1] rise steeply in males, reaching a peak at ages 65 to 74, and decreasing at older ages (Figure 2). In females, the mortality curve ascends gradually to a plateau at age 65, and levels off thereafter. However, Levin[3] in 1953 demonstrated by cohort analysis that the shape of the cross sectional age specific mortality curve for lung cancer in males is under the dual influence of an aging effect in the same birth cohort and a cohort effect at the same age in successive birth cohorts. When the aging effect exceeds the cohort effect over a specified time period, the age specific mortality curve at the end of the time period will show higher rates in older age groups. When the cohort effect exceeds the aging effect, the age specific curve turns down, even though mortality rates actually increased with age in the same birth cohort. That lung cancer death rates in U.S. men and women continue to increase with older age and at the same age in successive birth cohorts was well demonstrated by Wynder, Covey, and Mabuchi,[4] as illustrated in Figure 3. Each curve in the figure reveals higher lung cancer death rates in successive cohorts at the same age. The aging effect can be identified by following a vertical line through any year

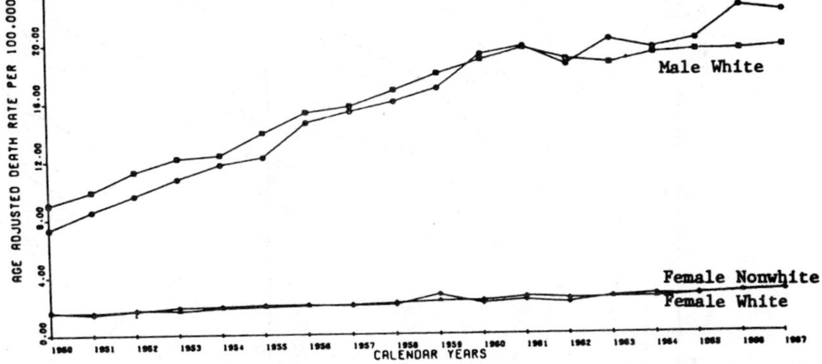


**Figure 1.** Age-adjusted lung cancer death rate per 100,000, U.S. 1950–1967, according to sex and race. (Burbank, F: National Cancer Inst Monograph, 1971.[1])

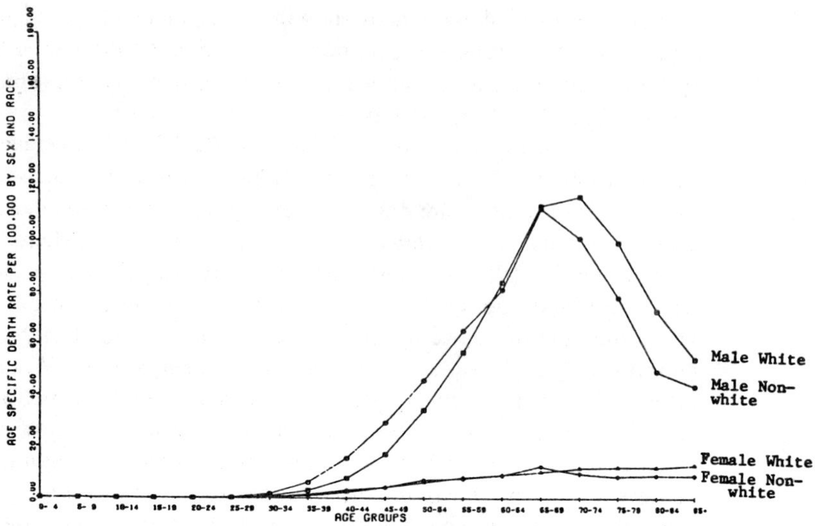


**Figure 2.** Age specific lung cancer death rate per 100,000, U.S. 1950–1967, according to sex and race. (Burbank, F: National Cancer Inst Monograph, 1971.[1])

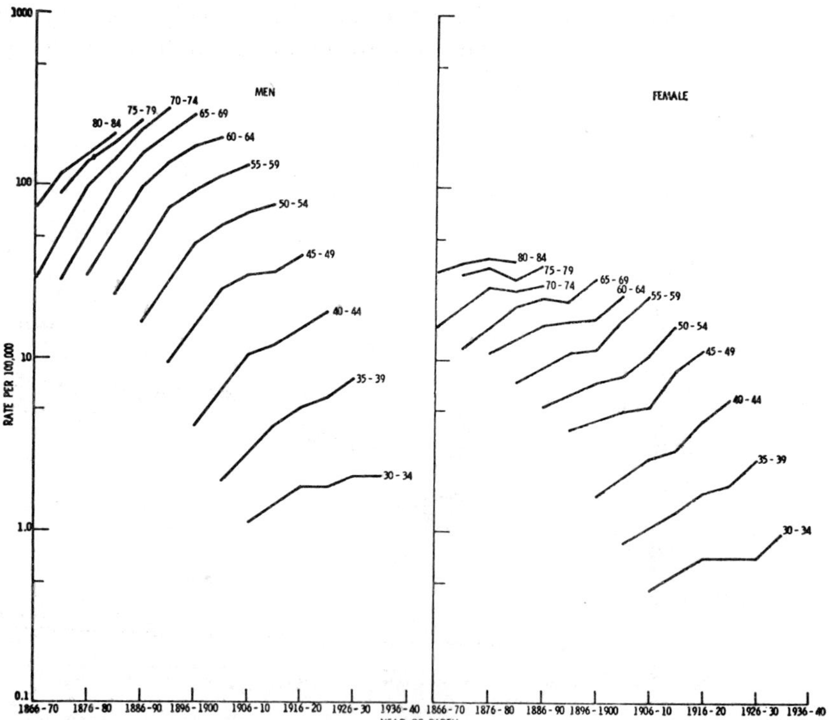


[Fi]gure 3. Lung and cancer death rates in U.S. men and women, by age and five-year birth [co]hort. (Wynder, Covey, Mabuchi: J Nat Cancer Inst, 1973.[4])

of birth. At younger ages, the aging effect on lung cancer mortality is greater than the cohort effect, while at the oldest ages, especially in males, the cohort effect is greater than the aging effect. The important points to be derived from Figure 3 are that lung cancer rates increase steadily from the youngest to the oldest ages in all birth cohorts and that successive birth cohorts experience higher lung cancer rates at the same attained age. Successive male cohorts generally show a declining rate of increase while successive female cohorts between ages 45 and 59 are showing an increasing rate of increase.

The lung cancer risk from cigarette smoking has been amply reviewed in the Report of the Advisory Committe to the Surgeon General of the Public Health Service[5] and in subsequent periodic Public Health Service reviews on the Health Consequences of Smoking.[6–9] Comparisons of lung cancer risks by smoking class reveal a range of relative risks, in various studies from a 2- to 9-fold greater rate of lung cancer in regular cigarette smokers of one pack or less daily, to a 6- to 30-fold greater rate in cigarette smokers of more than one pack per day. The mortality rate for lung cancer is estimated to be 3 to 5 per 100,000 in the nonsmoking U.S. population, age adjusted over all age groups.[10]

The impact of lung cancer on male survival and on work-years lost in the male labor force was well documented for the year 1968 by Murray and Axtell.[11] As shown in Figure 4, lung cancer deaths in 1968 accounted for nearly 700,000 person-years lost among males, nearly 3½ times greater

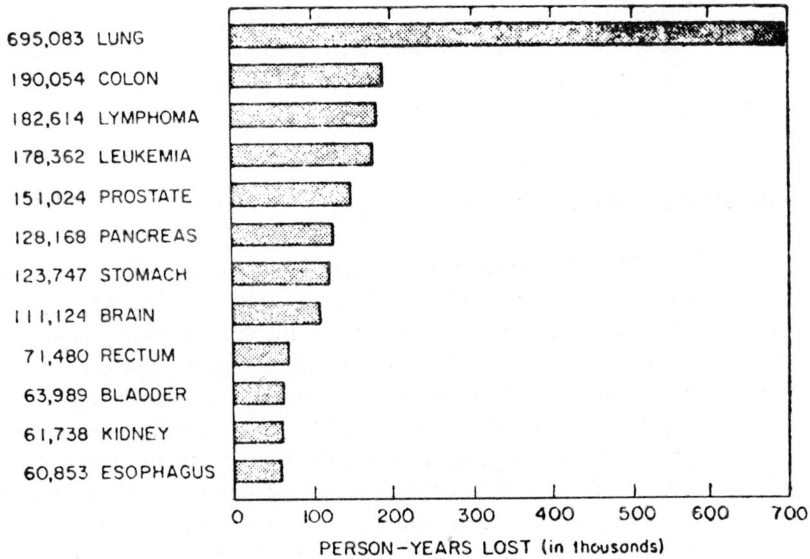


**Figure 4.** Person-years lost from leading types of cancer, 1968: U.S. males. (Murray and Axtell: J Nat Cancer Inst, 1974.[11])

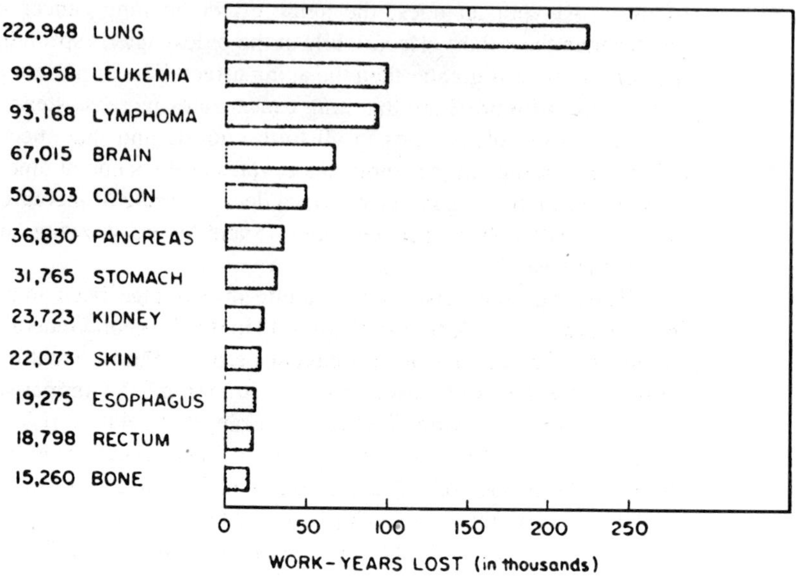


**Figure 5.** Work-years lost from leading types of cancer, 1968: U.S. males (Murray and Axtell: J Nat Cancer Inst, 1974.[11])

than the effect on survival of any other organ cancer. Person-years lost were calculated on the basis of life expectancy for each age group in 1968. In terms of the labor force (Figure 5), lung cancer was associated with 223,000 lost work-years (based on life expectancy for working years between ages 20 and 65), more than double the effect of any other organ cancer. These data illustrate the large magnitude of the lung cancer problem relative to the effect of other forms of cancer among males in the United States.

## EXPERIMENTAL STUDIES

Possible mechanisms of action for carcinogenesis by chemicals have been reviewed by Miller[12] and more recently by Heidelberger.[13] Chemical carcinogens generally require metabolic conversion to some form of "ultimate" carcinogen, which is thought to be an electrophilic reagent that reacts through formation of a covalent bond with reactive centers in the macromolecule. The result is an altered critical molecule that initiates the process of carcinogenesis at the cellular level. As shown in Figure 6, as derived from Heidelberger's review,[13] carcinogenic chemicals enter the body as precarcinogens and may be detoxified to noncarcinogenic products or activated through several stages into an ultimate carcinogen, capable of forming a covalent bond with DNA, RNA, or protein of target cells.

Heidelberger[14] notes that hydrocarbons, which are thought to comprise the carcinogenic fraction of atmospheric pollutants, are not chemically capable of forming covalent bonds with cellular DNA, RNA, or protein. They must, therefore, be metabolically activated to a chemically reactive form.

The issue of airborne carcinogens in the non-occupational environment was thoroughly reviewed by the Committee on Biologic Effects of Atmospheric Pollutants of the National Academy of Sciences.[15] The major thrust of this report was "the working hypothesis that there is an urban-rural difference in lung cancer rates that may stem from community air pollution and that may be attributable to polycyclic hydrocarbons known to be carcinogenic, such as benzo(a)pyrene.[16]

Bioassays on the skin, subcutaneous tissue, cervix, and trachea of experimental animals have shown that particulate matter of urban air can be carcinogenic in the intact animal.[17] Three classes of chemical carcinogens have been experimentally identified or suggested in urban air pollutants: polycyclic aromatic hydrocarbons, N-heterocyclic hydrocarbons, and oxygenated neutral compounds of unknown structure. Direct inhalation studies in animals have required high doses of agents, such as benzo(a)pyrene (BaP), far above concentrations measured in urban atmospheres, to produce a carcinogenic response. Induction of tumors by BaP

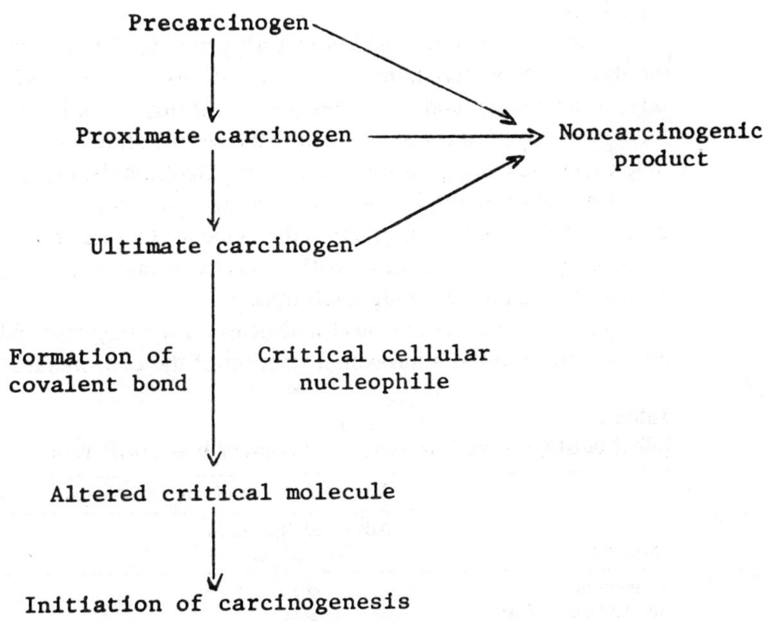


**Figure 6.** Schematic representation of the mode of action of chemical carcinogens. The carcinogen may be detoxified to an inactive state or activated to form an electrophilic reagent capable of covalently binding with components of the target cell.(Heidelberger, Fed Proc, 1973.[13])

usually requires addition of another physical or chemical factor, which probably inhibits tracheo-bronchial clearance and thereby prolongs the residence time of the carcinogen. No tumors were produced by injection of BaP alone, but adsorption of BaP on carbon particles, hematite or asbestos readily produced tumors in animals.[18] Saffiotti[19] mixed dust containing 3 mg BaP and 3 mg ferric oxide suspended in saline and administered the mixture by intratracheal instillation to three groups of hamsters; the first group was given 5 doses at weekly intervals; the second, 10 doses; and third, 15 doses. A dose-response relationship was observed, with 15 tumor-bearing animals out of 100 in the first group, 42 of 100 in the second, and 60 of 79 in the third group. By implanting 3-mg pellets ranging from 0.1% to 100% BaP or 3-methylcholanthrene in the trachea of rats, Kuschner and colleagues[20,21] were able to obtain a dose-response relation between BaP alone, or 3-methylcholanthrene alone, and the incidence of lung tumors of various types. This study is one of the few in which BaP induced malignant tumors in the lung without the addition of other physical or chemical agents. However, removal of BaP from the trachea was prevented by fixing the implanted pellet within the wall of the airway. Shabad[22] reported data supporting the conclusion that repeated injections of BaP, mixed with black india ink powder in casein solution, gave a greater tumorigenic effect for the same total dose than from single injections.

Irritant gases combined with BaP promote the tumorigenic effect of the hydrocarbon. Inhalation exposures of rats to 10 ppm $SO_2$ for 6 hours/day, 5 days/week and a combination of 10 mg/m$^3$ BaP and 3.5 ppm $SO_2$ for 1 hour/day, 5 days/week yielded the results shown in Table 1.[21] Thus, $SO_2$ exposures at high doses seemed to potentiate the effect of BaP.

Of possible importance with respect to childhood exposures and cancer rates in foreign migrants is the evidence that newborn mice are more susceptible to carcinogenesis with a variety of chemicals and routes of administration than are young adult mice.[23]

The enzyme system, aryl hydrocarbon hydroxylase, which is generally localized in the microsomal fraction of the cell, metabolizes a variety

**Table 1**
**Effect of $SO_2$ on the Carcinogenic Action of Benzo(a)Pyrene***

| | Ratio of Number of Animals Affected to Number Exposed | |
|---|---|---|
| Exposure | Advanced Squamous Metaplasia | Squamous Cell Carcinoma |
| Ambient air | 0/3 | 0/3 |
| Ambient air + BaP | 1/21 | 2/21 |
| $SO_2$ | 0/3 | 0/3 |
| $SO_2$ + BaP | 2/21 | 5/21 |

*Based on data from Laskin et al.[21]

of polycyclic hydrocarbons including BaP.[24] This enzyme catalyzes reactions *in vitro* that result in a covalent bond between BaP and DNA or protein. Inhibition of the enzyme reduces mouse skin tumorigenesis caused by repeated treatment with 7,12 dimethylbenz(a)anthracene. Pretreatment of rodents with chemicals that induce microsomal enzyme activity diminishes the incidence of lung and mammary gland tumors caused by 7,12 dimethylbenz(a)anthracene. Pretreatment with enzyme inducers may protect the animal by increasing the enzyme content in target tissue and thereby hastening metabolic detoxification of the carcinogen. These studies suggest the need to investigate the role of environmental factors such as pesticides, drugs, and nutrition in enhancing either the activating or the detoxifying action of the microsomal enzyme system on chemical carcinogens. Polycyclic aromatic hydrocarbons of weak carcinogenicity are present in polluted air and cigarette smoke. These compounds can exert an anticarcinogenic reaction when combined with more potent carcinogens.[25] Thus, a mixture of polycyclic aromatic hydrocarbons in polluted air may have a smaller net carcinogenic potential than would be expected from the action of the individual potent carcinogens in the mixture.

## ENVIRONMENTAL SOURCES AND LEVELS OF BENZO(A)PYRENE

Polycyclic organic matter can be formed from any combustion of fossil fuels or compounds containing carbon and hydrogen.[26] Inefficient combustion, which provides an oxygen deficient environment, and the presence of higher molecular weight molecules in fuel favors formation of larger amounts of polycyclic organic matter. In the 1960s the U.S. Public Health Service [27] conducted a survey to identify processes likely to produce emissions of polycyclic organic matter, with emphasis given to heat-generation and incineration, motor vehicles, petroleum refining, and other industrial processes. Ten polynuclear hydrocarbons were quantitatively analyzed in this survey but, for convenience and analytical simplicity, benzo(a)pyrene (BaP) was selected as an index of polycyclic organic matter. Sources were evaluated either by direct sampling of effluent gases from a given process or by atmospheric sampling in residential areas near the process.

Newer data compiled in the National Academy of Sciences report[28] indicate significantly greater BaP emissions from refuse burning in the United States than was accounted for in the original Public Health Service estimates. These data include as sources a large contribution from forest and agricultural fires and even greater emissions from coal refuse bank burning. Newer air sampling measurements also indicate that gaseous discharge from coke ovens contribute to higher BaP concentrations in areas adjacent to these processes.[29] These revisions lead to the summary of BaP emissions from stationary sources in the U.S.[38] given in Table 2. The major individual sources of BaP emissions within source categories are given in Table 3.

Ambient levels of BaP in urban and rural sites of the United States were determined by Sawicki in 1959.[33] In over 100 U.S. sampling sites, the median winter-spring value was found to be 6.6 ng/m$^3$ (=6.6 $\mu$g/1,000 m$^3$) in urban areas, and 0.4 ng/m$^3$ in nonurban areas (Figure 7). Ambient concentrations varied considerably by season, with peak levels occurring in winter months. The average BaP concentration in U.S. cities has decreased since 1950 by a factor of nearly three, probably due to a nationwide shift from combustion of coal for domestic heating to oil, gas, and electricity and to improved furnace design in stationary sources.[34] Recent unpublished data from the National Air Surveillance Network shows a median January-March 1967 BaP concentration of 2.5 ng/m$^3$ in urban sites, and 0.2 ng/m$^3$ in rural locations (Figure 8).[35] These values are considerably below those reported by Waller[36] for 12 consecutive months in 1949–1950 at eight different urban sites in Great Britain. BaP concentrations averaged 26 ng/m$^3$ over the eight cities and varied from an annual low of 13 ng/m$^3$ in Bristol to 46 ng/m$^3$ in London. From Waller's data, Blacklock[37] estimated that breathing 511,000 m$^3$ of air during a 70-year lifetime could result in accumulation of 16 mg BaP within the body.

Exposure to BaP, and to other polynuclear hydrocarbons, can result from cigarette smoke as well as from ambient air. One hundred cigarettes

**Table 2**
**Estimated Annual BaP Emissions in the U.S.**

| Source Category | BaP, Tons/Year | Percent of Total |
|---|---|---|
| Heat and power generation | ~ 475 | 36.5 |
| Refuse burning | ~ 600 | 46.0 |
| Industrial processes | 206 | 15.8 |
| Motor vehicles | 22 | 1.7 |
| Total | 1,303 | 100 |

**Table 3**
**Major Individual Sources of BaP Emissions in the U.S.**

| Major Sources | BaP, Tons/Year | Percent of Total |
|---|---|---|
| Residential hand stoked coal furnaces | 420 | 32.2 |
| Coal refuse fires | 340 | 26.1 |
| Wood combustion | 40 | 3.1 |
| Vehicle disposal | 50 | 3.8 |
| Forest and agricultural fires | 140 | 10.7 |
| Coke ovens | 200 | 15.3 |
| All other | 114 | 8.8 |
| Total | 1,303 | 100 |

yield 3.0 to 4.0 μg of BaP.[38] Sidestream smoke also contains BaP, and concentrations of 2 to 4 ng/m$^3$ have been measured in a medium-size room (40 m$^3$) in which three cigarette smokers were polluting the air with sidestream smoke.[32] A comparison of possible BaP exposures in polluted, average urban and average nonurban sites with mainstream and sidestream cigarette smoke yielded data presented in Table 4. In one year, a one-pack-per-day cigarette smoker is exposed to 2.5 times more BaP than a nonsmoker living in a moderately polluted city (Pittsburgh, average BaP 13.7 ng/m$^3$ in 1969).[38] Compared to the average of urban sites in 1969, the one-pack-a-day smoker receives a 14-fold greater BaP exposure. Nonsmokers exposed to sidestream smoke for 25% of the day could accumulate a total annual dose of 21 μg BaP rather than the 18 μg which would be calculated for 100% exposure to the ambient air of the median urban site in the National Air Surveillance Network, 1969. In nonurban areas, under similar assumptions nonsmokers exposed to sidestream smoke

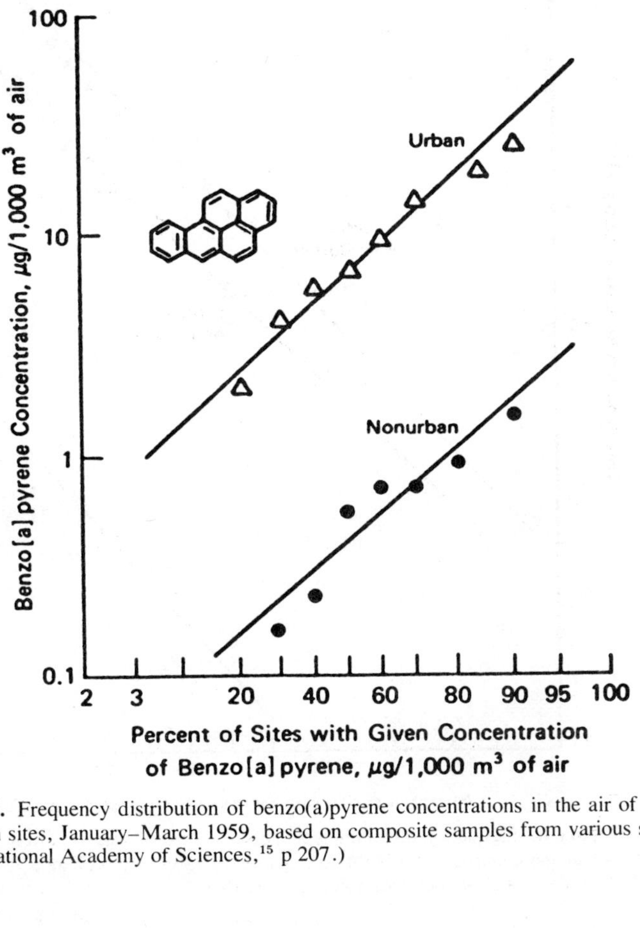


**Figure 7.** Frequency distribution of benzo(a)pyrene concentrations in the air of urban and nonurban sites, January–March 1959, based on composite samples from various sites in the U.S. (National Academy of Sciences,[15] p 207.)

over one year could accumulate a total BaP dose of 8 $\mu$g rather than 1 $\mu$g resulting from ambient nonurban exposure alone. A nonurban one-pack-a-day cigarette smoker would, of course, receive an overwhelming dose of BaP from mainstream smoke, compared with his BaP exposure from ambient nonurban air (ratio of 255:1).

Localized sources of BaP emissions may result in considerably higher exposures of workers and nearby residents than are found in the ordinary urban environment. Sawicki[39] reported the following range of BaP concentrations in the ambient air near various emission sources (Table 5).

Colucci and Begeman,[40] determined the relative contribution of motor vehicle exhaust to BaP concentrations in various sectors of Detroit

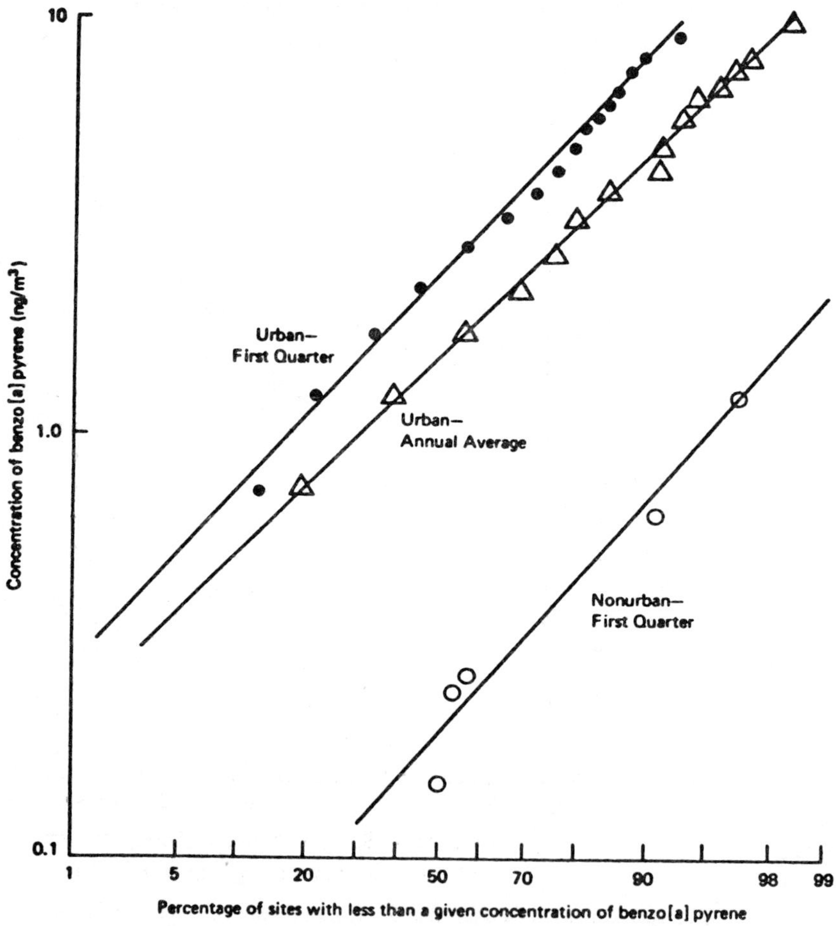


**Figure 8.** Frequency distribution of benzo(a)pyrene concentrations in the air for all U.S. urban and nonurban sites in the first quarter of 1969, and with annual average urban distribution. (National Academy of Sciences,[15] p. 49.)

in 1965 (Table 6). Thus, although the contribution of motor vehicles to total nationwide emission of BaP amounts to less than 2%, in the absence of other major sources as in some suburban locations, motor vehicles may contribute as much as 40% to 50% of the local BaP concentration. This contribution should be considerably diminished by hydrocarbon controls installed on vehicles since 1967. Uncontrolled vehicles prior to 1965 emitted 170 $\mu$g BaP per gallon of fuel consumed[41] while advanced control systems have brought down the emission rate to less than 10 $\mu$g per gallon.

**Table 4**
**Indoor and Outdoor Exposures to BaP**

| Exposure | BaP ng/m³ | Total BaP Exposure Per Year (in micrograms) |
|---|---|---|
| I. Ambient air | | |
| Polluted city, 1969* | 13.7 | 100 |
| Urban median, 1969* | 2.5 | 18 |
| Nonurban median, 1969* | 0.2 | 1 |
| II. Indoor air | | |
| Mainstream cigarette smoke† | 100,000 | 255 |
| Sidestream cigarette smoke‡ | 4 | |
| Median urban exposure§ | | 21 |
| Median nonurban exposure§ | | 8 |

*Assumes daily respiration of 20 m³/day.
†Assumes 35 ng BaP/cigarette[31] and 1 pack/day.
‡Three smokers in medium-size room.[32]
§Assumes exposure to sidestream cigarette smoke for 25% of day and exposure to urban average of 2.5 ng/m³ for 75% of day or to the nonurban average of 0.2 ng/m³.

**Table 5**
**Localized Sources of BaP Emissions**

| Sources | BaP ng/m³ |
|---|---|
| Sidewalk tarring | 78,000 |
| Coal tar pitch working area | 75,000 |
| Roof tarring area | 14,000 |
| Boston Sumner Tunnel | 69 |
| U.S. cities, 100 average (1958) | 6 |
| Grand Canyon Park, Arizona | 0.04 |

**Table 6**
**Relative Contribution of Motor Vehicles to BaP Concentrations in Detroit**

| Location | Relative Motor Vehicle Contribution (percent) |
|---|---|
| Downtown area | 5 |
| Freeway area | 18 |
| Suburb | 42 |

There is practically no information on the particle size or mass distribution of polycyclic hydrocarbons in atmospheric aerosols.[42] In one study, DeMaio and Corn[43] found 75% of the weight of selected polycyclic hydrocarbons to be associated with aerosol particles less than 5 micrometers diameter. Other aspects of the atmospheric physics and chemical reactivity of polycyclic hydrocarbons are reviewed in the National Academy of Sciences report.[44]

## STUDIES OF FOREIGN IMMIGRANTS

Lung cancer death rates of immigrants to various countries have been compared with rates in their home countries and in their country of settlement. Consistent changes in the lung cancer experience of the migrants suggest an influence of their early non-occupational environment.

In one of the first lung cancer studies of migrants, Eastcott[45] compared lung cancer deaths for the years 1949 to 1953 in New Zealand among native New Zealanders and immigrants from the United Kingdom (Table 7).

British born immigrants revealed a 32% excess over the New Zealand born population as a whole. The excess was accounted for by higher lung cancer death rates in British immigrants who had entered New Zealand after age 30 years (Table 8). Eastcott observed generally heavier cigarette smoking habits among New Zealanders and British born immigrants to New Zealand than among residents of Britain. Thus, the excess cancer deaths in British immigrants could not be explained by smoking habits alone. These findings suggested that environmental exposures in

**Table 7**
**Lung Cancer in Native New Zealanders and in British Immigrants**

| | Lung Cancer Deaths, Age 35+ in New Zealand | | |
|---|---|---|---|
| Country of Birth | Observed | Expected* | |
| New Zealand | 632 | 721.8 | P<.001 |
| United Kingdom | 369 | 279.2 | |

*Based on age specific rates for New Zealand population as a whole.

**Table 8**
**Lung Cancer in British Immigrants to New Zealand, According to Age at Entry**

| | Lung Cancer Deaths | | |
|---|---|---|---|
| Age at Entry | Observed | Expected | |
| Under 30 | 187 | 231.8 | P<.001 |
| 30+ | 168 | 123.2 | |

childhood and early adulthood exerted a lasting effect on the lung cancer experience of the immigrant cohort.

Dean[46] similarly compared white male lung cancer death rates of British immigrants to various sectors of South Africa with rates of native born residents in South Africa and of British born residents remaining in England and Wales (Table 9). Lung cancer mortality rates of immigrants fell between the lower rates of native South Africans and the higher rates of England and Wales. These differences could not be explained on the basis of cigarette consumption habits in the two countries, and Dean provides evidence that cigarette consumption per adult was greater among South African whites than among residents of England and Wales.[47] Daily cigarette consumption of British born immigrants to South Africa (24 cigarettes/day) was 75% higher than daily consumption in England and Wales (13.5 cigarettes/day). A statistically significant excess in lung cancer mortality occurred in immigrants age 45 to 64 but not in those age 65 and older. Dean speculates that this age difference may be explained by the fact that the majority of British immigrants age 45 to 64 dying of lung cancer came to South Africa after 1920, when lung cancer became common in England and Wales. Immigrants dying at ages 65 and over generally immigrated before 1910, when lung cancer was relatively uncommon in England and Wales. These data further support the hypothesis that environmental exposures early in life affect the lung cancer mortality experience of the migrating cohort.

Mancuso[48] estimated lung cancer death rates of white male foreign immigrants living in Ohio. Information on country of birth was obtained from death certificates, and estimates of the age distribution of the native

**Table 9**
**Lung Cancer Mortality in Native Born Residents and Immigrants to South Africa, 1947–1956, Rate per 100,000 White Males**

| Residence in South Africa | Age 45–64 | | Age 65+ | |
|---|---|---|---|---|
| | South African Born | British Immigrants to SA | South African Born | British Immigrants to SA |
| Johannesburg | 72 | 117 | 234 | 213 |
| Cape Town | 76 | 145 | 153 | 181 |
| Durban | 111 | 178 | 308 | 300 |
| Pretoria | 71 | 118 | 169 | 149 |
| Port Elizabeth | 53 | 125 | 175 | 181 |
| Other urban | 48 | 71 | 105 | 119 |
| Rural | 30 | 94 | 82 | 82 |
| Total | 50 | 112* | 112 | 172† |
| Residents of England & Wales | 135 | | 219 | |

*P<.001 (British immigrants vs. native born SA).
†P>.05 (British immigrants vs. native born SA).

and foreign born white male population in Ohio counties were derived from the age distribution of foreign born reported in the 1950 census of the Ohio Standard Metropolitan Statistical Areas. Migrant lung cancer rates were compared with those of native residents of Ohio and of the country of birth (Table 10), and were found generally to be intermediate to rates of their country of birth and the U.S.

Haenszel[49] compared lung cancer mortality among foreign born immigrants to the U.S. with the mortality experience of residents staying in the country of origin. All lung cancer deaths occurring in 1950 in 35 states of the northeastern, north central and western parts of the U.S. were used for this analysis. Table 11 shows that the lung cancer mortality rates of male foreign immigrants to the U.S. generally fell between those of the country of origin and those of native born U.S. residents. No consistent pattern was found for the lung cancer experience of female immigrants, whose absolute lung cancer death rates were much lower than that of males.

Reid and associates[50] compared mortality rates for lung cancer and for coronary heart disease among British and Norwegian immigrants to the U.S. with native born residents of Britain and Norway, respectively, and with native born U.S. residents. To obtain these data, the investigators utilized a multistage probability sample of the U.S. population and random subsamples of the British and Norwegian population. Deaths of persons age 35 to 74 occurring in 1960 to 1962 in Great Britain and Norway and in 1959 to 1961 in the U.S. were analyzed. As shown in Table 12, lung cancer mortality rates of immigrants to the U.S. were generally intermediate to those of the U.S. and their home country. On the other hand, coronary heart disease mortality rates of immigrants to the U.S. were much closer to U.S. rates than to rates prevalent in the country of origin. These results further support the body of evidence for an effect of early environment on lung cancer mortality. In contrast, coronary heart disease mortality rates did not strongly reflect early environmental influences. Likewise, immigrant rates for non-cancerous chronic respiratory disease morbidity and mortality were similar to those of coronary heart disease—that is, dis-

**Table 10**
**Lung Cancer Mortality in Natives and Foreign Immigrants Residing in Ohio (White Males, Age 25–64, 1947–51)**

| | Lung Cancer Mortality/100,000 | |
|---|---|---|
| | Country of Birth | Migrants to Ohio |
| U.S. (Ohio) | 21 | — |
| England and Wales | 55 | 33 |
| Germany | 36 | 45 |
| Italy | 16 | 18 |
| All foreign born in Ohio | — | 37 |

**Table 11**
**Standardized Mortality Ratios for Foreign Born White Immigrants to the U.S. and Residents in Country of Origin**

| | SMR for Males* | | SMR for Females† | |
|---|---|---|---|---|
| | Migrants to U.S. | Country of Origin | Migrants to U.S. | Country of Origin |
| England-Wales | 131 | 243 | 150 | 167 |
| Norway | 70 | 44 | 168 | 79 |
| Sweden | 90 | 56 | 108 | 108 |
| Germany | 112 | 124 | 126 | 101 |
| Italy | 90 | 53 | 106 | 82 |
| U.S. | — | 100 | — | 100 |

*SMR = 100 for native born white U.S. males.
†SMR = 100 for native born white U.S. females.

**Table 12**
**Age Adjusted Mortality Rates per 100,000, Ages 35–74**

| | Lung Cancer | | Coronary Heart Disease | |
|---|---|---|---|---|
| Population Group | Males | Females | Males | Females |
| Residents of Great Britain | 151.2 | 19.3 | 426.5 | 157.7 |
| British born immigrants to U.S. | 93.7 | 11.5 | 549.3 | 195.3 |
| Native born residents of U.S. | 72.2 | 9.8 | 578.8 | 196.5 |
| Residents of Norway | 30.5 | 5.6 | 315.5 | 112.0 |
| Norwegian born immigrants to U.S. | 47.5 | 10.7 | 510.0 | 180.3 |
| Native born residents of U.S. | 72.2 | 9.8 | 578.8 | 196.5 |

ease and death rates of immigrants were much closer to the rates of their adopted country than to their country of origin.

The migrant studies are important in providing consistent results that suggest a lasting effect of early environment on lung cancer mortality later in life. Consideration of smoking habits in New Zealand, South Africa, and Britain show that the lung cancer rates of immigrants could not be accounted for solely by cigarette smoking habits. Unfortunately, the studies could not point to any one component of early environment responsible for the residual effect on lung cancer mortality, and adequate measurements of variables of interest (such as occupation, air pollution exposure, and indoor sources of hydrocarbons) were lacking.

## URBAN-RURAL STUDIES

A considerable number of studies have compared lung cancer rates in urban and rural residents of the same country, or lung cancer rates for urban areas stratified according to population size.

Lung cancer death rates for the years 1946–1949 in England and

Wales were shown by Curwen[51] to increase with population size (Table 13). Differences in smoking habits were not assessed in this study.

Stocks[52] obtained information on cigarette smoking habits and air pollution exposure of male lung cancer decedents and non-cancer patients ages 45 to 74 for the years 1952–1954 in three areas of England: urban Liverpool, a mixed urban-rural district (Wrexham and Chester), and a purely rural area (Bangor and Llandudno). Smoking and environmental data were obtained from all patients at the time of hospital admission or from relatives of decedents. Information on smoking habits of 3,500 non-cancer patients was used to estimate the proportion of male smokers in the general population ages 45 to 74. Benzo(a)pyrene concentrations were measured at 10 sites for six months between October 1954 and April 1955. Results are shown in Table 14.

Both an urban effect and smoking effect were present. The urban effect diminished with intensity of cigarette smoking, while the contrast between nonsmokers and smokers was greatest among rural inhabitants. The number of lung cancer deaths within each combined smoking-residence category was extremely small, and the estimates of smoking habits for the general population based on information obtained from non-cancer patients is subject to significant bias in view of the general increase in morbidity associated with cigarette smoking.

Dean[53] obtained more precise smoking information from personal interviews with relatives of male lung cancer patients dying at age 35 and older in Northern Ireland, 1960–1962. Similar information was obtained from a 1 in 1,500 survey of households selected at random and yielded information on 12,307 persons. A total of 743 lung cancer deaths of males was available for analysis and was distributed by smoking category and area of residence. Table 15 indicates the computed lung cancer death rates per 100,000 population for males age 35 years and older. The results suggest increasing lung cancer death rates with increasing urbanization among nonsmokers, pipe and cigar smokers, and smokers of 1 to 10

**Table 13**
**Cancer of Lung, England and Wales 1946–1949**

| | Males | Females |
|---|---|---|
| I. England and Wales: Total mortality rate per 100,000 | 40.9 | 8.0 |
| SMR | 100 | 100 |
| II. SMRs for Population Sectors | | |
| Greater London | 137 | 132 |
| County boroughs* | 117 | 108 |
| Urban districts† | 86 | 88 |
| Rural districts | 62 | 74 |

*Regions with towns 100,000 or larger population, excluding Greater London.
†Regions with towns less than 100,000 populations.

cigarettes per day. Although rural residents of all smoking categories experienced lower death rates than smoking counterparts in urban areas, there was no evidence for a lung cancer gradient among heavier cigarette smokers with increasing size of urban area. These findings generally substantiate those reported by Stocks noted above.

Several studies of urban-rural differences in lung cancer rates in the U.S. were lacking in data on cigarette smoking habits. Age adjusted incidence rates for cancer of the bronchus and lung in Connecticut were calculated by Griswald[54] according to population size of area of residence (Table 16). Among males, mortality rates for cancer of the bronchus and lung were highest in residents of metropolitan cities exceeding 50,000 population and lowest in rural towns. These differences were less consistent and of lesser magnitude among females.

Mancuso[55] calculated standardized mortality ratios (SMR) for lung cancer deaths in Ohio among white males, age 25 to 64 years, from 1947 to 1951. Area of residence was stratified into three groups: eight metropolitan counties (cities exceeding 100,000 population), seven urban counties

**Table 14**
**Age Standardized Lung Cancer Death Rates per 100,000**
**Males, Age 54–74, 1952–1954**

| | Area of Residence | | |
|---|---|---|---|
| Smoking Category | Rural | Mixed | Urban Liverpool |
| Nonsmokers | 14 | 0 | 131 |
| Pipe smokers | 41 | 25 | 143 |
| Cigarette smokers | | | |
| Light | 87 | 153 | 297 |
| Medium | 183 | 132 | 287 |
| Heavy | 363 | 303 | 394 |
| Number of deaths | 68 | 118 | 539 |
| Avg. BaP ng/m³ | 7 | 27 | 77 |

**Table 15**
**Lung Cancer Mortality in Northern Ireland, 1960–1962: Rate per 100,000**

| | Area of Residence | | | | |
|---|---|---|---|---|---|
| Smoking Habit | Inner Belfast | Outer Belfast | Other Urban Districts | Small Towns | Rural Districts |
| Nonsmoker | 36 | 40 | 21* | — | 10* |
| Pipe & cigar only | 73 | 62 | 57 | 13* | 28 |
| 1–10 cigarettes/day | 138 | 140 | 121 | 71* | 25 |
| 11–22 cigarettes/day | 288 | 207 | 171 | 260* | 74 |
| 23+ cigarettes/day | 509 | 430 | 515 | 772 | 173 |
| Total number of deaths | 226 | 102 | 174 | 24 | 162 |

*Less than 10 deaths.

(cities between 50,000 and 100,000 population), and 73 nonurban counties (Table 17). Zimmer[56] reported data on urban-rural respiratory system cancer rates from cases diagnosed in Iowa during 1950 (Table 18). Urban areas were defined as towns with populations of 2,500 and above. The authors noted that urban-rural differences were widespread throughout Iowa and were not concentrated in a few selected counties. Respiratory system cancer rates for rural segments of metropolitan counties were not different than rates for rural areas in remote nonmetropolitan counties.

Levin compared age-adjusted lung cancer death rates in urban and rural areas of New York State, exclusive of New York City, subdivided into metropolitan counties (Standard Metropolitan Statistical Areas) and nonmetropolitan counties (Table 19). Among males, incidence rates in urban areas exceeded those in rural, and rates in metropolitan counties exceeded those in nonmetropolitan within urban and rural categories. The urban-rural ratios among females were not consistent.

Hoffman[57] studied the urban-rural difference in lung cancer death rates for the entire United States (Table 20). Lung cancer deaths in 1948–

**Table 16**
**Connecticut: Age Adjusted Cancer Incidence Rate/100,000**

| | Males | | Females | |
|---|---|---|---|---|
| Area of Residence | 1941–46 | 1947–51 | 1941–46 | 1947–51 |
| Metropolitan Areas | | | | |
| Cities 50,000+ | 21.7 | 31.2 | 4.1 | 4.7 |
| Other towns | 13.2 | 23.1 | 3.2 | 4.5 |
| Nonmetropolitan Areas | | | | |
| Urban towns* | 14.8 | 28.3 | 2.9 | 5.1 |
| Rural towns† | 12.9 | 17.8 | 2.9 | 3.8 |

*Urban towns: 2500+ population and density of 130+ persons/sq mi.
†Rural towns: <2500 population or density <130 persons/sq mi.

**Table 17**
**Ohio: SMR for Lung Cancer, White Males, Age 25–64, 1947–1951**

| Area of Residence | SMR | Urban-Rural Ratio |
|---|---|---|
| Metropolitan counties (a) | 123 | a/c = 1.8 |
| Other urban counties (b) | 82 | b/c = 1.2 |
| Nonurban counties (c) | 69 | |

**Table 18**
**Iowa: Respiratory System Cancer Rate Per 100,000, 1950**

| | Urban | Rural | U/R |
|---|---|---|---|
| Male | 32.8 | 12.1 | 2.7 |
| Female | 9.0 | 6.3 | 1.4 |

1949 and data from the population census of 1950 were utilized. Urban populations were defined as persons living in places of 2,500+ population.

Lung cancer mortality rates occurring in U.S. whites during 1949–1951 were categorized by Prindle[58] according to residence within the central county (having a city of 50,000+ population) of a Standard Metropolitan Statistical Area (SMSA), other counties within SMSAs, and non-SMSA counties (Table 21). Among the 163 SMSAs in the U.S. in 1950, lung cancer death rates of males increased proportionally to the size of the central city. Mortality ratios, where a value of 1.00 represents the mortality experience from lung cancer for the entire U.S. population, ranged from a low of 1.10 in SMSAs with central cities of 50,000 to 100,000 population to a high of 1.89 in SMSAs with central cities of 1 million or larger.

**Table 19**
**New York State: Age Adjusted Lung Cancer Incidence Rates Per 100,000 1949–1951**

| | Metropolitan Counties | Nonmetropolitan Counties |
|---|---|---|
| Males | | |
| Urban* | 29.2 | 20.8 |
| Rural | 23.9 | 15.2 |
| U/R ratio | 1.2 | 1.4 |
| Females | | |
| Urban* | 3.2 | 3.2 |
| Rural | 3.5 | 2.4 |
| U/R ratio | 0.9 | 1.3 |

*Urban: Incorporated places of 2,500+ population.

**Table 20**
**U.S. Age Adjusted Lung Cancer Death Rates/100,000, 1948–1949**

| | White | | | Nonwhite | | |
|---|---|---|---|---|---|---|
| | Urban | Rural | U/R | Urban | Rural | U/R |
| Males, all ages | 22.3 | 12.3 | 1.8 | 16.9 | 7.3 | 2.3 |
| 45–64 | 62.3 | 34.9 | 1.8 | 49.8 | 22.1 | 2.3 |
| 65–69 | 110.9 | 55.7 | 2.0 | 58.7 | 27.1 | 2.2 |
| Females, all ages | 4.7 | 3.7 | 1.3 | 4.2 | 2.1 | 2.0 |
| 45–64 | 10.3 | 7.9 | 1.3 | 10.7 | 5.3 | 2.0 |
| 65–69 | 22.8 | 18.5 | 1.2 | 14.3 | 8.9 | 1.6 |

**Table 21**
**U.S. Lung Cancer Mortality/100,000 Whites, 1949–1951**

| Residence | Males | Females |
|---|---|---|
| Central SMSA counties | 11.5 | 2.8 |
| Other SMSA counties | 9.0 | 1.4 |
| Non SMSA counties | 6.0 | 1.0 |

As might be expected from these data, lung cancer rates also increased in proportion to population density of the SMSA.

Several U.S. investigators were able to obtain information on smoking habits of the total population studied or of probability samples of the U.S. population, and were thereby able to compare urban-rural lung cancer differences adjusted for cigarette smoking habits.

Buell [59] conducted a prospective study of American Legion members who had resided in the same area of California for 10 or more years prior to the onset of the survey. Information on smoking habits was obtained directly from Legion members (Table 22). As in studies reported in Britain and Northern Ireland, larger urban-rural differences in lung cancer mortality were observed among nonsmokers than smokers. Buell stated that his data suggested an additive, not a multiplicative or synergestic, effect of smoking and urban residence on lung cancer mortality. The author also cautioned that differences in duration of smoking habits may account for some of the urban-rural differences, although no data were presented on this variable. Hammond and Horn [60] reported results of a 44-month prospective study of 187,783 U.S. white males age 50–69 years interviewed by American Cancer Society volunteers beginning in January 1952. Participants were contacted once yearly. A total of 11,870 men died during the 44 months of follow-up, 335 with well-established cases of bronchogenic carcinoma, exclusive of adenocarcinoma of the lung. These 335 cases were categorized according to area of residence (Table 23). The resi-

**Table 22**
**California American Legion Members, Lung Cancer Deaths Per 100,000**

<table>
<tr><th rowspan="2">Area of Residence</th><th rowspan="2">Age-Smoking Standardized</th><th rowspan="2">Age Standardized Nonsmokers, Only</th><th colspan="3">Age Standardized, Smokers</th></tr>
<tr><th>1/2 pack per day</th><th>1 pack per day</th><th>1 pack per day</th></tr>
<tr><td>A. Los Angeles</td><td>96</td><td>28</td><td rowspan="2">77</td><td rowspan="2">135</td><td rowspan="2">226</td></tr>
<tr><td>B. San Francisco Bay</td><td>106</td><td>44</td></tr>
<tr><td>C. All other counties</td><td>80</td><td>11</td><td>61</td><td>125</td><td>138</td></tr>
<tr><td>A&B/C</td><td>1.2–1.3</td><td>2.6–4</td><td>1.3</td><td>1.1</td><td>1.6</td></tr>
</table>

**Table 23**
**Hammond and Horn's Prospective Study**
**Lung Cancer Mortality/100,000 White Males Age 50–69**

<table>
<tr><th rowspan="2">Area of Residence</th><th rowspan="2">Age and Smoking Standardized</th><th colspan="2">Age Standardized</th></tr>
<tr><th>Nonsmokers</th><th>Cigarette Smokers</th></tr>
<tr><td>Rural</td><td>39</td><td>0</td><td>65.2</td></tr>
<tr><td>Towns <10,000</td><td>43</td><td>4.7</td><td>71.7</td></tr>
<tr><td>Towns 10,000–50,000</td><td>44</td><td>9.3</td><td>70.9</td></tr>
<tr><td>Cities 50,000+</td><td>52</td><td>14.7</td><td>85.2</td></tr>
</table>

dence gradient for lung cancer was again strongest among nonsmokers; urban residence appeared to be additive to smoking habits with respect to lung cancer rates in urban smokers.

Haenszel[61] calculated lung cancer mortality rates for U.S. white males ages 35 and older from a 10% sample of all lung cancer deaths in the U.S. Data on population at risk were obtained from a probability sample of U.S. households obtained under the Current Population Survey for 1958. Residence and smoking histories were obtained from a total of 2,191 lung cancer deaths and 25,000 households. The overall lung cancer rate per 100,000 in this population of white males, age 35 years and older, was 78.2. Lung cancer standardized mortality ratios (SMR) adjusted for age and smoking were computed for various residence categories. As shown in Table 24, the SMR urban-rural ratio increased with duration of residence in urban and rural areas. Lifetime residents were further categorized by smoking habits, and age adjusted SMRs were contrasted in lifetime urban and rural residents and in lifetime residents from communities of increasing urbanization (Table 25). Both the overall urban-rural gradient and the gradient with increasing urbanization are larger for nonsmokers than for cigarette smokers — findings that have been described consistently in most studies in which these variables were reported.

The effect of migration from communities of one size to another was also analyzed in Haenszel's report. Lung cancer SMRs were computed for persons with only one exposure residence, defined as residence for 10 or more years in the same population-size category. SMRs were also derived for persons with two exposure residences — that is, residence for 10 or more years in each of two categories of population-size. In Table 26, data for individuals having only one exposure residence appears within the parallel diagonals. Data for individuals who moved from a smaller to a larger place are shown in vertical rows to the right of the diagonals, and for migrants from larger to smaller places in vertical rows to the left of the diagonals. Sample sizes precluded adjusting for smoking habits, but more detailed analysis by Haenszel consistently indicated that adjustment for smoking history made no change in the broad pattern of results shown in

**Table 24**
**Lung Cancer SMR*, Age and Smoking Adjusted**
**U.S. White Males, Age 35+, 1958**

| | Duration of Residence (yrs.) | | | | |
|---|---|---|---|---|---|
| | 1 | 1–9 | 10–39 | 40+ | Lifetime |
| Urban | 166 | 107 | 117 | 117 | 100 |
| Rural | 154 | 88 | 83 | 75 | 50 |
| U/R | 1.08 | 1.22 | 1.41 | 1.56 | 2.0 |

*SMR = 100 for U.S. white males age 35+, 1958.

Table 26. In general, movement from smaller to larger places resulted in higher lung cancer risks, while movement to smaller places brought reductions in SMRs. The gradient from farm to large cities in lung cancer risks among persons with one exposure residence, as shown by data within diagonals, was less steep than the gradient in lung cancer risks for migrants from smaller to larger places, as shown by data arranged vertically to the right of the diagonals. These results suggest that migrants from small to large places experience an increased lung cancer risk that combines the adverse influences of urbanization and of migration. Unlike results from studies of foreign migrants, these data reveal that migrants from low cancer risk to higher risk areas in the same country assume a lung cancer mortality

**Table 25**
**Lung Cancer SMR*, Age Adjusted: Lifetime in Area of Birth**
**U.S. White Males, Age 35+, 1958**

| Area of Residence | Never Smoked | Exsmoker or Pipe† | Regular Cigarette Smoker <1 pack/d | Regular Cigarette Smoker >1 pack/d |
|---|---|---|---|---|
| Urban | 16‡ | 30 | 110 | 533 |
| Rural | 5‡ | 12 | 66 | 287 |
| U/R | 3.2 | 2.5 | 1.8 | 1.9 |
| Metropolitan Counties | | | | |
| Cities 10,000+ | 18‡ | 30 | 115 | 537 |
| Cities 10,000 | 14‡ | 29‡ | 62‡ | 356 |
| Nonmetropolitan Counties | | | | |
| Cities 2,500+ | 9‡ | 22‡ | 78 | 353 |
| Rural nonfarm and farm | 4‡ | 11‡ | 72 | 329 |

*SMR = 100 for U.S. white males, age 35+.
†Exsmoker or pipe: includes occasional and exsmokers, pipe and cigar smokers.
‡Less than 10 deaths in this category.

**Table 26**
**Lung Cancer SMRs* for Persons With Only One or With Only Two Exposure Residences**
**U.S. White Males, Age 35+, 1958**

| Second Exposure Residence \ One Exposure Residence | First Exposure Residence 500,000+ | 50,000–500,000 | 10,000–50,000 | 2,500–10,000 | Rural Nonfarm | Farm |
|---|---|---|---|---|---|---|
| 500,000+ | 133 | 136 | 133 | 120 | 167 | 169 |
| 50,000–500,000 | 121 | 119 | 98 | 123 | 108 | 156 |
| 10,000–50,000 | 145 | 118 | 102 | 72 | 89 | 119 |
| 2,500–10,000 | 69† | 73† | 71† | 94 | 48† | 101 |
| Rural nonfarm | 101 | 96 | 79 | 30† | 71 | 87 |
| Farm | 27† | 68† | 109† | 49† | 86 | 40 |

*SMR = 100 for U.S. white males, age 35+.
†Less than 10 deaths in this category.

ratio in excess of the high risk area. On the other hand, migrants from large to smaller places generally take on lung cancer ratios intermediate to risks in the larger and smaller place, and these results are similar to data from studies of foreign migrants.

Haenszel also analyzed the effect of migration between communities of various sizes on standardized lung cancer ratios in non-cigarette smokers and regular smokers of cigarettes (Table 27). These data reveal an adverse effect of migration on nonsmokers, independent of the size of the community of current residence. Conversely, lung cancer risks were lower among the less mobile nonsmoking population. Table 27 also shows high lung cancer risks for migrants from farm to metropolitan counties, when combined with regular smoking of cigarettes. Both nonsmokers and regular cigarette smokers tend to take on the lung cancer risk found in long-term inhabitants of their current residence.

As in the other studies reviewed here, Haenszel's data demonstrate the strong effect of cigarette smoking on lung cancer risks, and a lesser

**Table 27**
**Lung Cancer SMR* According to Birthplace and Current Residence**
**U.S. White Males, Age 35+, 1958**

| | Not Regular Cigarette Smoker | Regular Cigarette Smoker |
|---|---|---|
| I. Birthplace: farm | | |
| Current residence: | | |
| 1. Farm | 11 | 111 |
| 2. Non SMSA, rural nonfarm | 43 | 177 |
| 3. Non SMSA, urban | 46 | 240 |
| 4. SMSA, <50,000 | 36 | 343 |
| 5. SMSA, 50,000+ | 32 | 337 |
| II. Birthplace: rural nonfarm | | |
| Current residence: | | |
| 1. Farm | 36† | 150 |
| 2. Rural nonfarm | 11† | 130 |
| 3. Non SMSA, urban | 21† | 125 |
| 4. SMSA <50,000 | 19 | 139 |
| 5. SMSA, 50,000+ | 34 | 238 |
| III. Birthplace: towns 2,500–50,000 | | |
| Current residence: | | |
| 1. Farm and rural nonfarm | 34 | 117 |
| 2. Towns 2,500–50,000 | 22 | 163 |
| 3. SMSA, 50,000+ | 28 | 200 |
| IV. Birthplace: SMSA, 50,000+ | | |
| Current residence: | | |
| 1. Non SMSA | 10† | 164 |
| 2. SMSA, <50,000 | 12† | 184 |
| 3. SMSA, 50,000+ | 34 | 198 |

*SMR = 100 for U.S. white males, age 35+.
†Less than 10 deaths in this category.

**Table 28**
**Age Adjusted Lung Cancer Death Rates per 100,000 Nonsmokers U.S. White Males, Age 35+, 1958 and U.S. White Females, Age 35+, 1958–1959**

| Lifetime Residence | Male | Female |
|---|---|---|
| Urban areas | 12.5 | 8.4 |
| Rural areas | 3.9 | 5.0 |
| Metropolitan counties | 14.1 | 6.6 |
| Nonmetropolitan counties | 3.9 | 7.2 |

effect of urbanization. Important information is also provided on the effect of residence duration and migration between communities of different sizes. Age and smoking adjusted lung cancer risks decrease with longer duration in rural areas. Migration itself is associated with increased lung cancer risks, while migration from farm to metropolitan areas when combined with cigarette smoking yields the highest lung cancer risk. Conversely, migration from metropolitan to smaller places appears to lessen the lung cancer risk for smokers and nonsmokers alike.

A subsequent analysis of lung cancer rates in white females age 35 and over during 1958–1959[62] yielded essentially the same overall pattern of results as for males. The residence effect on lung cancer risk was apparent but was less than the effect of smoking, and farm-born females migrating to metropolitan areas showed the highest lung cancer risk of all categories. Male and female nonsmokers who were lifetime residents of communities of various sizes had absolute lung cancer rates that were numerically not very different between the two sexes, as shown in Table 28. Lowest lung cancer rates were found in lifetime rural nonsmokers, the rate per 100,000 being 3.9 for males and 5.0 for females. Dean[53] found higher rates — 10 per 100,000 — for male nonsmokers in Northern Ireland, ages 35 and older, and, in reviewing other studies, reported a small variation of 8 to 12 lung cancer deaths per 100,000 in nonsmoking native-born white South Africans aged 45–64 and in nonsmoking rural English physicians ages 35 and older. Thus the lung cancer risk among nonsmoking white U.S. rural inhabitants appears to be one-half the risk of their counterparts in Northern Ireland, England, and South Africa.

## AIR POLLUTION STUDIES

Very few investigators have obtained direct measurements of air pollution exposures and related these measurements to geographic differences in lung cancer rates. Difficulties in obtaining estimates of long-term air pollution exposures, especially for benzo(a)pyrene, largely account for this deficit. Until recently in the U.S., few regular measurements of atmospheric BaP were made. High population mobility has also rendered difficult estimates on long-term exposures, particularly for the large population sizes required to detect lung cancer rates as low as 5 per 100,000 in adult nonsmokers. The long latent period between exposure and development of lung cancer compounds the problem of associating measurements of ambient concentrations with lung cancer risks.

Stocks[63] computed correlation coefficients among (a) standardized mortality ratios for lung cancer in 53 county boroughs throughout England and Wales and (b) average amount of deposited matter in summer and winter at 207 sites in the 53 boroughs, and also (c) population density. Similar analyses were made for 74 administrative areas in Lancashire and West Riding. The results are given in Table 29. Associations were not consistent among the two study groups.

Winklestein[64] and Zeidberg[65] found no relationship between air pollution levels and lung cancer death rates of residents of census tracts stratified according to particulate concentrations or sulfation rates in Buffalo and Nashville, respectively. Likewise, Boucot[66] found no evidence for an association between sulfation rates and lung cancer incidence in 6,136 Philadelphia men followed prospectively with semi-annual chest roetgenograms for 10 years from 1951 to 1961. A total of only 162 lung cancer cases developed in this population of males age 45 years and older.

Menck[67] found higher lung cancer rates occurring in 1968–1969 among white males living in a contiguous group of highly industrialized census tracts within the south central area of Los Angeles than in the remainder of the city. No similar clustering was observed for mortality rates from other cigarette-smoking associated cancers, including larynx, esophagus, and bladder, or for other cancers including stomach, colon, pancreas, prostate, and hematopoietic system. Age adjusted male lung cancer rates in the contiguous industrialized geographic areas ranged from 70 to 75 per 100,000, while rates in other geographic areas of Los Angeles ranged from 43 to 63 per 100,000. These differences persisted when lung cancer mortality rates in census tracts with the three highest of five categories of socioeconomic status were analyzed or when data for census tracts in the lowest socioeconomic categories were considered. Higher

**Table 29**
**Correlation Coefficients Among Lung Cancer, Deposited Matter and Population Density**

| Variables | 58 County Boroughs | 79 Administrative Districts |
|---|---|---|
| Lung cancer and deposited matter | 0.51 | 0.326 |
| Lung cancer and population density | 0.65 | 0.625 |
| Lung cancer and deposited matter, Population density held constant | 0.50<br>$P < .002$ | 0.182<br>$P > .05$ |

benzo(a)pyrene concentrations were measured during 1971 and 1972 in the air and soil at a monitoring station situated in the center of the high lung cancer area than at three other stations in Los Angeles located to the north and west of the above area. The authors suggested that the BaP excess in the high lung cancer area resulted from effluents of petroleum and chemical industries concentrated in the area. A preliminary review of death certificates from members of the union serving the petrochemical industry did not reveal an excess of lung cancer. The authors suggested that the lung cancer excess might be explained by a synergistic action between smoking and neighborhood air pollution, although no data on smoking habits of decedents was available.

Prindle[58] compared lung cancer mortality rates in the U.S. white male population of 163 SMSAs during 1949–1951 with an industrialization index for the same SMSAs. The geographic distribution of lung cancer was dissimilar to that of industrialization. In particular, several metropolitan areas along the southeastern and Gulf coasts showed high lung cancer rates that failed to correspond with the industrialization index.

Carnow and Meier[68] performed additional regression analyses of data reported by Stocks[69] relating lung cancer mortality in 19 countries with national cigarette consumption and solid fuel consumption figures. Regression calculations gave results that indicated a 5.4% increase in lung cancer death rate per cigarette per day (comparing differences in per capita cigarette consumption among the 19 countries) and a 20% increase in lung cancer rate per metric ton of coal burned per year per capita. The authors suggest that the products of solid fuel consumption, or some variable highly related with such consumption, may be an important etiologic factor in lung cancer differences between various countries.

Carnow and Meier[70] performed regression analyses of lung cancer death rates in the 48 contiguous states of the U.S. as a function of cigarette sales per person over 15 years of age (1963 data) and benzo(a)pyrene concentrations measured quarterly in 1968 or 1969 in various urban and nonurban places of the U.S. Each state was divided into three categories of population: urban areas whose 1960 population exceeded 1 million, other urban areas, and rural areas. Benzo(a)pyrene measurements at several sites in each state were then averaged in each of the three population categories for a given state, and a weighted average, using the corresponding populations as weights, of the three averages was obtained to represent the average BaP exposure for that state. The regression of lung cancer rates on cigarette sales and weighted BaP average yielded results showing that a 1 ng/m$^3$ increase in benzo(a)pyrene across the 48 states was associated with a 5% increase in lung cancer death rates among white males, and a 15% increase among nonwhite males. Results for females were more variable and mostly nonsignificant. Acceptance of the quantitative relationship between increases in male lung cancer rates and in BaP concentrations across the 48 U.S. states is contingent on the validity of the weighted averaging

**Table 30**
**Osaka, Japan**
**Male Lung Cancer Death Rate Per 100,000**

| Cigarette Smoking Category | Pollution Level | | |
|---|---|---|---|
| | Low | Intermediate | High |
| Nonsmoker | 11.5* | 3.8* | 4.9* |
| Exsmoker | 26.2 | 42.6* | 61.7* |
| 1–14 | 10.6* | 14.2 | 23.5 |
| 15–24 | 14.7 | 19.1 | 27.0 |
| 25+ | 36.3 | 15.8* | 46.4* |
| Levels of Specific Air Pollutants | | | |
| Dustfall (tons/km²/mo) | 8.0 | 9.2 | 12.8 |
| Particulate Matter (mg/m³) | 0.19 | 0.22 | 0.39 |
| Max. BaP (ng/m³) | 26.0 | 31.0 | 79.0 |

*Less than 10 deaths in this category.

method to obtain a single BaP exposure estimate for the inhabitants of each state. The likelihood is very great that differences in BaP exposure between areas within the same state are considerably larger than differences between states. A single weighted BaP exposure estimate for all residents of a state appears to be a highly artificial construct, derived principally for ease of analysis but excessively simplying intra-city and intra-state differences in exposure. It is difficult to believe that a single exposure estimate for all residents of an entire state can be a valid measure of differences in air pollution exposure among the 48 states.

Lung cancer cases reported over a 7-year period in an area near Osaka, Japan were distributed by Hitosugi[71] according to smoking habits and residence in low, intermediate, or high pollution areas. Families of 259 lung cancer cases, and a random sample of 4,500 adults age 35 to 74 years were interviewed to obtain data on relevant factors. Results for males stratified by smoking habits and air pollution exposure reveal higher lung cancer rates in smokers, but not in nonsmokers, associated with higher levels of pollution (Table 30.) The number of cases was very small in most smoking-air pollution categories. Relationships between maximum BaP concentrations and long-term exposures may be quite inconstant. Whether the stated air pollution measurements can reasonably represent long-term differences in exposure to atmospheric pollution cannot be judged from the information presented.

## DISCUSSION AND CONCLUSIONS

Differences in smoking habits of men and women and of successive birth cohorts may well account for most, if not all, of the time trends and age and sex patterns in lung cancer rates. None of the data presented in this review is inconsistent with the evidence for an extremely dominant causal role of cigarette smoking in lung cancer risk. At the same time, cigarette smoking habits cannot account for all of the observed geographic and population differences in rates of lung cancer. The urban-rural lung cancer difference is almost as consistent, though not nearly as strong, as the contrast in lung cancer rates between smokers and nonsmokers. An urban excess in lung cancer persists when death rates or incidence rates for individuals of any given smoking category are considered. Lung cancer rates of communities distributed by population size yield a dose response curve of lesser slope but similar to that for smoking intensity. Change in residence from smaller to larger places is associated with higher lung cancer risks among migrants. Migration itself is apparently accompanied by an increase in lung cancer risks, although migrants from large metropolitan areas to small town or rural environments end up with lower lung cancer rates than their counterparts who remain in populous places.

Differences in diagnostic practices or medical treatment cannot reasonably explain the consistent lung cancer gradient from small to large population areas. Over the past 10 to 20 years, small town and rural inhabitants have generally had ready access to urban-level medical care in this country.

The pattern of lung cancer in foreign immigrants suggests a lasting effect of early environment on lung cancer risk. The immigrant does not experience the lung cancer rate prevalent in his country of settlement but tends to assume a risk intermediate to that of his country of birth and country of adoption. Here again, smoking habits cannot account for these observed findings. Strangely, migrants within a country, from small to large places, assume a lung cancer risk equal to or greater than that found in lifetime residents of larger places. The difference in lung cancer risk of foreign and native migrants cannot be readily explained. A change in environment from one country to another would appear to be a greater stress than a change from rural to urban living within the same country. It is possible that the ardor of foreign migration is more selective of hardy biological stock.

Can atmospheric pollution as indexed by benzo(a)pyrene account for the urban-rural differences in lung cancer risk? I believe that we have no substantial evidence to support this hypothesis. Between 1945 and 1965 there was a considerable decrease in urban emissions of benzo(a)pyrene, due primarily to conversion from coal fired furnaces to oil, natural gas, and electric sources of domestic space heating. The average benzo(a)pyrene concentration in the air of U.S. cities decreased from 1950 to 1969 by a factor of nearly three.[72] Over this same period, lung cancer rates in the U.S. increased by more than two-fold in males and nearly two-fold in females. Thus lung cancer rates have doubled while ambient levels of benzo(a)pyrene have been reduced three-fold. If BaP air pollution causes or contributes to lung cancer risk, it would be difficult to establish a relationship in the face of these strongly opposite time trends. Furthermore, geographic variations in lung cancer rates have not been correlated with atmospheric levels of sulfur oxides or particulate matter.[63–66]

Few investigators have attempted to make geographic comparisons of lung cancer rates and benzo(a)pyrene concentrations. This deficiency is largely due to a scarcity of BaP measurements in a sufficient number of cities. The effort of Carnow and Meier[70] to aggregate BaP data from scattered sites in each state into one air pollution exposure index for that state resulted in a highly artificial exposure estimate that, when analyzed by multiple regression techniques, yielded an association between state-wide BaP exposure and state-wide lung cancer rates. These results are unconvincing primarily because of the quality of the exposure estimates themselves.

Menck and associates[67] found above average lung cancer rates in a contiguous group of highly industrialized census tracts in Los Angeles. However, BaP measurements were extremely scanty in this report, and no information was obtained concerning smoking habits or migratory experience of the decedents. These results clearly require replication with additional data on lung cancer mortality in Los Angeles. Similar analyses of exposure differences within other large metropolitan areas might also verify these results.

Analyses of geographic patterns of mortality in the United States, by county or among urban areas, need to be undertaken with appropriate adjustments for population density. In Prindle's analyses[58] of lung cancer rates among 163 SMSAs in 1949–1951, a discrepancy between index of industrialization and lung cancer mortality was found. In particular, high lung cancer rates occurred in Gulf and southeastern Atlantic coast metropolitan areas, where the index of industrialization was not elevated. This regional localization of high lung cancer mortality should be verified and further characterized.

In considering the causes of urban-rural differences in lung cancer, we should give greater attention to indoor and occupational exposures of lung cancer victims. Evidence for excess cancer associated with several industrial chemicals has been accumulating. Asbestos, chloromethyl-methyl ether, coke oven gases, anesthetics, and propellants in aerosol cans have been variously implicated as carcinogens. Exposures to these materials should be evaluated in urban-rural comparisons. The challenge to replicate the major prospective studies of smoking-lung cancer relationships, this time with emphasis on other environmental carcinogens, ought to be accepted by the epidemiological community. Until these studies yield the quality of results that has been provided in investigations of cigarette smoking, we are unlikely to come forward with a convincing explanation for the urban excess in lung cancer.

## SUMMARY

Lung cancer rates in the U.S. have increased over time in the same age groups and in successive birth cohorts reaching the same age. Much of this increase can be explained by changes in cigarette smoking habits. However, a gradient of increasing lung cancer associated with increasing population size of area of residence persists even when cigarette smoking habits are taken into account. Migration from small to larger places is also associated with higher lung cancer risks, and foreign immigrants experience lung cancer rates intermediate to that of their country of origin and country of settlement. While levels of benzo(a)pyrene (BaP), which is frequently used as an index of atmospheric carcinogens, are higher in urban areas, few studies of the quantitative relationship between BaP and lung cancer rates have been conducted. BaP levels in urban areas have declined by a factor of three since 1950, while lung cancer rates have increased by two-fold over the same time. More complete measurements of indoor, occupational, and external ambient exposures to carcinogens are required to explain the environmental factors that contribute to the urban excess in lung cancer. At present, the available body of evidence does not convincingly support a relationship between exposure to benzo(a)pyrene and lung cancer in the urban environment.

## REFERENCES

1. Burbank F: Patterns in Cancer Mortality in the United States: 1950–1967. *Natl Cancer Inst Monogr* 33: 199–204, 1971

2. Higgins ITT: Trends in Respiratory Cancer Mortality in the United States and in England and Wales. *Arch Environ Health* 28:121–129, 1974.

3. Levin ML, Haenszel W, Carroll BE, et al: II Cancer Incidence in Urban & Rural Areas of New York State. *J Natl Cancer Inst* 24:1243–1257, 1960.

4. Wynder EL, Covey LS, Mabuchi K: Lung Cancer in Women: Present and Future Trends. *J Natl Cancer Inst* 51:391–401, 1973.

5. *Smoking and Health: Report of the Advisory Committee to the Surgeon General of the Public Health Service.* US Dept of Health, Education and Welfare. Public Health Service, Publication 1103. Washington, DC, US Government Printing Office, 1964, 387 pp.

6. *The Health Consequences of Smoking, a Public Health Service Review:* 1967. Public Health Service Publication 1696, revised 1968, US Dept of Health, Education and Welfare, Washington, DC, US Government Printing Office, 1968, 277 pp.

7. *The Health Consequences of Smoking: 1969 Supplement to the 1967 Public Health Service Review*. US Dept of Health, Education and Welfare, Washington, DC, US Government Printing Office, 1969, 98 pp.

8. *The Health Consequences of Smoking, a Report to the Surgeon General:* 1971. Public Health Service, DHEW Publication No. (HSM) 71–7513, US Dept of Health, Education and Welfare. Washington, DC, US Government Printing Office, 1971. 458 pp.

9. *The Health Consequences of Smoking; A Report of the Surgeon General: 1972*. Public Health Services, DHEW Publication No. (HSM) 72–7516. US Dept of Health, Education and Welfare. Washington, DC, US Government Printing Office, 1972. 158 pp.

10. Hammond EC, Horn D: The Relationship Between Human Smoking Habits and Death Rate. *JAMA* 155:1316–1328, 1954.

11. Murray JL, Axtell LM: Impact of Cancer: Years of Life Lost Due to Cancer Mortality. *J Natl Cancer Inst* 52:3–7, 1974.

12. Miller JA: Carcinogenesis by Chemicals: an Overview. *Cancer Res* 30:559–576, March 1970.

13. Heidelberger C: Current Trends in Chemical Carcinogenesis. *Fed Proc* 32:2154–2161, Dec 1973.

14. Heidelberger C: Studies on the Cellular and Molecular Mechanisms of Hydrocarbon Carcinogenesis. *Eur J Cancer* 6:161–172, 1970.

15. National Academy of Sciences, Committee on Biologic Effects of Air Pollution: *Particulate Polycyclic Organic Matter*. Washington, DC, 1972. 361 pp.

16. National Academy of Sciences, *op cit,* p 2.

17. National Academy of Sciences, *op cit,* Chapters 8–12, pp 95–150.

18. National Academy of Sciences, *op cit,* p 112.

19. Saffiotti V: Experimental Respiratory Tract Carcinogenesis and its Relation to Inhalation Exposures. In *Inhalation Carcinogenesis*. MG Hanna Jr, P Nettesheim, JR Gilbert (eds). US Atomic Energy Commission, 1970. pp 27–51.

20. Kuschner M: The J Burns Amberson Lecture. The Causes of Lung Cancer. *Am Rev Resp Dis* 98:573–590, 1968.

21. Laskin S, Kuschner M, Drew RT: Studies in Pulmonary Carcinogenesis. In *Inhalation Carcinogenesis, op cit,* pp 321–350.

22. Shabad LM: Dose Response Studies in Experimentally Induced Lung Tumors. *Environ Res* 4:305–315, 1971.

23. National Academy of Sciences, *op cit,* pp 122, 123.

24. National Academy of Sciences, *op cit,* pp 138–141.

25. National Academy of Sciences, *op cit,* pp 106, 107.

26. National Academy of Sciences, *op cit,* pp 13–35.

27. Hangebrauck RP, von Lehmden DJ, Meeker JE: *Sources of Polynuclear Aromatic Hydrocarbons in the Atmosphere.* US Dept of Health, Education and Welfare. National Air Pollution Control Administration Publication No. 999–AP–33, 1967.

28. National Academy of Sciences, *op cit,* p 25.

29. National Academy of Sciences, *op cit,* p 28.

30. National Academy of Sciences, *op cit,* pp 15–31.

31. Hoffman D, Wynder EL: Selective Reduction of the Tumorigenicity of Tobacco Smoke. Experimental Approaches. *Natl Cancer Inst Monogr* 28:151–172, 1968.

32. National Academy of Sciences, *op cit,* p 29.

33. Sawicki E: Airborne Carcinogens and Allied Compounds. *Arch Environ Health* 14:46–53, 1967.

34. National Academy of Sciences, *op cit,* pp 47, 48.

35. National Academy of Sciences, *op cit,* pp 49, 209.

36. Waller RE: The Benzpyrene Content of Town Air. *Br J Cancer* 6:8–21, 1952.

37. Blacklock J, Kennaway E, Lewis GM, et al: The Carbon Content of the Human Lungs and Bronchial Glands. *Br J Cancer* 8:40–55, 1954.

38. National Academy of Sciences, *op cit,* p 48.

39. Sawicki E: Airborne Carcinogens and Allied Compounds. *Arch Environ Health* 14:46–53, 1967.

40. Colucci JM, Begeman CR: The Automotive Contribution to Airborne Polynuclear Aromatic Hydrocarbons in Detroit. *J Air Pollut Control Assoc* 15:113–122, 1965.

41. National Academy of Sciences, *op cit,* p 17, Table 3–2.

42. National Academy of Sciences, *op cit,* p 52.

43. National Academy of Sciences, *op cit,* p 52 and reference 193.

44. National Academy of Sciences, *op cit,* Chapters 4 and 5, pp 36–81.

45. Eastcott DF: The Epidemiology of Lung Cancer in New Zealand. *Lancet* 1:37–39, 1956.

46. Dean G: Lung Cancer Among White South Africans. *Br Med J* 2:852–857, 1959.

47. Dean G: Lung Cancer in South Africans and British Immigrants. *Proc Roy Soc Med* 57:984–987, 1964.

48. Mancuso TF, Coulter EJ: Cancer Mortality Among Native White, Foreign-born White, and Nonwhite Male Residents of Ohio: Cancer of the Lung, Larynx, Bladder and Central Nervous System. *J Natl Cancer Inst* 20:79–105, 1958.

49. Haenszel W: Cancer Mortality Among the Foreign-born in the United States. *J Natl Cancer Inst.* 26:37–132, 1961.

50. Reid DC, Cornfield J, Markush RE, et al: Studies of Disease Among Migrants and Native Population in Great Britain, Norway, and the United States. III. Prevalence of cardio-respiratory symptoms among migrants and native-born in the United States. *Natl Cancer Inst Monogr* 19:321–346, 1966.

51. Curwen MP, Kennaway EL, Kennaway NM: The Incidence of Cancer of the Lung and Larynx in Urban and Rural Districts. *Br J Cancer* 8:181–198, 1954.

52. Stocks P, Campbell MJ: Lung Cancer Death Rates Among Non-smokers and Pipe and Cigarette Smokers, and Evaluation in Relation to Air Pollution by Benzpyrene and Other Substances. *Br Med J* 2:923–939, 1955.

53. Dean G: Lung Cancer and Bronchitis in Northern Ireland, 1960–1962. *Br Med J* 1:1506–1514, 1966.

54. Griswold MH, Wilder CS, Culter SJ: *Cancer in Connecticut, 1935–1951*. Hartford, Connecticut, State Dept of Health, 1955, 141 pp.

55. Mancuso TF, MacFarlane EM, Porterfield JD: Distribution of Cancer Mortality in Ohio. *Am J Public Health* 45:58–70, 1955.

56. Zimmer EG, Haenszel W: *Cancer in Iowa*. Public Health Service Publication No. 466. Washington, DC, US Govt. Printing Office, 1956.

57. Hoffman EF, Gilliam AG: Lung Cancer Mortality. Geographic distribution in the United States for 1948–1949. *Public Health Rep* 69:1033–1042, 1954.

58. Prindle RA: Some Considerations in the Interpretation of Air Pollution Health Effects Data. *J Air Pollut Contr Assoc* 9:12–19, 1959.

59. Buell P: Relative Impact of Smoking and Air Pollution on Lung Cancer. *Arch Environ Health* 15:291–297, 1967.

60. Hammond EC, Horn D: Smoking and Death Rates — Report on 44 Months of Follow-up of 187,783 Men. I. Total mortality. II. Death rates by cause. *JAMA* 166:1159–1172, 1294–1308, 1958.

61. Haenszel W, Loveland DB, Sirken MG: Lung Cancer Mortality as Related to Residence and Smoking Histories. I. White Males. *J Natl Cancer Inst* 28:947–1001, 1962.

62. Haenszel W, Loveland DB, Sirken MG: Lung Cancer Mortality as Related to Residence and Smoking Histories. II. White Females. *J Natl Cancer Inst* 32:803–838, 1964.

63. Stocks P: Cancer and Bronchitis Mortality in Relation to Atmospheric Deposit and Smoke. *Br Med J* 1:74–79, 1959.

64. Winkelstein W, Kantor S, Davis EW, et al: The Relationship of Air Pollution and Socioeconomic Status to Total Mortality and Selected Respiratory System Mortality in Men. I. Suspended particulates. *Arch Environ Health* 14:162–171, 1962.

65. Zeidberg L, Horton RJM, Landau E: The Nashville Air Pollution Study. V. Mortality From Diseases of the Respiratory System in Relation to Air Pollution. *Arch Environ Health* 15:214–224, 1967.

66. Boucot KR, Weiss W, Seidman J: The Philadelphia Pulmonary Neoplasm Project: Basic Risks Factors of Lung Cancer in Old Men. *Am J Epidemiol* 95:4–16, 1972.

67. Menck HR, Casagrande JT, Henderson BE: Industrial Air Pollution: Possible Effect on Lung Cancer. *Science* 183:210–211, 1974.

68. Carnow BW, Meier P: National Academy of Sciences, *op cit,* pp 222–224.

69. Stocks P: Recent Epidemiological Studies of Lung Cancer Mortality, Cigarette Smoking and Air Pollution, With Discussion of a New Hypothesis of Causation. *Br J Cancer* 20:595–623, 1966.

70. Carnow BW, Meier P: National Academy of Sciences, *op cit,* pp 224–226.

71. Hitosugi M: Epidemiological Study of Lung Cancer With Special Reference to the Effect of Air Pollution and Smoking Habits. *Inst Public Health Bull* 17:237–256, 1968. Cited in National Academy of Sciences, *op cit,* pp 229–233, reference 379.

72. National Academy of Sciences, *op cit,* p 47.

# 2 Metaplasias Induced by Benzo(a)pyrene in Human Respiratory Epithelium

T.T. Crocker, M.D.
E.R. Dirksen, Ph.D.

## INTRODUCTION

The purpose of this presentation is to describe findings from *in vitro* exposure of human tracheobronchial tissues to chemical carcinogens and to relate such findings to the larger question of *in vitro* assessment of carcinogenic risk for humans exposed to chemical carcinogens in the environment.

The data for *in vitro* studies of the action of carcinogens on human tracheobronchial mucosa have been derived from organ cultures. The originator of many of the methods was Dr. Ilse Lasnitzki of Strangeways Research Laboratory, Cambridge, England, who adapted methods developed in the 1930s by Dame Honor B. Fell, F.R.S., Emeritus Director of that laboratory. A small number of investigators have continued to work with organ cultures of animal and human respiratory tissue exposed to known or suspected carcinogens. The accumulated data will be reviewed for both animal and human tissue because the vital place of this method in cancer research is in comparing the response of animal and human tracheobronchial tissues to the same carcinogens under identical conditions. The object of such comparative study is to determine whether animal and human respiratory structures respond alike under comparable conditions of exposure to carcinogens *in vitro*. The organ culture system can also be used to test whether the animal tracheas and bronchi acquire the same lesions *in vitro* as appear *in vivo* during experimental lung carcinogenesis. By this means an equation can be developed to relate findings in cultures of human tissue to *in vitro* findings in animal tissues and to *in vivo* changes in animal tracheas and bronchi during experimental carcinogenesis.

## METHODS

Human tracheobronchial tissue for study in organ culture has been obtained by sterile autopsy of abortuses[1] and by dissection from lobes of lung removed surgically.[2] Biopsies of adult human bronchial mucosa have also been placed in organ culture, but these tissues are usually too damaged to survive in culture.[2] Tracheas from suckling rats,[3,4] suckling and adult hamsters, monkeys, and dogs[5] have been prepared in the same way as human tracheobronchial tissues. In most cases, the tubular organ is opened and cut into nearly flat plaques representing the full thickness of the wall of the organ. The plaques are laid, mucosa side upward, on a support or on the bottom of a culture dish.

Medium is added to moisten but not immerse the organ pieces. Oxygen is supplied at high concentrations, usually 45% to 50%,[5,6] because epithelial differentiation is better sustained from day 1 onward by high concentrations than by ambient levels. Media for support of organ cultures have included rich but undefined nutrients, as clotted plasma and chick embryo extract,[3] or defined nutrients free of serum.[6] Defined media that contain no vitamin A allow production of vitamin A deficient states with associated squamous metaplasia as a means to compare epithelial abnormalities due to avitaminosis A with those produced in response to carcinogens.[6]

| IN VITRO | Animal Tissue | Human Tissue | Type of Lesion |
|---|---|---|---|
| short-term cultures | lesion present | lesion present | squamous metaplasia<br>dysplasia<br>basement memb. breaks |
| IN VIVO | | | |
| short-term carcinogen exposure | lesions, as above | not possible to test | as above |
| carcinogenic exposure | lesions, as above; carcinoma | lesions, as above carcinoma | naturally occurring or experimental carcinoma |

## RESULTS: CONTROL TISSUES IN CULTURE

Full thickness segments of the walls of tracheas or bronchi can be maintained with a normal state of preservation of connective tissues, cartilage, smooth muscle, and epithelia for 15 or more days. Mucous secretion from goblet cells is retained and cilia are normal in appearance. Ciliary beating usually stops after 5 to 10 days but has been retained for as long as 4 months.[7] Multiple criteria of epithelial normality have been satisfied, including replicative DNA synthesis at rates comparable to respiratory epithelia *in vivo*[8] histochemical evidence of mucous differentiation and retention of fine structural details such as normal basement membranes, cellular organelles, intercellular junctions and replication of ciliary basal bodies.[1, 4]

## RESULTS: CARCINOGEN-TREATED TISSUES IN CULTURE

Against the background of adequate maintenance of normal tracheobronchial structures, the production of epithelial metaplasia and dysplasia is convincing evidence that the carcinogen enters the tissue by diffusion from the nutrient medium and that proximate carcinogens produce (or that protocarcinogens are converted metabolically to active compounds that induce) epithelial lesions comparable to the presumptively preneoplastic dysplasias and metaplasias found in association with bronchogenic carcinoma in man and in experimental animals (Table 1).[1,3–6] Many cellular characteristics described for human lung cancer in Table 1 have been produced in organ cultures of suckling rat trachea,[4] in adult hamster trachea, and in human fetal trachea.[1] The epithelial dysplasias require 7 to 10 days to appear and are associated with increased replicative activity. The concentration of benzo(a)pyrene required to produce epithelial metaplasias in human or rodent respiratory epithelia is about 10$\mu$g/ml.[1,4]

The epithelial lesions consist of a layer of cells four or more cells high without evidence of mucous or ciliary differentiation. Cells are rounded and suggest a preponderance of basal cells or are flattened to give a "typical squamous metaplasia" with keratin formation or to produce an "atypical squamous metaplasia" or dysplasia in which keratin is not produced. The flattened cells of both kinds of squamous metaplasia have deep interdigitations, frequent desmosomal connections, reduced rough endoplasmic reticulum, and many free ribosomes.[1,4]

## BASEMENT MEMBRANE DEFECTS

Basement membranes at the epithelio-mesenchymal junctions in human fetal tracheas are readily distinguished in control organ cultures but

**Table 1**
**Cellular Abnormalities of Epithelial Metaplasia in Organ Culture of Human and Hamster Trachea and Squamous Cell Carcinoma of Man**

| | Squamous Cell Carcinoma of the Bronchus in Man* | Epithelial Metaplasia Induced by Benzo(a)pyrene: Ferric Oxide in Hamsters† | Epithelial Metaplasia Induced by Benzo(a)pyrene in Human Tracheal Cultures‡ |
|---|---|---|---|
| Intercellular relationships and attachments | | | |
| Cellular morphology | Polygonal, interdigitations | Polygonal, interdigitations | Columnar and polygonal |
| Intercellular space | Widened | Widened | Widened |
| Desmosomes (at basement membrane) | Decreased | Decreased, abnormal | Decreased |
| Nucleus | | | |
| Nucleus/cytoplasm | Increased$^{+++}$§ | Increased$^{++}$ | Increased$^{+}$ |
| Shape | Polylobulated | Polylobulated | Irregular |
| Nucleoli | Enlarged, pleomorphic | Enlarged, pleomorphic | Enlarged, segregated |
| Cytoplasm | | | |
| Tonofilaments | Increased$^{+++}$, dispersed, and perinuclear bundles | Increased$^{+++}$, dispersed, and perinuclear bundles | Increased$^{+}$ |
| Lysosomes | Increased$^{++}$ | Increased$^{++}$ | Increased$^{+}$ |
| Mitochondria | Decreased$^{+++}$, abnormal | Decreased$^{+++}$, abnormal | Decreased$^{+}$, abnormal |
| Endoplasmic reticulum | Increased smooth$^{++}$<br>Decreased rough$^{+++}$ | Increased smooth$^{++}$<br>Decreased rough$^{++}$ | Increased smooth$^{+}$<br>Decreased rough$^{+}$ |
| Free ribosomes | Increased | Increased | Increased |
| Other organelles | Internalized cilia, rare | Filamentous granules | Autophagic vacuoles |

*Refs. 17, 18, 19, 20.
†Ref. 16.
‡Ref. 1.
§Gradation of change: $^{+++}$, greatest; $^{++}$, moderate; $^{+}$, least.

become discontinuous and adjacent epithelial cell processes extend as microinvasions into the connective tissues.[1] This defect is present at the same time as epithelial metaplasia and accompanies the cellular pleomorphism described above and in Table 1.[1]

Some states of squamous metaplasia induced by benzo(a)pyrene resemble, by light microscopy, the state induced by vitamin A deficiency. Electron microscopy reveals differences between the "typical" and the "atypical" squamous metaplasias associated with exposure to carcinogens and the squamous metaplasia of avitaminosis A.[9] In the latter state, nuclei and nucleoli are normal and basement membranes are intact (Table 2). Basement membrane defects occur in hamsters exposed to benzo(a)pyrene: ferric oxide on a regimen that leads to squamous carcinoma of trachea and bronchi.[10] The defect is noted in tissues sectioned one day after 10 intratracheal applications made in a total period of three weeks; hence, it is observed many weeks before overt invasion occurs. The basement membrane defects observed *in vivo* in hamsters are the same as those observed after 14 days of incubation of human fetal trachea with benzo(a)pyrene. Basement membrane defects and microinvasion of the connective tissue have

**Table 2**
**Histogenesis of Squamous Metaplasia***

| | Exposure to Carcinogen | | Vitamin A Deficiency |
|---|---|---|---|
| A. | Increase in lysosomes, dilation of endoplasmic reticulum, and abnormal nucleoli in ciliated and mucous cells | A. | Flattening of ciliated and mucous cells |
| B. | Abnormal basal cells<br>1. Pleomorphic nucleoli<br>2. Nucleolar dispersal | B. | Normal basal cells |
| C. | Basal cell hyperplasia | C. | Basal cell hyperplasia |
| D. | Sloughing of altered ciliated and mucous cells | D. | Sloughing of altered ciliated and mucous cells |
| E. | Basal cells differentiate into abnormal squamous metaplastic cells<br>1. Regular squamous metaplasia<br>a. Cornification<br>b. Keratohyaline granules<br>c. Abnormal nucleoli<br>d. Defects in basement membrane<br>2. Atypical squamous metaplasia<br>a. Enlarged nuclei indented by cytoplasmic invaginations<br>b. Nucleolar plaques and microsegregation<br>c. Defects in basement membrane<br>d. Widened intercellular spaces and abnormal cellular junctions | E. | Basal cells differentiate into normal squamous metaplastic cells<br>1. Cornification<br>2. Keratohyaline granules<br>3. Normal nucleoli<br>4. Normal basement membrane |

*Adapted from Harris et al.[9]

been observed in premalignant lesions and malignant neoplasms of skin,[11] mammary gland,[12,13] and larynx.[14] The defects are not specific to neoplasia, however, since they occur in psoriasis.[15]

## NUCLEAR ABNORMALITIES

The most characteristic abnormality of neoplastic cells in squamous carcinoma, and in metaplasias or dysplasias resulting from exposure to carcinogens *in vivo* and *in vitro,* is irregularity or multilobulation of enlarged nuclei and segregation of nucleoli into filamentous and granular components.[1,4,9,16] These are identified in Tables 1 and 2 and are similar to the nuclear abnormalities of human squamous cell carcinoma of the bronchus.[17–20]

Incubation of tracheas with tritiated benzo(a)pyrene has been followed by electron microscopic radioautography. Label has been observed in thin sections over heterochromatin rather than over echromatin. Harris, Kaufman, Sporn, Boren, et al.[21] have interpreted this localization as potentially associated with an increase in DNA-dependent RNA polymerase activity. According to a model proposed by Frenster,[22] agents that enhance DNA-dependent RNA polymerase activity would become associated preferentially with heterochromatin. Such agents include mercuric chloride,[23] phytohemagglutinin,[23] and perhaps hydrocortisone.[24] Evidence for increase in the polymerase has not been reported. Binding of labeled benzo(a)pyrene to DNA purified from carcinogen treated hamster respiratory epithelial cells has been reported by Kaufman et al.,[25] but binding to heterochromatin has not been confirmed directly by using DNA extracted from exposed tissue. The significance of localization over heterochromatin of label from tritiated benzo(a)pyrene arises from the conclusion that such labeling required conversion of the inactive compound to a metabolite that can bind to DNA in nuclei of the epithelial cells that undergo morphologic alteration during chemical carcinogenesis.

## CONCLUSIONS AND INTERPRETATION

The degree of similarity between cellular and histologic lesions in animals and in organ cultures supports the interpretation that the organ culture system provides a valid estimate of the interaction between carcinogen and tissue. Based on this interpretation, it is reasonable to propose that the human fetal trachea in organ culture undergoes responses that would occur in living subjects, should these be exposed to benzo(a)pyrene and other carcinogens. Data from fetal tissue cannot be applied directly to adult tissue, as experiments in human adult bronchial mucosa have demonstrated.[26] Thus it is only permissible to interpret these data as indicating that lesions that resemble, in cellular details, human and experimental animal squamous carcinoma of the respiratory epithelium can be produced at comparable concentrations and exposure times in human fetal and animal fetal and adult tracheas in culture. These conclusions suggest that human respiratory epithelium is as sensitive to the action of some chemical carcinogens as animal epithelium and studies in animals of the mechanism of induction of lesions and of methods for inhibition[27] or reversal[6] of lesions in animals will have relevance to human lung cancer.

## REFERENCES


1. Dirksen ER, Crocker TT: Fine Structural Changes Produced by Benzo(a)pyrene on Human Respiratory Epithelium in Organ Culture. *In 30th Annual Proc Electron Microscopy Soc Amer,* Los Angeles, Ca, 1972, Arceneaux CJ (ed).

2. O'Donnell TV, Crocker TT, Nunes LL: Maintenance of Normal, Metaplastic and Dysplastic States of Adult Human Bronchial Mucosa in Organ Culture. *Cancer Res* 33:78–87, Jan 1973.

3. Crocker TT, Nielsen BI, Lasnitzki E: Carcinogenic Hydrocarbons: Effect on Suckling Rat Trachea in Organ Culture. *Arch Environ Health* 10:240, 1965.

4. Dirksen ER, Crocker TT: Ultrastructural Alterations Produced by Polycyclic Aromatic Hydrocarbons on Rat Tracheal Epithelium in Organ Culture. *Cancer Res.* 28:906–923, 1968.

5. Crocker TT: Effect of Benzo(a)pyrene on Hamster, Rat, Dog and Monkey Respiratory Epithelia in Organ Culture. *Inhalation Carcinogenesis.* Proceedings of Biology Div, Oak Ridge National Laboratory, Conference held in Gatlinburg, Tenn, Oct. 8–11, 1969. AEC Symposium Series 18. Hanna MG Jr, Nettesheim P, Gilbert JR, (eds), AEC Division of Technical Information, April 1970 pp. 433–443.

6. Clamon GH, Sporn MB, Smith JM, et al: $\alpha$-and $\beta$ Retinyl Acetate Reverse Metaplasias of Vitamin A Deficiency in Hamster Trachea in Organ Culture. *Nature* 250:64–66, 1974.

7. Cailleau R, Crocker TT, Wood DA: Attempted Long-term Culture of Human Bronchial Mucosa and Bronchial Neoplasms. *J Natl Cancer Inst* 22:1027–1037, 1959.

8. Crocker TT, Nielsen BI: Chemical Carcinogenesis and Respiratory Epithelium. Proceedings of the Fifth Berkeley Symposium on Mathematical Statistics and Probability, Vol. IV: Biology and Problems of Health. Le Cam LM, Neyman J., (eds.) University of California Press, Berkeley, 1967.

9. Harris CC, Kaufman DG, Sporn MB, et al: Histogenesis of Squamous Metaplasia and Squamous Cell Carcinoma of the Respiratory

Epithelium in an Animal Model. *Cancer Chemother Rep* Part 3, No. 4:43–54, 1973.

10. Saffiotti U: Experimental Respiratory Tract Carcinogenesis and Its Relation to Inhalation Exposures. *Inhalation Carcinogenesis,* Atomic Energy Commission Symposium, Ser. 18:27–54, Hanna M, Nettesheim P, Gilbert J (eds), Springfield, Va: National Technical Information Service, U.S. Department of Commerce, 1970.

11. Tarin D: Sequential Electron Microscopial Study of Experimental Mouse Skin Carcinogenesis. *Int J Cancer* 2:195-211, 1967.

12. Ozzello L, Sanpitak P: Epithelial-stromal Junction of Intraductal Carcinoma of the Breast. *Cancer* 26:1186–1198, 1970.

13. Tarin D: Fine Structure of Murine Mammary Tumors: The Relationship Between Epithelium and Connective Tissue in Neoplasms Induced by Various Agents. *Br J Cancer* 21:417–425, 1969.

14. Sugar J: A Gege Praecancerosisok Es Rakok Korszovettana. *Magy Onkol* 13:201–209, 1969.

15. Cox A: The Dermal-epidermal Junction in Psoriasis. *J Invest Dermatol* 53:428–435, 1969.

16. Harris CC, Sporn MB, Kaufman DG, et al: Acute Ultrastructural Effects of Benzo(a)pyrene and Ferric Oxide on the Hamster Tracheobronchial Epithelium. *Cancer Res* 31:1977–1989, 1971.

17. Nagaishi C, Okada Y, Diado S, et al: Electron Microscopic Observations of the Human Lung. *Cancer Exptl Med Surg* 28:177–202, 1965.

18. Obiditsch-Mayer I, Breitfellner G: Electron Microscopy in Cancer of the Lung. *Cancer* 21:945–951, 1968.

19. Razzuk M, Race G, Lynn J, et al: Observations on Ultrastructural Morphology of Bronchogenic Carcinoma. *J Thorac Cardiovasc Surg* 59:581–587, 1970.

20. Watson J, Bryant V, Brinkman G: The Ultrastructure of Epithelium of Human Bronchus in Epidermoid Carcinoma. Sixth International Congress for Electron Microscopy, pp. 759–760. Uyeda R, (ed), Tokyo, Maruzen Co Ltd, 1966.

21. Harris CC, Kaufman DG, Sporn MB, et al: Localization of Benzo(a)pyrene-$^3$H and Alterations in Nuclear Chromatin Caused by Benzo(a)pyrene-ferric Oxide in the Hamster Respiratory Epithelium. *Cancer Res* 33:2842–2848, 1973.

22. Frenster JH: Biochemistry and Molecular Biophysics of Heterochromatin and Enchromatin. *Handbook of Molecular Cytology* A Lima-de-Faria (ed), pp. 251–276, Amsterdam: North Holland Publishing Co, 1969.

23. Frenster JH: Electron Microscopic Localization of Nuclear Binding Within Human Leukocytes. *Blood* 34:847, 1969.

24. Csaba G, Kiss J, Olah J: Mechanism of Formation of Mast-cell Granules. IV. Electron Microscopic Radioautographic Studies With $^{3}$H-hydrocortisone. *Acta Biol Budapest* 21:85-90, 1970.

25. Kaufman DG, Genta VM, Harris CC, et al: The Binding of $^{3}$H-labeled Benzo(a)pyrene to DNA in Hamster Tracheal Epithelial Cells. *Cancer Res* 31:2837–2841, 1973.

26. Crocker TT, O'Donnell TV, Nunes LL: Toxicity of Benzo(a)-pyrene and Air Pollution Composite for Adult Human Bronchial Mucosa in Organ Culture. *Cancer Res* 33:88–93, 1973.

27. Crocker TT, Sanders LL: Influence of Vitamin A and 3,7-dimethyl-2,6-Octadienal (citral) on the Effects of Benzo(a)pyrene on Hamster Trachea in Organ Culture. *Cancer Res* 30:1312–1318, 1970.

# 3 Environmental Cancer Associated with Inorganic Microparticulate Air Pollution*

Irving J. Selikoff, M.D.

## INTRODUCTION

It has been known for 40 years, since polychlorinated biphenyl chloracne was found in family contacts of an exposed worker,[1] that toxic substances might leech into the community from the factory. This possibility, and the potential importance of air pollution as a route for dissemination, was later emphasized by the identification of beryllium lung disease among individuals living close by a beryllium factory.[2]

With occupational hazards now recognized to sometimes include significant cancer risk, it has been perhaps natural to inquire whether neoplastic disease also extends beyond the workplace.

*Research supported by Grant ES 00358 of the National Institute of Environmental Health Sciences and by the Environmental Cancer Research Project (Grant R-53B) of the American Cancer Society.

## CANCER HAZARD WITH OCCUPATIONAL EXPOSURE TO INORGANIC DUSTS

There is considerable interest in mineral dusts since mining, milling, smelting, or processing operations have been found to be associated with significant cancer risk with a number of such dusts, including talc,[3] hematite,[4] chromates,[5] nickel ores,[6] and a variety of asbestiform fibers, including chrysotile,[7] amosite,[8] crocidolite,[9] and anthophyllite.[10] Lung cancer has been most frequently found, as with other hazardous dusty environments (uranium mining[11]), but neoplasms have occurred at different sites as well, including bucco-pharyngeal,[12] laryngeal,[13,14] gastrointestinal,[12] and pleural and peritoneal mesotheliomas.[9,12,15]

### Asbestos

Of the inorganic dusts, cancer risk has been most extensively studied with asbestos. Investigation of the mortality experience of several cohorts of asbestos workers have shown cancer incidence to be significantly increased in this population. Thus, while variations occurred with intensity and duration of exposure, duration of observation, concomitant cigarette smoking and perhaps other factors, lung cancer was often found to be increased fivefold or more, gastrointestinal cancer often twice that expected, and mesothelioma of the pleura or peritoneum was frequently seen, sometimes accounting for 5% of all deaths.[12]

**Prolonged period of clinical latency** Where data have been available concerning the lapsed period between onset of exposure and clinical appearance of cancer, there has been the consistent finding that many years, generally decades, are required for the carcinogenic effect to be observed. Usually, the span has been greater than 20 years, and latency of 30 or 40 or more years has not been unusual. (Parenthetically, it follows that neoplasms diagnosed in recent years often had their origin in the 1930s and 1940s, during a period in which there may have been inadequate appreciation of the necessity for rigorous control of dust exposures.) The importance of prolonged observation for the evaluation of cancer risk with inorganic dust exposure can be appreciated by review of data obtained in one ongoing study in our laboratory. We have been following all members of the insulation workers' union (International Association of Heat and Frost Insulators and Asbestos Workers, AFL-CIO) in the New York metropolitan area who were registered on the Union's rolls on January 1, 1943. There were 632 men who held membership on that date.[16] Nine of these died before reaching 20 years from the date of onset of their work. The remaining 623 men have been kept under observation and their mortality experience compared with that expected, age, year, and sex being taken into account. For the first 9 years of observation (1943–1951), 73 deaths occurred, with 76.4 expected. There was only a modest increase in cancer, with 11.3 expected deaths and 28 observed. The next two decades, however, with further passage of time from onset of exposure, saw a worsening of their experience. From 1952 through 1961, 18.3 cancer deaths were expected and 57 occurred, while from 1962 through 1971, the cancer deaths were 17.6 expected and 104 observed (Table 1).

We have also followed the mortality experience of another cohort of asbestos insulation workers, composed of the entire registered membership of the Union throughout the United States and Canada on January 1, 1967.

Through December 31, 1972, 1,358 men had died although only 960.02 deaths were expected. There were 575 cancer deaths, approximately a threefold increase (Table 2). Analysis showed, however, that by far the major portion of the increase occurred among those who had reached more than 20 years from onset of their work (Table 3). Thus, of 12,681 workers with less than 20 years from onset of exposure in 1967, only 28 lung cancer deaths occurred, against 8.40 expected. In contrast, among the men with more than 20 years from onset, there were 240 lung cancer deaths, compared with 47.47 expected. As indicated in Table 4, death from bronchogenic carcinoma was uncommon in less than 15 years from onset of work.

**Fiber variety** Apart from asbestos, little is known concerning variations in carcinogenic potential that might be associated with differences in the nature of the inorganic dusts associated with neoplastic risk. The possibility that there may be such differences among the several commercial varieties of asbestos has, however, attracted considerable attention. While all types of fibers are known to increase cancer risk among exposed workers, it has been proposed that the level of risk might vary and that the types of neoplasms that could result might differ. Thus, while cancer is significantly increased among chrysotile miner and millers,[7,17] it has been proposed that this fiber variety is less likely to produce mesothelioma.[7] Some investigators hold the opinion that crocidolite has a greater propensity for the development of such neoplasms.[18] Anthophyllite exposure has been shown to sharply increase the risk of lung cancer but not of mesothelioma.[19] On the other hand, extensive laboratory studies with animals exposed to the inhalation of the various types of asbestos have failed to show any marked difference among them.[20]

**Multiple factor etiology** In 1968, data became available indicating that asbestos exposure greatly increased the lung cancer risk of cigarette smoking.[21] In 1963, 370 New York metropolitan area asbestos insulation

**Table 1***

**Expected† and Observed Number of Deaths Among 623 New York-New Jersey Asbestos Insulation Workers January 1, 1943 to December 31, 1971, 20 or More Years After Onset of First Exposure to Asbestos**

| | 1943–51 | | 1952–61 | | 1962–71 | | Total 1943–71 | |
|---|---|---|---|---|---|---|---|---|
| | Expected | Observed | Expected | Observed | Expected | Observed | Expected | Observed |
| Total cancer: all sites | 11.3 | 28 | 18.3 | 57 | 17.6 | 104 | 47.2 | 189 |
| Lung cancer | 1.5 | 13 | 3.5 | 23 | 5.1 | 48 | 10.1 | 84 |
| Pleural mesothelioma | n.a.‡ | 1 | n.a. | 2 | n.a. | 5 | n.a. | 8 |
| Peritoneal mesothelioma | n.a. | 1 | n.a. | 3 | n.a. | 20 | n.a. | 24 |
| Cancer of stomach, colon, rectum, esophagus | 4.1 | 7 | 5.0 | 18 | 3.9 | 16 | 13.0 | 41 |
| Cancer all other sites | 5.7 | 6 | 9.8 | 11 | 8.6 | 15 | 24.1 | 32 |
| Asbestosis | n.a. | 1 | n.a. | 10 | n.a. | 22 | n.a. | 33 |
| All other causes | 65.1 | 44 | 89.4 | 94 | 78.3 | 61 | 232.8 | 199 |
| Total all causes | 76.4 | 73 | 107.7 | 161 | 95.9 | 187 | 280.0 | 421 |
| Person-years of observation | 3726 | | 4406 | | 2898 | | 11,030 | |

*632 members were on the union's rolls on January 1, 1943. Nine died before reaching 20 years from first employment. All others entered these calculations upon reaching the 20-year point from onset of first exposure point.

†Expected rates are based upon age-specific death rate data of the U.S. National Office of Vital Statistics from 1949–67. Rates were extrapolated 1943–48 from rates for 1949–55, and for 1968–71 from rates for 1961–67.

‡U.S. death rates not available, but these are rare causes of death in the general population.

workers, the survivors of the 1943 cohort, were enrolled in a prospective study, with smoking habits recorded. By 1967, among the 87 men with no history of cigarette smoking, no death of lung cancer occurred (fewer than 1 was expected). On the other hand, of the 238 asbestos workers with a history of cigarette smoking, approximately 3 deaths of lung cancer were expected; but 24 occurred. It was considered that the combination of the two carcinogens had a multiplying effect and it was calculated that an asbestos worker who smoked cigarettes had approximately 92 times the risk

**Table 2**
**Expected and Observed Deaths Among 17,800 Asbestos Insulation Workers in the United States and Canada—January 1, 1967 to December 31, 1972**

| | Expected* | Observed |
|---|---|---|
| Total deaths—all causes | 960.02 | 1,358 |
| Cancer all sites | 175.55 | 575 |
| Lung cancer | 55.87 | 275 |
| Pleural mesothelioma | † | 29 |
| Peritoneal mesothelioma | † | 63 |
| G.I. cancer | 37.79 | 61 |
| All other cancer | 81.89 | 147 |
| Asbestosis | † | 101 |
| All other causes | 784.47 | 682 |
| Number of persons | 17,800 | |
| Person-years of observation | 102,864 | |

*Expected rates are based upon age-specific white male death rate data of the U.S. National Office of Vital Statistics. Rates for 1972 were extrapolated from rates for 1967–1971.
†U.S. rates are not available, but these are rare causes of death in the general population.

**Table 3**
**Expected and Observed Deaths Among 17,800 Asbestos Insulation Workers in the United States and Canada—January 1, 1967 to December 31, 1972**

| | Duration from Onset of Exposure | | | |
|---|---|---|---|---|
| | Less than 20 Years | | More than 20 Years | |
| | Expected* | Observed | Expected* | Observed |
| Total deaths—all causes | 203.90 | 249 | 756.12 | 1,109 |
| Cancer all sites | 30.42 | 64 | 145.13 | 511 |
| Lung cancer | 8.40 | 28 | 47.47 | 247 |
| Pleural mesothelioma | † | 2 | † | 27 |
| Peritoneal mesothelioma | † | 3 | † | 60 |
| G.I. cancer | 4.64 | 5 | 33.15 | 56 |
| All other cancer | 17.38 | 26 | 64.51 | 121 |
| Asbestosis | † | 7 | † | 94 |
| All other causes | 173.48 | 178 | 610.99 | 504 |
| Number of persons | 12,681 | | 5,119 | |

*Expected rates are based upon age-specific white male death rate data of the U.S. National Office of Vital Statistics. Rates for 1972 were extrapolated from rates for 1967–1971.
†U.S. rates are not available, but these are rare causes of death in the general population.

of dying of lung cancer, compared to an individual of the same age who neither smoked nor worked with asbestos.

The conclusions have now been confirmed.[22] Included in the prospective study of the 17,800 members of the Insulation Workers' Union on January 1, 1967, was opportunity for evaluation of smoking habits. On establishment of the cohort, it was ascertained that 9,590 workers had a history of regular cigarette smoking; 2,066 had no such history, 609 having smoked pipe and/or cigar only and 1,457 having no history of any type of smoking. By December 31, 1972, it was clear that lung cancer risk was increased only among those workers with a history of cigarette smoking. Among the pipe and cigar smokers, 3.11 deaths from lung cancer were expected, but only 1 occurred. Among the non-smokers, 4.40 deaths were expected; 1 was observed. On the other hand, while 31.60 deaths of lung cancer were expected among those with a history of cigarette smoking, 179 deaths of this disease occurred (Table 5).

**Environmental asbestos disease** As with investigations of the nature and extent of cancer risk following exposures to inorganic microparticulates in general, associated environmental hazards have been more extensively studied and identified with asbestos exposure. This has been facilitated by two fortuitous circumstances. First, the characteristic, well-defined morphology, structure, and chemical composition of the several types of asbestos fibers have permitted the demonstration, characterization and quantitation of asbestos fibers both in tissues[23] and in air.[24] This has permitted study of asbestos air pollution from various sources[25] and has

**Table 4**
**Deaths from Lung Cancer Among 17,800 Asbestos Insulation Workers in the U.S. and Canada—January 1, 1967 to December 31, 1972—Relation to Elapsed Period from Onset of Work Exposure**

| | Lung Cancer | | |
|---|---|---|---|
| Years from Onset | Expected Deaths | Observed Deaths | Ratio |
| <10 | 0.56 | 0 | — |
| 10–14 | 1.97 | 5 | 2.5 |
| 15–19 | 5.87 | 23 | 3.9 |
| 20–24 | 9.55 | 34 | 3.6 |
| 25–29 | 10.70 | 56 | 5.2 |
| 30–34 | 8.20 | 60 | 7.3 |
| 35–39 | 4.68 | 29 | 6.2 |
| 40–44 | 4.84 | 27 | 5.6 |
| 45–49 | 4.51 | 19 | 4.2 |
| 50+ | 4.97 | 22 | 4.4 |
| Total | 55.87 | 275 | 4.9 |

Note: Expected deaths are based upon age-specific white male death rate data of the U.S. National Office of Vital Statistics. Rates for 1972 were extrapolated from data for 1967–1971.

also encouraged quantitative (and qualitative) estimation of the asbestos burden of several tissues, especially the lung. With experience, reasonably reliable estimates can be made concerning the asbestos lung burdens typically associated with occupational exposures, and those following less intense exposure. Similarly, data are now becoming available that relate to levels usually seen with a range of occupational exposures, as well as those that obtain in environmental circumstances, such as in the neighborhood of asbestos facilities or in the ambient air of our large cities.[24] It should be noted that a variety of analytical techniques are essential for such studies, ranging from optical microscopy to electron beam techniques, and include electron microprobe analysis, selected area diffraction, and transmission electron microscopy.[23]

Using these techniques, Langer and his colleagues studied lung tissue obtained at autopsy of New York City residents.[25] He found chrysotile asbestos regularly, a result that indicated that inhalation of these fibers had occurred, albeit at levels much below those seen in occupational circumstances. Similar observations were made in London.[26]

A second finding that facilitated investigation of disease with environmental asbestos contamination has been the curious susceptibility of the mesothelial surfaces, especially the parietal pleura, to changes following the inhalation of asbestos. Although the predilection for pleural disease had previously been described with occupational exposure to asbestos,[27,28] Kiviluoto made brilliant use of the phenomenon to identify 499 cases of calcification among residents of a county in Eastern Karelia, among 6,312 people x-rayed in a tuberculosis survey. Having found none in a neighboring county in 7,012 individuals similarly studied, he observed that the first

**Table 5**
**Expected and Observed Deaths of Lung Cancer Among 17,800 U.S. and Canada Asbestos Insulation Workers—January 1, 1967 to December 31, 1972—Relation of Cigarette Smoking**

| | No. of Persons | Deaths from Lung Cancer | | |
|---|---|---|---|---|
| | | Expected* | Observed | Ratio |
| Smoking habits not known | 6,144 | 16.76 | 94 | 5.6 |
| History of cigarette smoking | 9,590 | 31.60 | 179 | 5.7 |
| No history of cigarette smoking | 2,066 | 7.51 | 2 | 0.3 |
| Never smoked | 1,457 | 4.40 | 1 | 0.2 |
| History of pipe and/or cigar only | 609 | 3.11 | 1 | 0.3 |

*Expected deaths based upon age-specific U.S. mortality rates for white males disregarding smoking. Lung cancer estimates based upon U.S. rates for cancer of lung, pleura, bronchus, and trachea, categories 162 and 163 of the International Classification of Diseases and Causes of Deaths, 7th Revision.

county was the site of an asbestos mine and mill.[29] Air samples taken up to 27 km from the mill showed the specific type of asbestos (anthophyllite) produced in the mine.[30] This was the first clear demonstration that asbestos air pollution could result in at least one of the biological effects previously associated with occupational exposure to the dust.

**Mesothelioma as an environmental asbestos cancer** Until 1960, only occasional cases of mesothelioma of the pleura or mesothelioma of the peritoneum had been reported among asbestos workers, and the relation of this neoplasm to asbestos exposure was tenuous and uncertain. This situation changed with the report by Wagner, Sleggs, and Marchand[9] of 47 cases of mesothelioma, all seen in the northwestern section of Cape Province of South Africa, an area rich in crocidolite asbestos mines and mills; in 45 of the cases investigation had shown that prior asbestos contact has been possible. Since this report, numerous other studies have confirmed and extended the remarkable association between asbestos exposure and subsequent occurrence of mesothelioma.[12,14,31–34] It is evident that mesothelioma can occur without asbestos exposure; nevertheless, when such exposure has occurred, there is concern for the possibility that mesothelioma will be found later. On the other hand, when mesothelioma is diagnosed, the possibility that asbestos exposure had occurred many years earlier is considered and investigated. The frequency of association between asbestos exposure and mesothelioma can be appreciated by review of recent reports from South Africa and Great Britain. Among 210 cases of mesothelioma in the South African Mesothelioma Registry 1955–1970, for whom adequate history was available, 85% had definite or possible exposure to asbestos.[35] Greenberg and Lloyd Davies analyzed the asbestos exposure history of 246 cases of mesothelioma entered by the Department of Employment in the Mesothelioma Registry, 1967–1968, from England, Wales, and Scotland. Appropriate histories were available for 234. Careful inquiry failed to elicit definite or possible asbestos exposure in only 38 (15%).[34]

The initial descriptions of the important association between asbestos exposure and the occurrence of mesothelioma were marked by a second observation, that such exposure need not be intense or prolonged. Thus, Wagner and his colleagues reported that approximately half their cases had contact, many years before, in environmental circumstances. Newhouse took this observation an important step further. Reviewing the 76 cases of established mesothelioma in the files of the London Hospital, she found that while 31 had had history of occupational exposure, in 9 cases exposure had occurred by virtue of residence within the household of an asbestos worker and in 11 there had been only that exposure that might have occurred with residence within a half mile of an asbestos plant in London. This report established the importance of conjugal or family contact as a vector for mesothelioma, as well as the equally significant hazard that might result, presumably from asbestos air pollution, in the immediate environs of asbestos facilities, thus extending the import of Kiviluoto's observations. Newhouse's findings have been confirmed and extended by other investigators[36,37] and have been of great value in evaluating approaches to the control of environmental asbestos disease.[38]

**Neighborhood exposure and mesothelioma** While the entity of neighborhood asbestos mesothelioma has been well established, [9,29,33,36,37] the incidence of this neoplasm among residents living about asbestos facilities is not known, owing to the difficulty of defining population bases from which the observed cases are derived. It is not known, for example, how many people lived near the London asbestos factory implicated in Newhouse's description of 11 "neighborhood" cases. Indeed, it is not known whether the 11 cases described are the only cases that occurred among those exposed.

Studies are under way in our laboratory (Hammond, E.C., Selikoff, I.J. and Churg, J., unpublished data) that seek quantitative estimates of the neoplastic risk associated with residence in the vicinity of an asbestos factory. We are tracing all 1,672 adult male residents living near an amosite asbestos factory in an eastern U.S. city, 1942–1943, and are comparing their mortality experience with that of a similar cohort of residents of the same city, in a neighborhood with no asbestos works. Results of this investigation will provide quantitative information concerning the level of risk associated with residence about an asbestos plant, at least as of 30 years from onset of exposure.

Simple geographical distribution of cases of mesothelioma, in relation to population density, is an uncertain guide to the incidence of neighborhood mesothelioma. The possibility of intimate admixture of occupational and environmental exposures provides an important constraint. In their study of the British mesothelioma cases, 1967–1968, Greenberg and Lloyd Davies noted that southwestern England, with a population of 3,652,000 had a total of 22 cases. One of its cities, Plymouth, with a population of 250,000, had 13 cases, all with histories of occupational exposure to asbestos (shipyard work being common). If it was assumed that Plymouth hospitals served a population of 500,000, the rate was calculated to be 26 cases per million per year. In the remainder of the region, with nearly 3,000,000 population, there were 9 mesotheliomas, of which only 4 had a history of occupational exposure to asbestos, for a rate of 1.5 cases per million per year. Areas with heavy asbestos-using industries may provide greater opportunity for neighborhood exposure; they also have more cases with occupational exposure, and, of the two, it is likely that occupational cases would be more numerous. Of the 234 definite mesotheliomas in the British study, there were 167 with definite occupational exposure and 16 with possible occupational exposure. Ten cases had neighborhood exposure, but two of these also had had household contact.[34]

The intensity and duration of exposure needed to cause neighborhood asbestos mesothelioma is not known. It does not seem to be very great, at least in comparison with exposures in occupational or family circumstances. W.J. Nicholson in our laboratory is investigating settled dust in houses standing during 1942–1943 in the asbestos neighborhood that we are studying, as well as the asbestos lung burden of neighborhood residents who have died and have been examined at autopsy. His initial impression is that the asbestos content of both settled dust and lung tissue is quite low. However, any definitive statement must await completion of his study.

A further reason for considering that low levels of exposure might suffice to cause neighborhood asbestos mesothelioma depends on scattered observations of mesothelioma occurring with low-level or brief exposure. Greenberg and Lloyd Davies report a case of mesothelioma whose asbestos exposure consisted of sawing up asbestos cement sheets to construct two sheds during one day. (It should be noted, however, that the day's exposure could have been quite intense!) Further, they report that "in this study the briefest occupational exposure to asbestos associated with a mesothelial tumor was 3 weeks, but if asbestos was a cause of mesothelioma, it cannot be assumed that lesser exposures are safe." Here, too, it should not be assumed that "brief exposure" is synonymous with "minimal exposure."

The distinction may not be unimportant from the point of view of control of environmental asbestos disease. Sources for intense, albeit brief, exposures may be identified and properly controlled. More pervasive, truly low-level, contamination of the ambient air from multiple sources may be more difficult to control. By the same token, however, it is not known whether this has the same pathogenic significance.[25] This critical question must still be answered.

**Conjugal disease; household contact** The same difficulty that attends neighborhood disease blurs the significance of conjugal mesothelioma (wives of asbestos workers who had cleaned, dusted, and washed their husbands' work clothes, for example) or other mesothelioma associated with residence in households of asbestos workers. Such neoplasms have been and continue to be reported, but population bases against which to judge the incidence have not been available. The likelihood that the numbers of individuals exposed in such circumstances have been much smaller than "neighborhood" groups supports the prediction that the incidence of such disease will be found to be greater than that of neighborhood mesothelioma. It is not known whether Newhouse's ratio of 9:11 family:neighborhood will represent common experience.

In the British Mesothelioma Registry series, there were 3 cases with history of family exposure to asbestos, but 2 also lived close to an asbestos factory.[34]

Current studies in our laboratory seek to provide information concerning the incidence of mesothelioma associated with family contact and household residence. We are tracing and inviting for examination relatives (wives, children, siblings, parents) who lived in the households of a group of 1,664 amosite asbestos factory workers employed for shorter or longer periods at some time between 1941 and 1954, when the plant closed. Pleural changes and parenchymal infiltration with the appearance of diffuse interstitial fibrosis are being noted. Somewhat to our surprise, such changes, reminiscent of those seen among asbestos workers, were found in 38.6% of the first 210 family contacts examined (Table 6) (Anderson, H.A., Daum, S., Lilis, R., Fischbein, A.S., Nicholson, W.J. and Selikoff, I.J., unpublished data.) The large percentage of individuals with pleural changes (diffuse pleural thickening, calcified and non-calcified plaques) is of particular concern, in view of the report by Edge that many such individuals being followed by him (the pleural changes having been presumably caused by indirect occupational "bystander" exposure in Barrow shipyards before and during World War II) have died of mesothelioma.[39] Nevertheless, with present knowledge, the temptation to extrapolate from observed disease in neighborhood or family contact circumstances should be resisted despite the fact that under the conditions of their experiment, Wagner and his colleagues were unable to detect a lower threshold level for the induction of mesothelioma by asbestos.[20] The relevance of this observation to the human situation is uncertain.

The question of disease with low-level exposure is an important one, in view of the ubiquity of asbestos contamination of ambient urban air[24] and the concomitant presence of asbestos fibers in the lungs of urban dwellers at this time.[25] Obviously, quantitation is the crux of the matter. Generally, the amount of asbestos in the ambient air is orders of magnitude less than that found in occupational circumstances and considerably less — so far as meager current data are concerned — than that found in households of asbestos workers. Similarly, the asbestos lung burden of unexposed dwellers at this time is very much less than that of an asbestos worker. Several cautions may be suggested however. First, data on asbestos air pollution and tissue burdens are comparatively scant and conclusions based upon them must necessarily be tentative. Second, in view of the long period of clinical latency between onset of exposure and evidence of disease, the significance of current findings must be evaluated with reserve since ambient air levels 20 to 40 years ago are unknown and the carcinogenic potential of asbestos recently retained in tissues will not be known for some decades. In an attempt to seek information concerning asbestos tissue levels 30 to 50 years ago, which would be more relevant to

**Table 6**
**X-ray Abnormalities* (1973–1974) Among 210 Family Members of Amosite Asbestos Workers employed 1941–1954**

| | Total Examined | Pleural Thickening Only | Pleural Calcification Only | Pleural Thickening and Pleural Calcification | Irregular Opacities Only | Irregular Opacities, Pleural Thickening and/or Pleural Calcification | Total Abnormalities |
|---|---|---|---|---|---|---|---|
| Wives | 63 | 8(12.7%) | 4(6.4%) | 12(19.1%) | 11(17.5%) | 10(16%) | 33(52.4%) |
| Daughters | 73 | 5(6.8%) | 2(2.8%) | 7(9.6%) | 5(6.8%) | 1(1.4%) | 13(17.8%) |
| Sons | 41 | 11(27%) | 1(2.4%) | 12(29.4%) | 3(7.3%) | 6(14.6%) | 21(51.2%) |
| Siblings | 20 | 1(5%) | 1(5%) | 2(10%) | 3(15%) | 2(10%) | 7(35%) |
| Other | 13 | 0 | 1(7.7%) | 1(7.7%) | 3(23%) | 3(23%) | 7(53.7%) |
| Total | 210 | 25(12%) | 9(4.3%) | 34(16%) | 25(12%) | 22(10.6%) | 81(38.6%) |

*ILO–U/C Classification

current experience, asbestos (''ferruginous'') bodies were sought in lung tissues of individuals who died in the 1930s and were examined at postmortem[40] and their presence compared with those in lungs of individuals coming to autopsy 1966–1968. No significant difference was found. However, it is known that ''ferruginous bodies'' may not reflect the asbestos content of lung tissue.[25] Therefore, current studies are designed directly to measure asbestos fibers in tissues over the past 50 years (Langer, A.M., personal communication). In an approach along the same lines, information is being sought concerning previous levels of asbestos air pollution by studying the asbestos content of arctic ice cap samples, carefully identified as to time of deposition. In this way, secular changes in ambient asbestos content will be evaluated, 1920–1970 (Langer, A.M., unpublished data).

The fact that there have been reports of increasing incidence of pleural and peritoneal mesothelioma in general, does not suffice as proof of cancer hazard associated with general community asbestos air pollution, although it does not rule it out. First, attention has been called to the possibility of background asbestos air contamination from natural sources,[38,41,42] especially that which might be due to natural erosion of outcroppings of asbestos minerals and from asbestos soils.[43]. Second, as noted, the majority of cases of mesothelioma in collected series have had indentifiable exposure under occupational, neighborhood, or family contact circumstances. Careful lifetime occupational and environmental histories are necessary before one can feel confident that well-defined exposure from known sources has not taken place.[44] This is particularly true when one considers the growing importance of ''indirect'' occupational (''bystander'') exposure. Opportunity for such exposure was, in the past, common in at least two important industries, shipbuilding and ship repair and building and construction work. Thus, among cases of mesothelioma in the British Mesothelioma Register 1967–1968, there were only 52 listed cases in ''regular'' asbestos work, of 167 mesotheliomas with definite occupational exposure to asbestos.[34] ''Bystander'' exposure has been of particular importance in shipbuilding and ship repair.[45,46,47]

**Other environmental cancer with asbestos air pollution** Although other neoplasms than mesothelioma, especially bronchogenic carcinoma, are increased in incidence after occupational exposure to asbestos, it has been difficult to determine whether or not such increased incidence, at least in part, also occurs among individuals exposed in environmental circumstances. The principal difficulty has been that while mesothelioma is so uncommon in general that its occurrence immediately attracts attention and raises the question of asbestos exposure, other asbestos-associated neoplasms are not at all uncommon even without asbestos exposure and their occurrence among individuals exposed in occupational or environmental circumstances requires comparison of the number observed with what would have been expected without asbestos dust exposure, i.e., comparison of rates of incidence among exposed and non-exposed groups. For this reason, it is not known whether bronchogenic carcinoma is increased among those exposed in neighborhood or family contact circumstances, and evaluation of the risk of gastrointestinal, pharyngeal, and other cancers suffers the same constraint. Therefore, it may be noted at this time that the only neoplasm that can be said to have been definitely identified as related to environmental exposure to asbestos is mesothelioma. If it should be established that lung cancer is only found with increased incidence in association with higher exposures,[48] then it will be unlikely that lung cancer will result from environmental exposures, since high exposure has been found unlikely in environmental circumstances[38] and risk of lung cancer will remain restricted to occupational exposure, or more intensive and prolonged neighborhood and family contact exposure.

## ENVIRONMENTAL CANCER HAZARD WITH OTHER AIRBORNE INORGANIC MICROPARTICLES

There is no convincing evidence of a recognized cancer risk associated with airborne environmental contamination by inorganic dust other than asbestos, despite the recognition of such hazard in the occupational setting. In large part, this statement should be evaluated with the knowledge that few studies have been attempted that would provide reliable information on the question, one way or the other.

### Talc

Perhaps the greatest temptation to extrapolate from known occupational risk has been associated with the industrial use of talc minerals and with the use of consumer talc products. First, increased cancer risk among talc workers has been well established. Second, many talc minerals contain asbestos fibers (chrysotile, tremolite, anthophyllite) as naturally occurring "contaminants." Environmental contamination with talc would therefore be associated with environmental dissemination of asbestos. Since it has been proposed that the neoplasms seen among talc workers, including bronchogenic carcinoma and mesothelioma, may be related to the asbestos content of talc,[3] it would seem reasonable to believe that environmental dissemination of asbestos in talc would be associated with potential environmental cancer risk. This may indeed be so, but appropriate epidemiological data that would support this hypothesis have not been reported.

### Fibrous Glass

The observation by Stanton[49] of experimental mesothelioma with several varieties of fine fibrous glass has raised the possibility that air contamination with these inorganic particles might result in environmental cancer risk. Once again, it should be pointed out that while this remains a reasonable hypothesis, such risk in humans has not been reported to date.

## CONTROL OF ENVIRONMENTAL CANCER FROM INORGANIC MICROPARTICLE AIR POLLUTION

The identification of the principal sources of inorganic microparticle air pollution carries with it the simultaneous opportunity for the establishment of appropriate control measures. Obviously, since most such pollution is derived from the wide use of these inorganic materials in industrial processes, source identification is not only feasible but likely, in view of the necessity for control of the occupational hazard in any case. However, this will have to be true for indirect (''bystander'') occupational exposure as well as that in the usually recognized mineral trades. Similarly, exhaust controls will prevent neighborhood exposures and appropriate measures are available to prevent household contamination. A problem will be the necessity to control ultramicroscopic particles.[50,51] Another problem will be the difficulty in evaluating the success of measures taken, before some time, perhaps decades,[52,53] has elapsed.

## REFERENCES

1. Schwartz L: Dermatitis From Synthetic Resins and Waxes. *Am J Public Health* 26:586–592, 1936

2. Chesner C: Chronic Pulmonary Granulomatosis in Residents of a Community Near a Beryllium Plant; Three Autopsied Cases. *Ann Intern* Med 32:1028–1048, 1950.

3. Kleinfeld M, Messite J, Zaki MH: Mortality Experiences Among Talc Workers: A Follow-up Study. *J Occup Med* 16(5)345–349, 1974.

4. Bonser GM, Faulds JS, Stewart MJ: Occupational Cancer of the Urinary Bladder in Dyestuffs Operatives and of the Lung in Asbestos Textile Workers and Iron-ore Miners. *Am J Clin Pathol* 25(2):126–134, 1955.

5. Health of Workers in Chromate Industry, Public Health Service Publication #192. Fed Security Agency, Public Health Service, Wash DC, 1953.

6a. Sunderman FW Jr: Nickel Carcinogenesis. Epidemiology of Respiratory Cancer Among Nickel Workers. *Dis Chest* 54 (6) :41–48, 1968.

6b. Pedersen E, Hogetveit A, Andersen A: Cancer of Respiratory Organs Among Workers at a Nickel Refinery in Norway. *Int'l J Cancer* 12:32–41, 1973.

7. McDonald JC, McDonald A, Gibbs GW: Mortality in the Chrysotile Asbestos Mines and Mills of Quebec. *Arch Environ Health* 22:677–686, 1971.

8. Selikoff IJ, Hammond EC, Churg J: Carcinogenicity of Amosite Asbestos. *Arch Environ Health* 25:183–186, 1972.

9. Wagner JC, Sleggs CA, Marchand P: Diffuse Pleural Mesothelioma and Asbestos Exposure in the North Western Cape Province. *Br J Ind Med* 17:260–271, 1960.

10. Kiviluoto R, Meurman M: Results of Asbestos Exposure in Finland. In: Pneumoconiosis Proc Int Conf, Johannesburg, 1969. Shapiro HA (ed), Oxford University Press, Capetown, 1970, pp. 190–196.

11. Lundin FE Jr, Lloyd JW, Smith EM: Mortality of Uranium Miners in Relation to Radiation Exposure, Hard-rock Mining and Cigarette Smoking — 1950 Through Sept 1967. *Health Phys* 16:571–578, 1969.

12. Selikoff IJ, Hammond EC, Seidman H: Cancer Risk of Insulation Workers in the United States. In: *Biological Effects of Asbestos,* Bogovski P, et al (eds), IARC Sc Pub No 8, Lyon, France, 1973, pp 209–216.

13. Stell PM, McGill T: Asbestos and Laryngeal Carcinoma. *Lancet* 2:416–417, 1973.

14. Musti M, Vimercati F: Neoplasia Laringea e Polmonare in Asbestosico. *Med Lavoro* 64:423–431, 1973.

15. Enticknap JB, Smither WJ: Peritoneal Tumours in Asbestosis. *Br J Ind Med* 21:20–30, 1964.

16. Selikoff IJ, Churg J, Hammond EC: Asbestos Exposure and Neoplasia. *JAMA* 188:22-26, 1964.

17. Kogan FM, Guselnikova NA, Gulevskaja MR: Cancer Mortality Amongst Ural Asbestos Workers. *Gigiena i Sanitarija* 37(7):29–32, 1972.

18. Wagner JC: Asbestos Cancer (guest editorial). *J Natl Cancer Inst* 46:v-ix, 1971.

19. Meurman M, Kiviluoto R, Hakama M: Mortality and Morbidity Among the Working Population of Anthophyllite Miners in Finland. *Br J Ind Med* 31:105–112, 1974.

20. Wagner JC, Berry G, Skidmore JW: The Effects of the Inhalation of Asbestos in Rats. *Br J Cancer* 29:252-269, 1974.

21. Selikoff IJ, Hammond EC, Churg J: Asbestos Exposure, Smoking and Neoplasia. *JAMA* 204(2):106–112, 1968.

22. Hammond EC, Selikoff IJ, Seidman H: Multiple Interaction Effects of Cigarette Smoking: Extrapulmonary cancer. Proc. XI Int. Cancer Congress of UICC, Florence, Italy, 1974 (To be published).

23. Langer AM, Ashley R, Baden V: Identification of Asbestos in Human Tissues. *J Occup Med* 15(3):287–295, 1973.

24. Nicholson WJ, Rohl AN, Ferrand EF: Proc Second Int Clean Air Cong. Englund HM, Beery WT (eds), Academic Press, New York, 1971, pp. 136–139.

25. Langer AM, Selikoff IJ, Sastre A: Chrysotile Asbestos in the Lungs of Persons in New York City. *Arch Environ Health* 22:348–361, 1971.

26. Pooley FD, Oldham PD, Chang-Hyun U: The Detection of Asbestos in Tissues. In: Pneumoconiosis Proc Int Conf, Johannesburg, 1969. Shapiro HA (ed), Oxford University Press, Capetown, 1970, pp. 108–116.

27. Weiss A: Pleurakrebs Bei Lungenasbestose, in Vivo Morphologisch Gesichert. *Mediz* (Stuttgart) 3:83–94, 1953.

28. Jacob G, Bohlig H: Roentgenological Complications in Pulmonary Asbestosis. *Fortschr Röntgenstr* 83:515–525, 1955.

29. Kiviluoto R: Pleural Calcification as a Roentgenologic Sign of Non-occupational Endemic Anthophyllite-asbestosis. *Acta Radiol* 194:1–67, 1960.

30. Laamenen A, Noro L, Raunio V: Observations of Atmospheric Air Pollution Caused by Asbestos *Ann NY Acad Sc* 132:240–245, 1965.

31. Hourihane DO'B: The Pathology of Mesotheliomata and an Analysis of the Association with Asbestos Exposure. *Thorax* 19:268–278, 1964.

32. Selikoff IJ, Churg J, Hammond EC: Relation Between Exposure to Asbestos and Mesothelioma. *New Eng J Med* 272:560–565, 1965.

33. Newhouse ML, Thompson H: Mesothelioma of Pleura and Peritoneum Following Exposure to Asbestos in the London Area. *Br J Ind Med* 22:261–269, 1965.

34. Greenberg M, Lloyd Davies TA: Mesothelioma Register 1967–68. *Br J Ind Med* 31:91–104, 1974.

35. Webster I: Asbestos and Malignancy. *S Afr Med J* 47(5):165–171, 1973.

36. Lieben J, Pistawka H: Mesothelioma and Asbestos Exposure. *Arch Environ Health* 14:559–563, 1967.

37. Bohlig H, Dabbert AF, Dalquen P: Epidemiology of Malignant Mesothelioma in Hamburg, A preliminary report. *Environ Res* 3:365–372, 1970.

38. Committee on Biologic Effects of Atmospheric Pollutants Airborne Asbestos NRC, National Academy of Sciences, Wash, DC 1971.

39. Edge JR: Asbestos Related Disease in Barrow-in-Furness. *Environ Res* (In press).

40. Selikoff IJ, Hammond EC: Asbestos Bodies in the New York City Population in Two Periods of Time. In: Pneumoconiosis Proc Int Conf, Johannesburg, 1969. Shapiro HA (ed), Oxford University Press, Capetown, 1970 pp.99–105.

41. Selikoff IJ, Nicholson WJ, Langer AM: Asbestos Air Pollution. *Arch Environ Health* 25:1–13, 1972.

42. Wright GW: Asbestos and Health in 1969. *Am Rev Resp Dis* 100: 467–469, 1969.

43. Burilkov T, Babadjov L: Ein Beitrag Zum Endemischen Auftreten Doppelseitiger Pleuraverkalkungen (Endemic occurrence of bilateral pleural calcifications). *Prax Pneumol* 24:433–438, 1970.

44. Selikoff IJ, Bader RA, Bader ME: Editorial: Asbestosis and Neoplasia. *Am J Med* 42(4):487–496, 1967.

45. Harries PG: Asbestos Hazards in Naval Dockyards. *Ann Occup Hyg* 11:135–145, 1968.

46. McEwen J, Finlayson A, Mair A: Mesothelioma in Scotland. *Br Med J* 4:575–578, 1970.

47. Whitwell F, Rawcliff RM: Diffuse Malignant Pleural Mesothelioma and Asbestos Exposure. *Thorax* 26:6–22, 1971.

48. Gilson JC: Asbestos Cancer; Past and Future Hazards. *Proc Roy Soc Med* 66:395–403, 1973.

49. Stanton MF, Wrench C: Mechanisms of Mesothelioma Induction With Asbestos and Fibrous Glass. *J Natl Cancer Inst* 48:797–821, 1972.

50. Natusch DFS, Wallace JR: Urban Aerosol Toxicity: The Influence of Particle Size. *Science* 186 (4165):695–699, 1974.

51. Langer AM: Inorganic Particles in Human Tissues and Their Association With Neoplastic Disease. Conf, Biol Effects of Ingested Asbestos, November 1973, *Environ Hlth Perspec* (In press).

52. Selikoff IJ: Recent Perspectives in Occupational Cancer, *Ambio* (In press).

53. Selikoff IJ, Hammond EC: Environmental Cancer in the Year 2000. Proc 7th National Cancer Conf, Los Angeles, 1973 pp. 687–696.

# 4 Carcinogenicity of Airborne Particulates*

P. Shubik, D.M., D.Phil.

Compounds classified as chemical carcinogens occur commonly in analyses of air, primarily from urban environments. Although a wide range of polycyclic aromatic hydrocarbons occur in polluted atmospheres, the current weight of evidence is against an association with bronchogenic carcinoma in man. This is in contradistinction to the association between cigarette smoking and lung cancer, which is well established and where, once again, a range of polycyclic and heterocyclic hydrocarbons occurs. The widespread assumption that the cigarette tars are of prime importance in human lung carcinogenicity would implicate these polycyclic and heterocyclic compounds. There is experimental evidence to be discussed that two of the compounds present in cigarette tars, dibenzo(a,i)pyrene and 7-H-dibenzo(c,g)carbazole, are singularly carcinogenic to hamster lung as compared with other related compounds. It seems possible that minor qualitative differences might create very different human hazards. Again, it is known that in the hamster lung test system certain polycyclic hydrocarbon carcinogens manifest overt carcinogenicity only when administered with

*Abstract of paper presented at the Conference.

certain particulates. It is possible that similar associations may be of considerable importance in determining carcinogenicity from different sources. It is clearly essential that we be in possession of chemical and animal test methods that enable us to predict carcinogenicity to man. By comparing situations that are currently known to involve varied levels of hazard it is possible to establish confidence levels for extrapolation. These problems and others concerned with the bioassay of potential airborne carcinogens will be discussed.

# 5 Epidemiology of Lung Cancer Based on Population Studies

Takeshi Hirayama, M.D.

## DESCRIPTIVE EPIDEMIOLOGY

Up until a few years ago, the number of deaths in Japan from pulmonary tuberculosis and lung cancer had been on a par, but in 1972 the number of deaths from lung cancer was on the upswing.[1] The death toll of pulmonary tuberculosis registered in 1947 (121,912) — 159 times more than that of cancer cases (768) — has decreased year by year. In 1972 the number of tuberculosis deaths (11,968) was lower than lung cancer deaths (12,968). This change indicates that respiratory diseases have now entered a new era.

The drastic reduction in the death toll from tuberculosis can be attributed to various anti-tuberculosis measures that were established in an effort to come to grips with the causes of this dreadful disease. The question here is whether similar countermeasures against lung cancer are being carried out today.

It is imperative, in the first place, to carry out epidemiological research on the actual condition of lung cancer and its causes. As shown in Table 1, deaths from lung cancer in Japan have been on a steady increase since after World War II, with the total in 1972 showing 16 times as many as that in 1947. If the current pace continues, the death rate due to lung cancer is expected to catch up soon with that of cancer of the stomach.

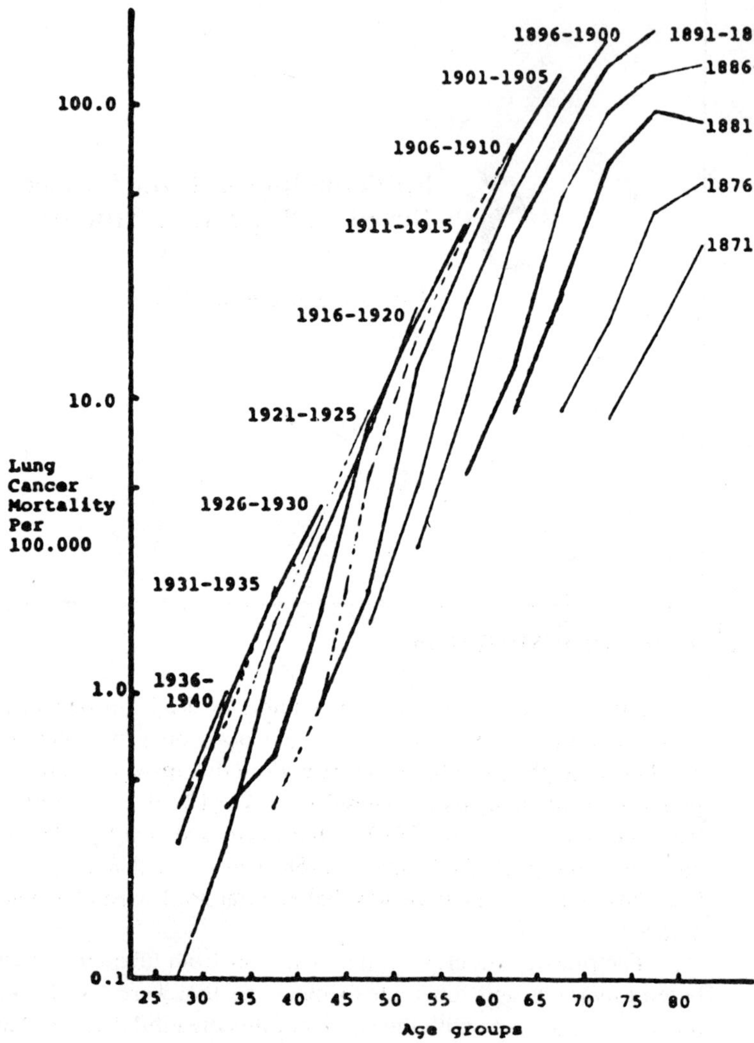


**Figure 1.** Cohort mortality rate for lung cancer in Japan.

Cohort mortality rate of lung cancer in Japan shows a great similarity to that of the U.S. (Figures 1 and 2).

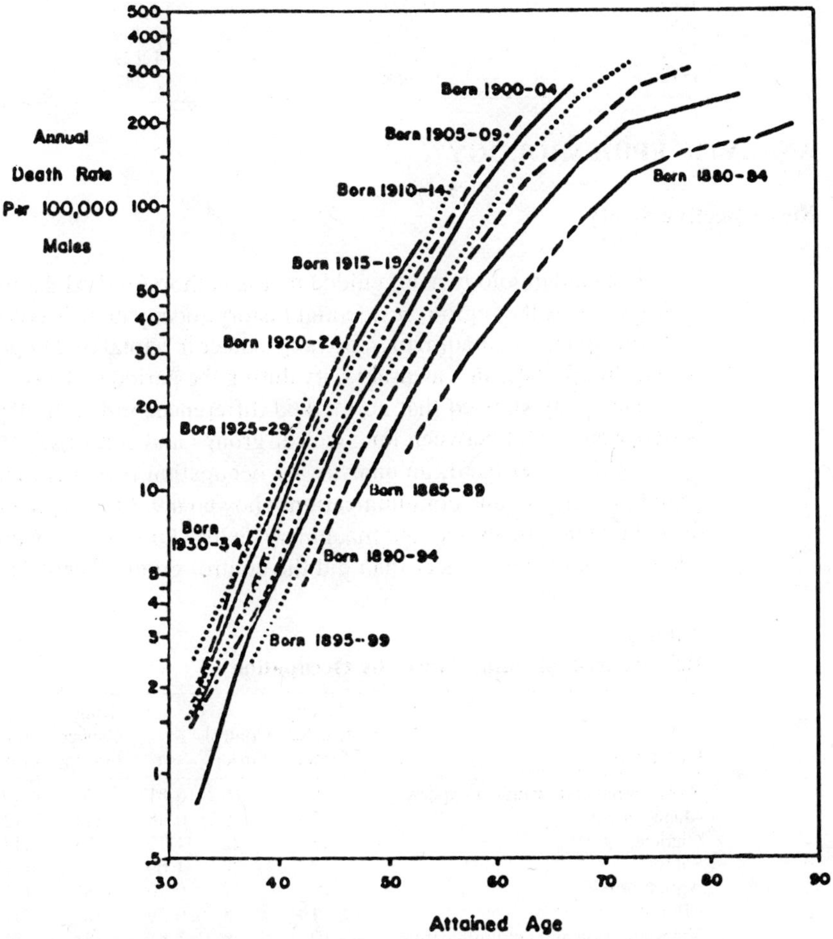


**Figure 2.** White male lung cancer death rates for year-of-birth cohorts by attained age U.S. 1934–1969. Source: U.S. National Center for Health Statistics Dept. of Epidemiology and Statistics, American Cancer Society, 8–73

**Table 1**
**Number of Deaths from Respiratory Tuberculosis and Lung Cancer, 1947 to 1972, Japan**

| Year | Respiratory Tuberculosis | Lung Cancer | Ratio |
|---|---|---|---|
| 1947 | 121,912 | 768 | 158.74 |
| 1950 | 101,355 | 1,119 | 90.58 |
| 1955 | 40,115 | 2,711 | 14.80 |
| 1960 | 29,176 | 5,171 | 5.64 |
| 1965 | 21,033 | 7,725 | 2.72 |
| 1970 | 15,132 | 10,489 | 1.44 |
| 1971 | 12,946 | 11,129 | 1.16 |
| 1972 | 11,968 | 12,287 | 0.97 |

## ANALYTIC EPIDEMIOLOGY

### Retrospective Study

The epidemiology team guided by the authors studied the influence of such factors as the region, residential history, occupation, smoking habits, and atmospheric contamination on lung cancer in a total of 239 persons who died of lung cancer in Kawasaki City during the period of 1964–1970.[2]

The study showed that no marked differences in length of residential period were found between lung cancer groups and controls.

On the other hand, an analysis by occupation revealed that those engaged in transport and communications showed a 9.3 times greater risk and those in metal processing and machinery fields showed a 12.8 times higher relative risk of lung cancer than did the control group (Table 2).

**Table 2**
**Relative Risk of Lung Cancer by Occupation**

| Occupation | Lung Cancer Group | Control Group | r.r | Lung Cancer Group | Control Group | r.r |
|---|---|---|---|---|---|---|
| Professional and technical experts | 9 | 16 | 0.81 | (5) | (7) | (0.83) |
| Management | 2 | 6 | 0.48 | (1) | (2) | (0.58) |
| Clerical | 18 | 22 | 1.18 | (7) | (11) | (0.74) |
| Sales | 28 | 24 | 1.69 | (15) | (12) | (1.63) |
| Agriculture | 18 | 37 | 0.70 | (18) | (27) | (0.73) |
| Mining | 1 | 5 | 0.29 | (0) | (2) | (—) |
| Transportation and communication | 10 | 8 | 1.81 | (8) | (1) | (9.30) |
| Factory workers | 62 | 106 | 0.85 | (25) | (46) | (0.63) |
| Service industry | 8 | 6 | 1.93 | (5) | (2) | (2.91) |
| Unclassified | 4 | 2 | 2.89 | (2) | (0) | (—) |
| Unknown | 10 | 12 | 1.21 | (0) | (0) | (—) |
| Metal processing Machine assembly workers | 15 | 10 | 1.50 | (11) | (1) | (12.79) |

(—) those in the occupation over 25 years.

In relation to smoking, an excess percentage was noted in the lung cancer group, both in male and female cigarette smokers. Male smokers showed a 2.5 times higher relative risk of lung cancer than non-smokers; those smoking more than 30 cigarettes a day, a 6.6 times higher risk. If atmospheric contamination is added to cigarette smoking, the risk of lung cancer increases to about 10 times (Figure 3).

A check on the regional occurrence of the disease in the whole city showed that there were 10 places of significantly higher rates of occurrence. When regional patterns of the degree of atmospheric contamination were examined, there proved to be a higher concentration of cases of epidermoid carcinoma and undifferentiated carcinoma patients in the highly contaminated areas, while the lesser contaminated areas showed an aggregation of adenocarcinoma patients.

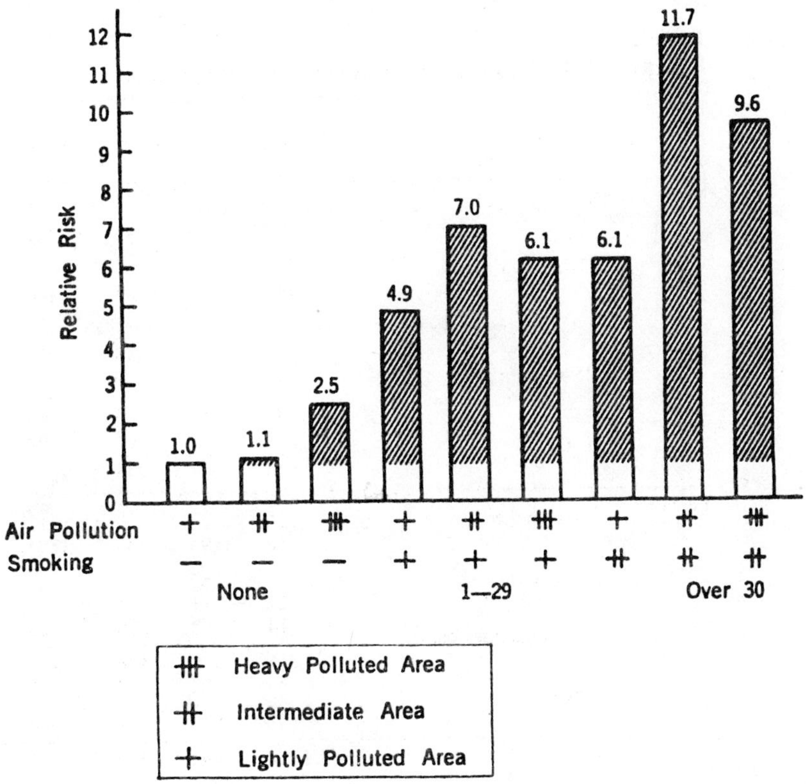


**Figure 3.** Lung cancer relative risk by extent of air pollution and number of cigarettes per day.

## Population Cross-Sectional Study

With regard to relations between air pollution and lung cancer, surveys in Amagasaki and Nishinomiya cities, Hyogo Prefecture, showed results similar to the one conducted by Kawasaki City, Kanagawa Prefecture — that air pollution, if coupled with smoking, increases the risk of getting lung cancer (Table 3).[3]

It is fair to say, therefore, that smokers living in air polluted areas and smokers with certain types of occupational exposures, in particular, belong to high-risk groups. For these reasons, prime efforts should be directed at smoking in a lung cancer prevention drive.

## Population Prospective Study

Smoking showed up as the most important cause of lung cancer when a survey was conducted on about 265,000 people by the National Cancer

**Table 3**
**Effect of Air Pollution on the Frequency of Lung Cancer**

| | Levels of Air Pollution | + | ++ | +++ | Total | $X^2$ for Linear Regression |
|---|---|---|---|---|---|---|
| Male | Non-smoker | 5/118 | 1/69 | 1/55 | 7/242 | 1.0777 |
| | Ex-smoker | 11/114 | 7/45 | 7/31 | 25/190 | 3.8542* |
| | 1–24 cigarettes per day | 27/563 | 27/433 | 31/336 | 85/1332 | 6.6482* |
| | 25 or more cigarettes per day | 19/143 | 4/69 | 9/53 | 32/265 | 0.04277 |
| | Total | 62/938 | 39/616 | 48/475 | 149/2029 | 4.4638* |
| Female | Non-smoker | 15/950 | 12/495 | 6/448 | 33/1893 | 0.0028 |
| | Ex-smoker | 1/22 | 2/11 | 3/7 | 6/40 | 6.0605* |
| | 1–24 cigarettes per day | 14/215 | 8/130 | 6/106 | 28/451 | 0.0884 |
| | Total | 30/1187 | 22/636 | 15/561 | 67/2384 | 0.1545 |
| | | Low | Intermediate | High | | |
| | | Air pollution area Amagasaki-Nishiomiya | | | | |
| | Age group | 35–74 | | | | |

Center Research Institute in collaboration with Miyagi and five other prefectures in 1965. In this survey, the people were followed for five years (Figure 4).[4,5]

If the death rate from lung cancer in non-smokers is set at 1, people who started smoking at 25 years of age or older and who smoked less than 190,000 cigarettes are faced with a risk 1.97 times greater than the base. However, the risk becomes 8 times greater if people started smoking at age 19 or younger and smoked 400,000 cigarettes, clearly indicating that there is a cause-effect relation between smoking and lung cancer (Figure 5).

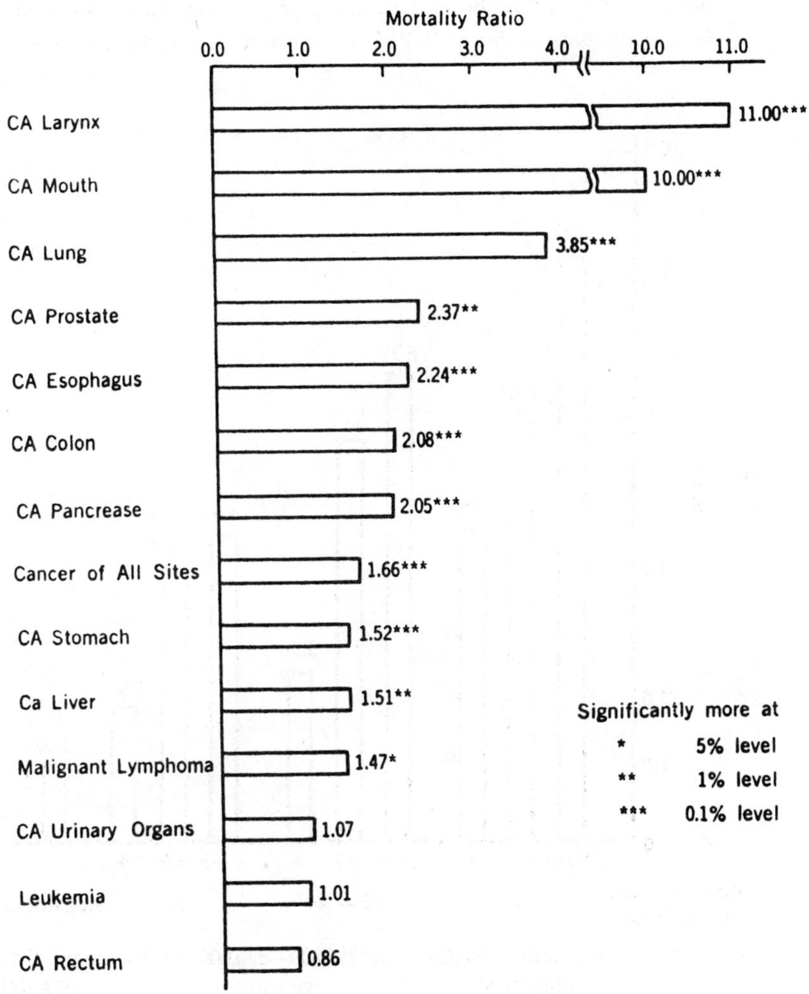


**Figure 4.** Effect of cigarette smoking on cancer death in men (excluding cases with onset prior to start of follow-up). Result of five-year follow-up, 1966–1970. CA means cancer.

The survey suggests that in order to prevent lung cancer measures should be worked out to reduce the number of cigarettes smoked by individuals and to discourage young people from smoking.

## OPERATIONAL EPIDEMIOLOGY

### Measures Against Teen-Age Smoking

A law exists in Japan prohibiting teen-agers from smoking. Not everybody observes this law, and those who started smoking before age 20 were 14.4% in the above-mentioned population study. This percentage is far lower than in the U.S. (66.1%), where a similar law does not exist.

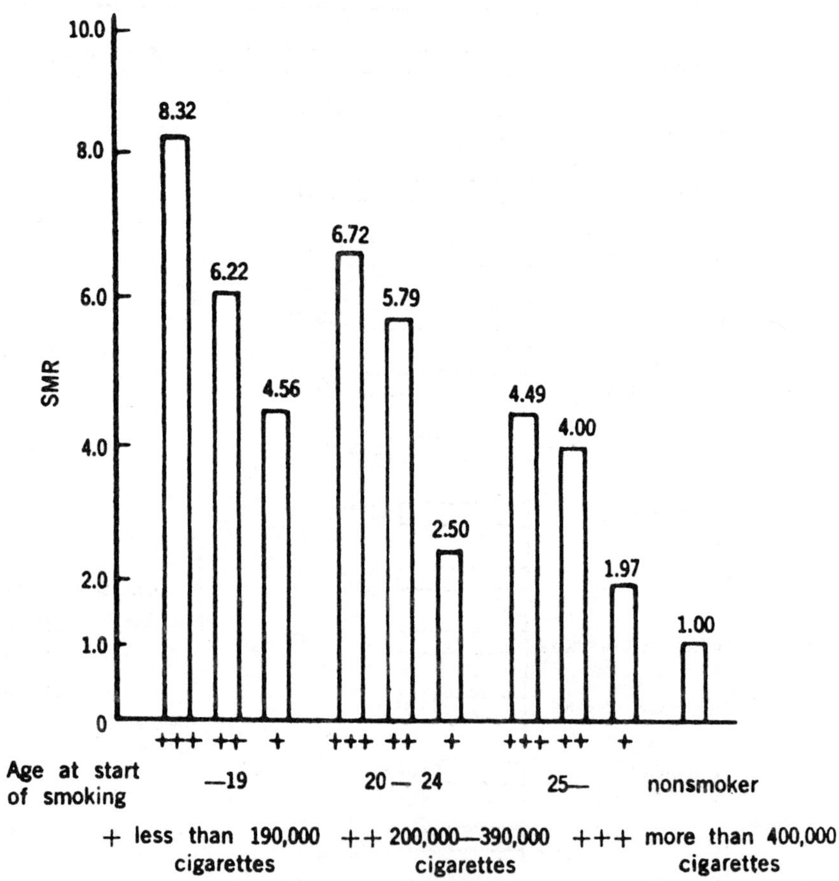


**Figure 5.** Standardized mortality ratio for cancer of the lung by age at start of smoking and by total amount of cigarettes smoked in Japan. Prospective study (1966–70) (From Hirayama et al.).

Therefore, efforts are currently focused on encouraging parents and teachers as well as teen-agers to observe this law.

## Measures Reducing Amount of Smoking

Production and sales of tobacco are a government monopoly in Japan. Consequently, there is practically no advertisement of cigarettes in Japan. Since 1972, a "caution" has been printed on each package of cigarettes sold in Japan that reads "refrain from excess smoking for your health." Our survey revealed an unexpected effect in that almost half of the smokers decided either to stop smoking or to reduce the amount of smoking because of this message. In survey 1, such response to the caution printed on cigarette packages was 51.5% (95 out of 186), and in survey 2, 48.2% (3,246 out of 6,737). The percentage was lowest in psychologically intoxicated smokers (24.8%, or 65 out of 262), next in positive smokers (42.8%, or 391 out of 914), then in habitual smokers (50.7%, or 2,352 out of 4,638), and highest in negative smokers (51.9%, or 280 out of 540).

## Measures Against Air Pollution

Environmental monitoring is quite important to reduce the effect of air pollution on lung cancer. Heavy smokers in highly polluted areas must be carefully examined as they are the highest risk group.

## Mass Screening Program for Early Detection of Lung Cancer

A detailed study by mass screening for lung cancer was conducted by the Japan Cancer Society in Kawasaki City, a city noted for severe air pollution. Out of 83,998 (male 15,976, female 68,022) screened, 16 cases of lung cancer were found (male 9, female 7). The detection rate per 100,000 was 12.19 for males under 40 and 23.27 for females over 40.[2] Such screens should be steadily expanded by full use of up-to-date diagnostic tests and measures.

## Identification of High Risk Group

Retrospective studies in Tokyo and Kawasaki clearly showed that the presence of family history is one of the most important risk factors for lung cancer. Out of 331 lung cancer patients, 11 had lung cancer cases among their blood relatives, as compared to only 1 out of 375 controls.[2] All of these were found in moderate or heavy cigarette smokers, a result suggesting the combined effect of host factors and cigarette smoking. High risk group identification such as AHH must seriously be considered.

**REFERENCES**

1. Health and Welfare Statistics Department, Ministry of Health and Welfare, Vital statistics, Japan 1948–72.

2. Japan Cancer Society: Fight Against Lung Cancer in Japan, Japan Cancer Society Conference on Lung Cancer Prevention and Control, 1974.

3. Hitosugi M: Epidemiology of lung cancer with special reference to the effect of air pollution and smoking habit: *Bulletin of Institute of Public Health* (Tokyo) 17:237–256, 1968.

4. Hirayama T: Strategy for cancer epidemiology. In: *Analytic and Experimental Epidemiology of Cancer,* Nakahara, W (ed), 393–420. Tokyo, University of Tokyo Press, 1973.

5. Hirayama T: Prospective Studies on Cancer of Epidemiology Based on Census Population in Japan. Proceedings XI International Cancer Congress, Florence, 1974, 3:26–35, 1975, Elsevier Excerpta Medica, Amsterdam.

# 6 Clinical Implications of Air Pollution Research for Respiratory Diseases

Benjamin Burrows, M.D.

While I have been asked to summarize the "clinical significance" of the session, I will really be talking about the significance of the session to a clinician. "Clinical significance" implies that I will take home the information I heard this morning and use it to modify diagnosis and treatment in my next clinic session. I am not quite sure that this is the case.

To a practicing chest physician like me, the most significant comments may have been those of Dr. Selikoff. He assured me that I will not be out of a job as a chest physician in the next 20 years. In fact, what I heard about the effects of air pollution and the potential effects of a number of occupational exposures suggests that I may get even busier.

In summarizing how clinicians may use the kind of information presented in this session, it is useful to separate the studies into two types. The first type is epidemiological, represented by the papers of Drs. Shy, Selikoff, and Hirayama. The second type consists of laboratory or experimental investigations, such as those presented by Drs. Crocker and Shubik.

How does a clinician apply epidemiological information in general and, more particularly, the kind of information we heard this morning? In considering this, one might first ask what the difference is between an epidemiologist and a clinician. The best distinction I have heard is that a clinician differs from an epidemiologist in that the former can only count to two. This makes the clinician like a bit in a computer, basically an on-off switch.

Epidemiological information finds its place in clinical practice in a variety of ways, the most spectacular being when epidemiological studies reveal a new diagnostic entity. One of the best examples, asbestosis, was discussed by Dr. Selikoff. Certain pleural manifestations of asbestosis are distinctive enough so that they represent a definite clinical disease. As a result of epidemiological investigations, we now have a new entity in clinical medicine.

The second way in which epidemiology affects clinical practice relates to considerations of diagnostic probabilities. Dr. Hirayama's talk is an excellent example. The obvious rapid decline in tuberculosis in Japan means that tuberculosis will no longer be the first diagnosis that is considered when a physician finds an indeterminate pulmonary lesion in a younger person. During my early medical career, an indeterminate atypical lesion in a relatively young person would automatically put him in a tuberculosis sanitarium, even if no further confirmation of the diagnosis was obtained. Such a procedure was justifiable only when tuberculosis was overwhelmingly the most common cause of that kind of lesion.

Epidemiological information also helps in making a differential diagnosis. When very striking relationships are demonstrable in a population, these become useful to the clinician in making his decisions. The information on carcinoma of the lungs is a good example. The effects of smoking, age, sex, and perhaps even place of residence on frequency of disease are very striking. In a place like Tucson, Arizona, where a very large portion of the population has little nodules in the lungs as a result of coccidioidomycosis, a nodular lesion occurring in a young female who does not smoke is likely to be ignored, while an identical appearing lesion on the x-ray of an older, smoking man will very likely lead to a thoracotomy for presumed carcinoma. This kind of application of epidemiological data to clinical practice requires that relationships be very well established and very close. A 1½- or 2-fold difference in prevalence, which may be extremely significant in terms of risk factors for populations as a whole, is relatively useless in considering the probability of diagnosis in an individual case. Perhaps it is here that the clinician and the epidemiologist differ most significantly. Since the clinician has to make a yes-no decision, he is only willing to make that kind of guess if the probabilities of being correct are very strong indeed.

Epidemiological studies may also point out methods for intervening in the course of the clinical illness. I was rather disappointed that Dr. Hirayama did not discuss one aspect of his study that was in the written presentation. His data on results of screening are important in evaluating the usefulness of screening procedures. In the case of early detection by screening of populations it is sometimes difficult to distinguish epidemiological studies from large scale clinical trials. Unfortunately, data thus far collected do not justify screening studies in detecting early cancer of the lung. For certain other diseases, such as high blood pressure, however, it is becoming clear that it may be useful to detect subjects with elevated blood pressure because effective and acceptable therapy can be offered.

Epidemiology also affects clinical practice in the area of prevention of disease. If risk factors that are controllable or modifiable can be uncovered in epidemiological studies, these studies will have a direct impact on clinical practice. For example, there are probably few physicians who do not recommend as strongly as they can that their patients stop smoking. This is certainly a modifiable habit and one only wishes that physicians were more effective in changing the smoking habits of their patients. On the other hand, it is very difficult to evaluate how effective they really are. If physicians were to say that smoking was fine, would there be even more people smoking than there are now?

One of the entrees to modifying personal habits that affects risk of disease is through clinical practice. I would also like to think that physicians do more than see sick patients in an attempt to cure them or relieve their symptoms. I would like to think that physicians have some impact on community attitudes and in influencing a variety of agencies to take action to reduce risk factors. Convincing data provided to physicians which indicate significant risks can and do lead to community action.

Finally, epidemiological information influences the doctor's conception of disease. The concept of multifactorial causes of disease is reaching the practicing physician because of a variety of epidemiological studies of the type discussed this morning. The clinician is finally becoming concerned with the kinds of exposures — occupation, air pollution, personal habits — and the kinds of foods and living conditions people are experiencing. He is slowly becoming aware that all of these things added together greatly influence the likelihood of various diseases. Direct clinical application is seen in the case of coronary artery disease, in which many cardiologists now do a formal assessment of the risk factors in their patients. They also make a concerted effort to change those factors that are modifiable on a personal level. Some day we hope to be in a position to do a similar analysis for chronic respiratory disease.

The second type of study I mentioned was laboratory investigation. It is extremely difficult for me to take the information from the experimental studies that were described this morning back to the clinic and immediately apply it. But because of the hope that it gives us for the future, I suspect that everyone — clinicians, epidemiologists, and interested bystanders —

recognizes the overwhelming importance of these studies. Specific carcinogenic agents able to induce metaplastic and malignant neoplastic change can be identified only by this kind of study. These investigative techniques provide an opportunity to study the specific mechanisms by which metaplasia and carcinogenesis occur. By uncovering the mechanisms, we may learn how to intervene in or prevent the process. As a clinician, however, I am concerned about the danger of premature application of certain observations. I am especially concerned about the possibility of confusing metaplastic change with neoplastic change. Drastic therapeutic procedures have been recommended at times to eliminate metaplastic changes when the certainty that these changes would have proceeded to malignant neoplasia had not been demonstrated. In terms of the preceding chapters, there could be widespread misapplication of vitamin A as a cancer preventive if studies of this agent are not presented in a careful way.

I have one final comment about this session. My own primary interests are in chronic bronchitis and emphysema, and I can hardly be regarded as an expert, even as a clinician, in the field of cancer. I was a bit disappointed that there were not more studies concerning other respiratory diseases. Air pollution research may have more direct clinical application to the non-neoplastic pulmonary disorders than in the case of cancer.

# DISCUSSION: PART ONE

**Dr. Theodore Sterling, Burnaby, B.C.:** In your presentation, Dr. Shy, you make certain statements that run, I believe, counter to the opinions and published documentation in literature. Your statement that there is no decline in lung cancer rates is not in agreement with the findings published by some of the very well-known people in the field. There is a large body of literature that shows that male lung cancer rates are indeed beginning to level off and decline quite noticeably by now. Furthermore, it is well known that there is a lag period between insult and the appearance of the disease. The lag period is known to be 15 to 20 years. I find it incongruous to hear someone say today that benzo(a)pyrene has been decreasing since 1950 (actually the decrease began in 1945 not in 1950) while cancer rates are going up when (1) they are beginning to level off today and (2) the lag period, which needs to be considered, is ignored.

**Dr. Shy:** Let me comment on the change on cancer rates. The absolute rate of lung cancer is still increasing in males and females, but the rate of increase is less steep than it used to be in males and more steep than it used to be in females. And much of this Higgins has recently shown can be related to the change of smoking habits of males and females. Male smoking habits are declining a little bit and the rate of increase of lung cancer is not as great. Female smoking habits are on the increase and their rate of lung cancer is on the increase. So I still think that it is true that the lung cancer rates are increasing both in males and females but the rate of increase is not as great in males as it used to be 15 years ago, or 10 years ago.

**Dr. Sterling:** I must disagree with you again. According to both Doll's and Schneiderman's data, lung cancer up to the age of 54 is declining. Now, it is not a question of what you say or what I say — it is in the literature.

**Dr. Goldsmith:** Rates for lung cancer have leveled off and this has been shown by Higgins for London, England. The total lung cancer mortality rates for males in the United Kingdom is beginning to plateau, and Sir Richard Doll thought that perhaps 1973 would be the first year in which it would not have increased. I think that it did continue to increase by a couple of deaths but in 1974, if these trends continue, there will surely be a decrease. What is very important to watch, although it is too early to draw causal interpretations, is the trend in mortality among the young age groups. The crude data between 1968 and 1972 have just been examined, and they appear to show that for males under 45 there has begun to be a leveling of, and probably a decrease in, lung cancer mortality. But this is not occurring among females. Unfortunately, in the United States age-, sex-, and race-specific smoking histories are not obtained in sufficient detail, and these are required in order to provide an adequate analysis of the non-smoking factors.

One other study that I think is very informative is the somewhat obscurely published report by Hammond of the large-scale prospective study of the American Cancer Society. Dr. Hammond divided the male population, with a 10-year period of stable residence in the same community, into those who said they had a history of exposure in their place of work to dust, fumes, and radiation, and those who gave no such history. This is a very coarse division, but it provided two sub-populations throughout that study. When the population that had no occupational exposure was analyzed for cities of different sizes and for cities with different levels of air pollution, there were no consistent urban-rural gradients. By contrast, the urban gradients that were present appeared in those who said they had an occupational exposure. So, I think that we now have to place the as yet not well characterized, occupational contribution to urban-rural differences fairly high in the list of suspicious factors. I fully agree with the need for establishing new prospective studies, and I think that the specification of occupational exposure, smoking, and migration, as you stressed, will be very important to the success of those studies.

**Dr. Shy:** We need to examine urban-rural differences and the contribution of occupational exposure and other indoor sources of carcinogens. Benzo(a)pyrene as an index of pollution has not offered us a good explanation for rural-urban differences.

**Dr. Allan J. Ebbin, Los Angeles:** How about genetic influences in addition to the environment as possibilities for the change in rates?

**Dr. Shy:** It is always difficult to separate out genetic and early environmental influences because you get into family aggregations, and so forth. How do you explain, though, the fact that the British immigrants to the United States have an intermediate rate between natives of Britain and the US natives if it is a matter of genetics?

**Dr. Ebbin:** It could be explained by multifactorial inheritance (determined by multiple genetic and nongenetic factors). The actual expression of the condition would require a genetic predisposition that pushes the individual beyond the "threshold" of risk at which point environmental influences will determine to what extent the individual is affected.

**Dr. Shy:** There certainly may be an ethnic influence, but there must also be an environmental influence, because the rates have gone off in the same ethnic group that move to this country. You can't rule out the ethnic influence in those comparisons.

**Question:** Is there any study being made on pesticide exposures, especially those in tobacco fields, in order to compare agricultural worker's lung and cancer rates? Are there any other studies on the relationship between pesticides and lung cancer or any other cancer in agricultural workers?

**Dr. Shy:** There have been many pesticide studies, but they have not investigated cancer or pre-carcinogenic changes. There is an interest in doing this now in view of the fact that animal studies have demonstrated an association of cancer with DDT, aldrin, and dieldrin. The Environmental Protection Agency is interested in investigating whether areas of the country in which a lot of pesticide is used show any geographic pattern of high mortality rates. But these cancer studies have not been done and have hardly been considered. The problem is that we have not had enough lung cancer data by counties all over the United States; we have only information by states.

**Dr. Kenneth Hayes, San Jose:** I have two questions. One, when you talk about lung cancer, do you include all types of cancer or just bronchiogenic carcinoma? The second question is related to clusters of leukemia in a population that suggests a viral etiology and contamination. How do you know your statistics on urban populations don't represent a possible viral spread as a source of cancer rather than industrial exposure or air pollution?

**Dr. Shy:** I said that I could not explain the urban-rural difference by atmospheric pollution, and so it can or must be explained by something — one possibility is the density of population and viral spread. There is no answer to that. In answer to the first question, the lung cancer considered here is specified by the international classification codes 162 and 163, which are primary lung cancers of the trachea and bronchus. Some authors have looked at the problem more specifically, by excluding adenocarcinomas. In general, investigators limit themselves to primary lung cancers of the trachea and bronchus and exclude the larynx.

**Dr. Leonard A. Sagan, Palo Alto, CA:** Perhaps you would comment on the epidemiology of chronic pulmonary disease other than cancer. Is the relationship between air pollution exposure and bronchitis well established now? Does the epidemiology of cancer correlate with other chronic pulmonary diseases or are there significant differences?

**Dr. Shy:** I believe that the relationship between non-cancerous chronic respiratory diseases and air pollution is established. Relationships are more definite between specific measures of air pollution, such as sulfur oxides and particulate matter, and excess chronic bronchitis than for lung cancer and benzo(a)pyrene. Again, cigarette smoking is a very dominant factor and, I believe, that the data suggest a very additive effect of air pollution to cigarette smoking.

PART TWO

# CARDIOVASCULAR DISEASES

# 7 Relationship Between Air Pollution and Cardiovascular Disease; A Review

Jack D. Hackney, M.D.

## I. INTRODUCTION

This state-of-the-art review of relationships between atmospheric pollutants and cardiovascular disease is generally limited to common community air pollutants and does not include heat as a factor. Its purpose is to provide a current supplement to other existing reports and reviews.[11,31,38,45] The emphasis is on results of human studies. Among the usual atmospheric pollutants found in ambient air, oxidants and sulfur oxides have been suggested as potentially harmful to the cardiovascular system, but evidence of the adverse effect is limited. Carbon monoxide (CO), however, is definitely capable of impairing cardiovascular function.

CO is a gas that is not readily preceived by human senses, but it is taken up from inspired air, combines with hemoglobin, and interferes with oxygenation in the body. Evidence indicates that exposures to high concentrations of CO impair the function of the heart, brain, and exercising skeletal muscles by compromising $O_2$ transport and delivery. At low levels of CO, a decreased exercise tolerance in patients with arteriosclerotic heart disease has been demonstrated by controlled exposure. Vascular disease in other organs probably increases their susceptibility to CO toxicity. Even in normals, the ability to exercise maximally appears to be limited by exposure to relatively low CO concentrations. Therefore, the major emphasis of this review will be CO and its effects on the cardiovascular system.

## II. CARBON MONOXIDE (CO)

## COMBINATION OF CO WITH HEMOGLOBIN AND SOURCES OF CO EXPOSURE

### Formation of Carboxyhemoglobin (COHb)

A relationship can be calculated for COHb at equilibrium with inspired CO concentration (see Table 1).[25] Notice that the starting point is not at zero, but at 0.36% COHb. This level represents the small amount of CO that is produced metabolically within the body and found even in the blood of persons not exposed to CO.[27,67]

Table 1 implies that COHb can be considered as an indicator of CO exposure, and several studies have used this approach.[2,46,54,70,71]

Table 1 also suggests that CO in a sample of alveolar air in equilibrium with blood COHb can serve as a useful estimator of COHb. Methods for indirect estimation of COHb have been described elsewhere.[48,53,68]

### Levels of COHb in Populations

Because COHb can serve as an indicator of exposure to CO, it has been used to investigate the extent to which CO is absorbed by people living or working in or passing through a downtown urban area. There is a problem in answering this question because there are multiple sources of CO exposure.

**Table 1**
**Carboxyhemoglobin (COHb) at Equilibrium with Inspired CO in Non-Smokers[25]**

| Inspired CO (ppm) | % COHb |
|---|---|
| 0 | 0.36 |
| 5 | 1.11 |
| 8.7 | 1.66 |
| 10 | 1.85 |
| 15 | 2.57 |
| 20 | 3.29 |
| 30 | 4.69 |
| 40 | 6.05 |
| 50 | 7.36 |

Sources of CO include community activities, tobacco smoking, and occupational operations. An unusual occupational source was reported by Stewart and associates.[72] Their evidence suggested that CO may be a metabolite of methylene chloride in that deliberate exposure to methylene chloride was followed promptly by a sustained elevation of COHb in all 11 human subjects studied.

Among the problems encountered in using population COHb levels to investigate CO exposure is that the concentration of carboxyhemoglobin in the blood is generally low,[54,71] and measurement methods for COHb must be especially carefully calibrated to obtain reliable results. Also, many factors other than errors in measurement can modify results. A person at rest exposed to sudden change in inspired CO may reach only about half the expected equilibrium blood concentration in three to four hours, and it may take ten or more hours to approach the equilibrium value.[41] Thus, short-term increases or decreases of CO levels will be reflected by small percentage increases in COHb levels in the blood.[46]

Despite the limitations in COHb as an indicator of CO exposure in populations, a number of interesting studies have been done. One investigative approach uses volunteer blood donors as an approximate stratified random sampling of "healthy" residents of a particular geographic area. Stewart and his associates[70,71] conducted the largest of these studies by obtaining venous blood samples from 29,000 blood donors in urban, suburban, and rural communities and analyzing them for COHb. They found that tobacco smoking was the most important factor in producing elevated COHb. Other significant factors were geographic location and occupational and meteorological conditions. They reported that 45% of all the non-smoking blood donors tested had COHb levels greater than 1.5%. The authors interpreted this to indicate that exposure to CO in excess of the Air Quality Standard was widespread and occurred regularly.

Kahn and associates[54] obtained blood samples from 10,157 non-smokers and 6,492 smokers in mobile blood donor units in St. Louis in a study that gave special attention to occupationally related sources of CO. The mean COHb for the 1,523 non-smoking subjects classified as "industrial workers" was 1.38 as contrasted to 0.75 for 8,634 designated "other than industrial." The study also showed for non-smoking, non-industrial workers, a small urban-rural gradient of COHb: "highly urban," "urban," and "rural," with a respective rating of about 0.8%, 0.7%, and 0.6%. These small differences do not support the importance of uptake of CO from ambient air. The results of Kahn et al.[54] indicated that less than 10% of non-smoking non-industrial subjects had COHb values above 2%; this differs from Stewart's findings that 35% of the non-smokers in St. Louis had exposures leading to COHb levels of greater than 1.5%. Both studies emphasize the contribution of smoking and occupational exposures, but Kahn concludes the CO level in St. Louis is a negligible factor in CO uptake, while Stewart's group concludes that in cities with widely divergent meteorological conditions and population densities, one-fourth to three-fourths of the non-smoking population is exposed to CO levels that result in COHb levels above 1.5%. (See Table 1)

In other studies, Aronow et al.[2,3,4] in Los Angeles found control values of about 1.0% that rose to 5.1% when 25 subjects drove on freeways for 1½ hours. In Santa Barbara, Dahms, and Horvath[33] measured a mean of 0.6% COHb in about 150 non-smokers. Ayres et al.[9] in New York City reported a mean of 1.0% COHb in 26 non-smoking hospitalized patients. If COHb concentrations in the general population are less than 1.5% in urban areas such as New York, Baltimore, St. Louis and Los Angeles, then transient exposure to high concentrations of CO on the streets and freeways is not usually reflected in significant accumulative net absorption. On the other hand, Stewart's findings are not consistent with this conclusion. Better standardization of analytic methods between laboratories must be accomplished before existing conflicts are resolved and convincing overall conclusions can be stated. Hopefully an inter-laboratory standards program will be established prior to further studies. Additional large scale studies should be collaborative and include analyses by more than one laboratory and by more than one method.

## CLINICAL IMPLICATION OF CO PHYSIOLOGY AND CHEMISTRY

### CO and Human Senses

CO is not readily perceived by the human senses and this fact helps explain the subtle hazard associated with this non-irritating gas.

### Effects of CO on $O_2$ Transport

The respiratory, cardiovascular, and blood systems transport $O_2$ from ambient air to the various tissues to support and maintain aerobic metabolism. Most of the $O_2$ in blood is bound reversibly to the red blood cell hemoglobin. CO competes with $O_2$ for binding sites on the hemoglobin molecule and the affinity of hemoglobin for CO is over 200 times that for $O_2$. The hemoglobin sites bound with CO are unavailable for $O_2$ binding, and the binding affinity of remaining hemoglobin sites for oxygen is changed by the precence of CO binding. Thus, the effect of CO poisoning is similar, although not identical, to that of anemia. The $O_2$-carrying capacity of the blood is compromised in both.

Prolonged or repeated exposure to CO may produce other changes, including compensatory responses. In animal studies,[7,15,17,24,34,60,73] the increase in hemoglobin concentrations has been found to be approximately proportional to the severity of CO exposure. Less detailed information is available on the response in man.[52,71] Other reactions that would tend to compensate for the decreased affinity of hemoglobin for $O_2$ in CO exposed individuals have been investigated, but the findings do not confirm them as important adaptive responses.[5,36,59]

There is a possibility, as yet not proven, that CO combines with intracellular compounds with resulting impairment of $O_2$ delivery to the mitochondria.[20,28,40,79,80]

### $O_2$ Supply to Tissues and CO Hazard

The hazard from CO may be increased by any condition in which the oxygen supply to tissues has already been compromised. Included are: (1) limitation of blood flow to the tissues, especially to the heart or brain, in patients with coronary or cerebral vascular disease, (2) increased amounts of CO already in the blood from cigarette smoking or from increased endogenous CO production, (3) reduced oxygen transport capacity in the blood in patients with anemia or methemoglobinemia; and (4) oxygenation of the blood that is less efficient than normal, leading to lower arterial oxygen saturation in newborn infants with respiratory distress or patients with pulmonary emphysema.

## ARTERIAL DISEASE AND CO—BLOOD FLOW PROBLEMS

### Coronary Artery Disease and Clinical Heart Disease

There are two possible relationships between environmental exposure to CO and the pathogenesis of arteriosclerotic heart disease: First, in association with other risk factors, such as elevated serum cholesterol, exposure to CO may enhance the development of underlying arteriosclerosis and subsequent coronary artery stenosis. Laboratory investigations using animal models[6,7] suggest an association between atherosclerosis and CO exposure. In general, possible association between CO and the development of the underlying coronary atherosclerosis is speculative at present. Second, in the presence of coronary atherosclerosis, CO may precipitate myocardial infarction, angina pectoris, or sudden death.

Evidence suggestive of an association between clinical coronary disease and CO exposure includes the following:

1. Because the binding of CO and hemoglobin can result in interference in tissue oxygenation, myocardial ischemia,[11] clinical symptomatology and disease could result, especially in those individuals with severe coronary artery stenosis due to arteriosclerosis.[10,11]

2. Experimental studies show a relationship between the exposure to CO and a decrease in exercise tolerance of individuals with angina pectoris.[1,2,3]
3. There is a possible association between the case-fatality percentage in patients admitted to the hospital with myocardial infarction and ambient CO levels of heavily polluted areas.[29,45]
4. On the assumption (without solid evidence) that CO is a major factor in cigarette smoke that predisposes heart problems, there is considerable evidence describing a relationship between cigarette smoking and
   (a) new clinical coronary artery disease
   (b) sudden deaths due to arteriosclerotic heart disease
   (c) lower survival rate of angina pectoris patients who are cigarette smokers.[13]

**Cardiovascular responses to CO exposures** The heart muscle requires adequate delivery of $O_2$ to maintain its function. CO can interfere with myocardial oxygen delivery. In normal man, a reduction in myocardial oxygen tension produces an increase in coronary blood flow that restores myocardial oxygen delivery toward normal levels. The normal heart can increase coronary flow by up to 300% during exercise in patients with severe anemia[47] or individuals exposed to arterial hypoxemia.[12] This is fortunate, because, unlike other vascular beds, the coronary circulation delivers almost 70% of its available oxygen to the perfused tissue, and increased oxygen requirements cannot be satisfactorily met by increasing oxygen extraction. Unfortunately, however, the development of coronary atherosclerosis reduces the capacity of the vascular bed to adapt to hypoxic stress.

Ayres et al.[9] studied coronary flow and myocardial metabolic response to acute elevation of COHb in man during diagnostic cardiac catheterization. COHb saturations between 6% and 12% were achieved. The groups with coronary artery and cardiopulmonary disease showed a decreased myocardial arteriovenous oxygen difference of 25% and 30.5%, respectively. Statistically significant changes in coronary blood flow occurred in all but two of these studies. Failure of the increase in coronary blood flow to compensate adequately for the CO produced decrease in oxygen delivery was indicated by a decrease in coronary sinus oxygen tension in most of the patients in this group. Nine of the 14 patients with coronary artery disease decreased lactate extraction with increasing COHb, a finding suggestive of anaerobic metabolism.

**Experimental studies that describe a relationship between exposure to CO and decreased exercise tolerance in angina subjects** Anderson et al.[1] studied 10 men with stable angina who breathed air with 50 or 100 ppm CO intermittently for four hours on five successive days in a double-blind study. The men exercised on a treadmill after each exposure and control period. COHb levels rose to 2.9% after exposure to 50 ppm and to 4.5% after exposure to 100 ppm. The time of exercise before the onset of pain decreased with either exposure when compared to air ($P < 0.005$). The duration of pain was prolonged with the 100 ppm exposure when compared to the control period.

Aronow studied the effects of driving in heavy Los Angeles freeway traffic in 10 subjects with angina pectoris.[2] They were driven for 90 minutes with the window open and then traveled the freeway while breathing compressed air. Control mean COHb levels were 1.42%, reached 5.08% after freeway travel, and decreased to 2.91% two hours after driving. After breathing compressed air, COHb averaged 0.65%. After exercising immediately following the drive and again two hours later, the men exhibited a significant decrease in time to the onset of angina. Electrocardiograms during exercise reflected no changes.

A second study in Los Angeles by Aronow[3] is very similar to Anderson's work previously cited. Ten non-smoking male patients, with clinically documented angina pectoris, were studied. Two mornings the men breathed 50 ppm COHb, and two mornings they breathed compressed air for two hours in a double-blind random trial. COHb levels after breathing 50 ppm for two hours were elevated from 1.03% to 2.68%, and decreased after breathing compressed air from 1.07% to 0.77%. The exercise period was similar to the previous study. The length of time that an individual could exercise before onset of angina pectoris was reduced after breathing the CO ($P<0.001$). Symptoms also occurred sooner and with lesser amounts of exercise. Again, there were no electrocardiographic changes. The reproducibility of treadmill tests for evaluating exercise tolerance in angina pectoris patients has been studied by others[69] and reported to be acceptable.

**Studies describing the relationship between ambient CO levels and myocardial infarction**[29,45,58] The location of Los Angeles hospitals was classified as being situated in high or low pollution areas, based on air pollution data. A review of 2,484 myocardial infarction admissions was made in 1958 at selected hospitals in high- and low-pollution areas. The results reflected no differences in admissions for myocardial infarction in relation to areas of high or low pollution. The case-fatality percentages, however, were higher in the highly polluted areas and rose to a still higher level during periods of elevated CO readings. A relationship was also indicated between ambient CO levels and case-fatality percentages. As pointed out by the authors, however, a causal relationship was not established because enough uncertainties existed so that differences in case-fatality between high and low pollution areas could be due to many factors other than CO concentrations. Areas of uncertainty included: the accuracy of the diagnosis; the small number of admissions to hospitals in the low pollution areas, which could cause wide variations in the case-fatality percentages from one week to another;[29] factors such as smoking history, sex, age, race, and socioeconomic differences that could affect the results, as well as differences in the severity of the disease on admission; and the unaccounted number of deaths occurring prior to hospitalization.

A 10-month study in Baltimore, Maryland[58] has some design similarities to the Los Angeles study. The Baltimore study had a relatively small base, but diagnosis and population characterization were given additional attention. However, levels of ambient CO in Baltimore were probably lower than found in Los Angeles. The Baltimore study found no relationship between either the incidence of myocardial infarction or sudden deaths due to Arteriosclerotic Heart Disease (ASHD) and the average 24-hour ambient CO levels. Nor was there a relationship between total deaths due to ASHD and ambient CO levels. However, the Baltimore study does not rule out the possibility that there may be an association between ambient CO levels and the incidence of myocardial infarction or sudden death, or the case-fatality percentages in areas with higher levels of ambient CO.

**Smoking, CO, and heart disease** The extent to which CO contributes to adverse cardiovascular effects from cigarette smoking is a question of major importance, but the answer is not known. Several points about cigarette smoking and clinical coronary disease can be noted: (1) The relationship with cigarette smoking is stronger for myocardial infarction and sudden death than for angina pectoris.[55] (2) The relationship between cigarette smoking and clinical coronary disease is apparently also dependent on other risk factors, such as a high fat diet and increased serum cholesterol.[56] The relationship between cigarette smoking and clinical coronary disease is relatively weak[43,56] in populations with low serum cholesterol. (3) In patients with pre-existing angina pectoris,[75] cigarette smoking apparently increases the risk of subsequent sudden death from myocardial infarction; but whether or not smoking after a myocardial infarction adversely affects subsequent survival is less clear.[13,76]

Evidence against a causal association between CO exposure and cardiovascular disease includes: (1) studies where no relationship was found between the number of new cases of myocardial infarction on a particular day and the ambient CO levels;[29,32,45,58] and (2) one study that, after being adjusted for cigarette smoking, failed to find an association between post-mortem blood COHb levels and causes of sudden death.[58]

**Opportunities for further studies** Types of data that can further clarify the relationship between ambient CO levels and clinical heart disease are: (1) studies on the prevalence of heart disease in those occupationally exposed to high CO levels; (2) additional studies to determine if patients with ASHD have increased risk of heart attacks, changes in symptomatology, or disability when ambient CO levels are elevated; and (3) clarification of whether CO is a major factor in cigarette smoke that contributes to the increased risk of heart disease.

### Effect of CO Exposure on Intermittent Claudication

Aronow, Stemmer, and Isbell[4] performed a double-blind study evaluating the effect of breathing 50 ppm of CO for two hours versus compressed purified air for two hours in intermittent claudication. Ten men with angiographically documented occlusive arterial disease were studied. Exercise was performed on a constant load bicycle ergometer. The mean exercise time until the onset of claudication decreased from 174.3 seconds in the control period to 144.0 seconds after CO exposure ($P< 0.001$). In control experiments, mean exercise time until claudication did not change significantly (168.8 to 173 seconds).

### Cerebrovascular Disease

Individuals with cerebral vascular disease and potential reduction in the supply of oxygen to higher cerebral centers may be another group susceptible to high levels of CO. In such cases, small amounts of COHb could have disproportionately severe effects on higher cerebral function. Studies are needed among older age groups, in whom cerebrovascular arteriosclerosis is also prevalent, to test the effects of low concentrations of COHb on cerebral activity, vigilance, or other criteria of functional capability. Adequate information does not exist on whether CO exposure might enhance the development of cerebral arteriosclerosis or, in the presence of pre-existing disease, be a major precipitant of significant cerebral ischemia and strokes.

## OTHER CLINICAL CONDITIONS AND CO

### High Blood COHb; Smokers, Hemolysers

Additive effects of ambient CO exposure would be expected in individuals with abnormal amounts of CO already in the blood. Cigarette smoking is the primary source, but CO can be added from breakdown of hemoglobin. The rate of production of CO in the body has been reported up to ten times more than normal in patients with hemolytic disease.[25,27]

### Reduced Oxygen Capacity, Anemia, Methemoglobinemia

The pathophysiology of anemia is similar to that of CO poisoning. The adverse effects of CO and anemia are qualitatively the same, so that CO exposure in an anemic person would aggravate the anemia. Thus, it is reasonable to believe that anemic persons are at greater risk than normal people to effects of CO exposure. The extent to which the hemoglobin concentration must fall before an increased risk of adverse effects of CO can be detected is not known.

The rare patient with methemoglobinemia may also be at special risk, since methemoglobin would act physiologically much like COHb to impair oxygen transport and delivery.

### Low Arterial $O_2Hb$; High Altitude, Newborns with Respiratory Distress, Chronic Lung Patients

The effects of CO would be expected to be greater at high altitude than at sea level because at altitude the mean capillary oxygen pressure would be reduced. Experimental work is needed to quantitate this possibility. Infants have been suggested as a potentially sensitive group, but in infants adaptation to low tissue oxygen may make the brain and other tissues less sensitive to hypoxia, and thus exposures to low levels of CO may be less significant. Adults with chronic pulmonary disease comprise another group of concern. Venous admixture may be aggravated by COHb because of the lower $pO_2$ of mixed venous blood.[14] This in turn can contribute to the low $pO_2$ of arterial blood. On the other hand adaptive processes do occur with chronic hypoxemia that may reduce the sensitivity of the patient to small amounts of COHb from ambient sources.

Residence at high altitudes or the presence of chronic lung disease added to other conditions discussed could make some individuals and their cardiovascular system more susceptible to CO than either condition alone.

### Critically Ill Patients with Impaired Blood Flow to the Tissues

This group would include individuals with impaired circulation resulting from surgical procedures or with reduced blood volume due to hemorrhage or acute trauma. These patients are generally under close observation and extensive treatment, including oxygen therapy, for any recognized signs of tissue hypoxia. Thus, they may not represent as important a susceptible population as otherwise could be the case.

## EXERCISE LIMITATIONS BY CO

COHb levels have been shown to decrease the maximal $O_2$ consumption during exercise in healthy young males,[23,30,37,39,51,61,63,64,77,78] whereas efforts of up to 75% maximum can be carried out with little change in maximum $O_2$ consumption.[21,39,57,61,62,77,78] However, only a relatively small number of young, healthy males have been studied.

Within the range of 5% to 35% COHb, there is approximately a linear relationship between the decrement in maximum $O_2$ uptake and the level of COHb. Maximum $O_2$ uptake was not significantly decreased in young non-smokers until a blood level of about 5% of COHb was attained.

## AMBIENT CO STANDARDS AND CARDIOVASCULAR DISEASE

The present federal statutory standards are: 9 ppm CO for an averaging time of eight hours, and 35 ppm for an averaging time of one hour. At 9 ppm for eight hours, a normal resting man will register a COHb of approximately 1.5%. It is important in protecting public health to know whether any population group would be adversely affected by CO exposure at these statutory levels. Data now indicate that there are measurable changes in human performance at levels of 5% and also that COHb levels as low as 2.5% to 3% can have measurable effects on patients with angina pectoris.

Most studies of CO effects have administered CO at relatively high concentrations for short periods of time. Studies in which CO exposure more closely simulates that experienced by the general population would be useful. Studies to date have included relatively small numbers of patients and subjects, and it is not known how the results obtained can be projected to large population groups. Concentrations of COHb associated with current standards need to be tested for effects in other susceptible groups.

Adverse health effects have not been found at COHb levels that result from long-term exposure to the eight-hour federal standard of 9 ppm CO. Current standards express ambient air CO in terms of one-hour and eight-hour averaging times. Although exposure to CO is usually defined by measurement of concentrations in outdoor air, and standards are expressed in these terms, recent data[54] suggests that an individual's exposure may be determined to a great extent by occupational and other personal CO exposure conditions. Cigarette smoking produces a much greater CO exposure than does community air pollution, but each exposure is partially additive in determining an individual's COHb concentration at any moment.

There is evidence that present monitoring systems may not adequately reflect CO exposure in the general population and it would be desirable to re-evaluate locations of monitoring stations, standardization, and operating techniques. In theory, CO exposure can be determined with blood COHb measurements, and further epidemiological studies are needed to clarify the relationship between ambient CO concentrations and the COHb levels in people living and working in the community.

## III. OTHER POLLUTANTS

### OXIDANTS

In one study with adult rabbits and mice of long-term ozone exposure at concentrations of 0.2 ppm for five hours daily for three weeks, structural changes were reported in the nuclei of heart muscle fibers.

These changes reverted to normal one month after exposure. Apparently this work has not been pursued by other investigators.[16]

### SULFUR OXIDES AND PARTICULATES

In a Chicago study[18] a model was used to estimate the $SO_2$ levels in each square mile of the city at 15-minute intervals; however, details of the model were not included in the report. The city was divided into high, medium, and low levels of $SO_2$ exposure, with results indicating a higher mortality rate in older males in high pollution area. Of course, other pollutants, especially particulates, were also present. A one-month study compared mortality rates at 10 days of low, 10 days of intermediate and 10 days of high pollution. A significant increase in deaths due to cardiovascular disease, particularly coronary arterial disease, was revealed in males 55 and older in comparisons between low and high pollution periods. Objective assessment is limited because the study report did not take into consideration other variables such as socio-economic factors, and subject differences.

Another study correlated frequency and severity of cardiopulmonary symptoms with air pollution in New York.[44] The study set up four panels of elderly citizens (60 years or older) from intermediate and low pollution areas, subjectively classified as ''well'' (no known symptoms), ''heart'' (subjects reported symptoms of angina, leg swelling, shortness of breath), ''lung'' (cough, phlegm, wheezing, shortness of breath), and ''heart and lung'' (combination of two prior groups). Subject responses and interviews determined the panel membership. Panelists rated symptoms on a five point scale: (1) never present, (2) better, (3) same, (4) worse, or (5) much worse than usual.

Among "heart" panelists, symptoms of angina and shortness of breath were significantly correlated with pollutants, although firm conclusions are difficult because of the subjective nature of responses and methodology of panelist selection and assignment.

## CADMIUM

Cadmium can enter the body either through the lungs or by being ingested and is excreted through the kidneys, possibly causing proteinuria.[42] Cadmium accumulation in the kidney may produce hypertension.[66] R.E. Carroll reported that communities with high cadmium levels in the air (above 0.007 $\mu g/m^3$) tended to have higher death rates due to cardiovascular disease.[19] On the other hand, weaknesses in existing evidence implicating cadmium as a cause of hypertension in humans are discussed in papers by D.I. Hammer et al.[49,50] Because of the public health significance, more work is indicated to assess these results.

## IV. SUMMARY

In this review of air pollution and cardiovascular disease, the major emphasis is on carbon monoxide and cardiovascular disease.

Evidence indicates that exposure to high concentrations of CO impairs the function of the heart, brain, and exercising skeletal muscle by mechanisms that are fairly well understood. Controlled exposure to low concentrations of CO provides evidence of decreased exercise tolerance in patients with arteriosclerotic heart disease. Vascular disease in other organs may increase their susceptibility to CO toxicity. Even in normals, ability to exercise maximally appears to be limited by exposure to relatively low levels of CO.

Other atmospheric pollutants such as oxidants and sulfur oxides may produce adverse cardiovascular effects, but the evidence is limited.

## REFERENCES

1. Anderson EW, Andelman RJ, Strauch JM, et al: Effect of Low-level Carbon Monoxide Exposure on Onset and Duration of Angina Pectoris. A Study in Ten Patients With Ischemic Heart Disease. *Ann Intern Med* 79:46–50, 1973.
2. Aronow WS, Harris CN, Isbell MW, et al: Effect of Freeway Travel on Angina Pectoris. *Ann Intern Med* 77:669–676, 1972.
3. Aronow WS, Isbell MW: Carbon Monoxide Effect on Exercise-induced Angina Pectoris. *Ann Intern Med* 79:392–395, 1973.

4. Aronow WS, Stemmer EA, Isbell MW: Effect of Carbon Monoxide Exposure on Intermittent Claudication. *Circulation* 49:415–417, 1974.

5. Astrup P: Intraerythrocytic 2, 3-diphosphoglycerate and Carbon Monoxide Exposure. *Ann NY Acad Sci* 174:252–254, 1970.

6. Astrup P, Kjeldsen K: Carbon Monoxide, Smoking, and Atherosclerosis. *Med Clin North Am* 58:323–350, 1973.

7. Astrup P, Kjeldsen K:, Wanstrup J: Enhancing Influence of Carbon Monoxide on the Development of Atheromatosis in Cholesterol-fed Rabbits. *J Atherosclerosis Res* 7:343–354, 1967.

8. Ayres SM, Giannelli S, Armstrong RG: Carboxyhemoglobin: Haemodynamic and Respiratory Responses to Small Concentrations. *Science* 149:193–194, 1965.

9. Ayres SM, Giannelli S Jr, Mueller H: Effects of Low Concentrations of Carbon Monoxide. Part IV. Myocardial and systemic responses to carboxyhemoglobin *Ann NY Acad Sci* 174:268–293, 1970.

10. Ayres SM, Mueller HS, Gregory JJ, et al: Systemic and Myocardial Hemodynamic Responses to Relatively Small Concentrations of Carboxyhemoglobin (COHb). *Arch Environ Health* 18:699–709, 1969.

11. Bartlett D Jr: Pathophysiology of Exposure to Low Concentrations of Carbon Monoxide. *Arch Environ Health* 16:719–727, 1968.

12. Bates DR, Witherspoon AE: The Photochemistry of Some Minor Constituents of the Earth's Atmosphere ($CO_2$, CO, $CH_4$ $N_2O$). Monthly Notices. *Royal Astron Soc* (London) 112:101–124, 1952.

13. Blackburn H, Canner P, Krol W, et al: The Natural History of Myocardial Infarction in the Coronary Drug Project. Prognostic indicators following infarction, pp 54–64. In: *Preventive Cardiology*. G Tibblin, A Keys, L Werko (eds). Proceedings of an International Symposium held at Billingebus, Skovde, Sweden, August 21, 1971, under the sponsorship of the Planning Group for Preventive Cardiology of the Swedish Medical Research Council. Stockholm: Almquist and Wiksell, 1972.

14. Brady JS, Coburn RF: Effects of Elevated Carbon Monoxide on Gas Exchange in the Lung. *Ann NY Acad Sci* 174:255–260, 1970.

15. Brieger H: Carbon Monoxide Polycythemia. *J Ind Hyg Toxicol* 26:321–327, 1944.

16. Brinkman R, Lamberts HB, Veninga TS: Radiomimetic Toxicity of Ozonized Air. *Lancet* 1(7325):133–136, Jan. 1974.

17. Campbell JA: Tissue Oxygen Tension and Carbon Monoxide Poisoning. *J Physiol* (London) 68:81–96, 1929.

18. Carnow BW: Sulfur Oxides and Particles: Effects on Health. Proceedings of the National Academy of Science Conference on Health Effects of Air Pollution. Washington, DC, US Government Printing Office, 1973. Stock No. 5270-02105.

19. Carroll RE: The Relationship of Cadmium in the Air to Cardiovascular Disease Death Rates. *JAMA* 198:267–269, 1966.

20. Chance B, Erecinska M, Wagner M: Mitochondrial Responses to Carbon Monoxide Toxicity. *Ann NY Acad Sci* 174:193–204, 1970.

21. Chevalier RB, Krumholz RA, Ross JC: Reaction of Nonsmokers to Carbon Monoxide Inhalation. *JAMA* 198:135–138, 1966.

22. Chovin P: Carbon Monoxide: Analysis of Exhaust Gas Investigations in Paris. *Environ Res* 1:198–216, 1967.

23. Chiodi H, Dill DB, Consolazio F, et al: Respiratory and Circulatory Responses to Acute Carbon Monoxide Poisoning. *Am J Physiol* 134:683–693, 1941.

24. Clark RT Jr, Otis AB: Comparative Studies on Acclimatization of Mice to Carbon Monoxide and to Low Oxygen. *Am J Physiol* 169:285–294, 1952

25. Coburn RF, Forster RE, Kane PB: Consideration of the Physiologic Variables that Determine Blood Carbon Monoxide Hemoglobin Concentrations in Man. *J Clin Invest* 44:1899–1910, 1965.

26. Coburn RF (ed): Biological Effects of Carbon Monoxide. *Ann NY Acad Sci* 174(1): 1–430, 1970.

27. Coburn RF, Blakemore WS, Forster RE: Endogenous Carbon Monoxide Production in Man. *J Clin Invest* 42:1172–1178, 1969.

28. Coburn RF, Gondrie P, Abbound F, et al: Myocardial Myoglobin Oxygen Tension. *Am J Physiol* 224:870–876, 1973.

29. Cohen SI, Deane M, Goldsmith JR: Carbon Monoxide and Survival from Myocardial Infarction. *Arch Environ Health* 19:510–517, 1969.

30. Collier CR, Workman JM, Mohler JG, et al: Physiological Adaptations to Carbon Monoxide Levels and Exercise in Normal Men. Final Report. US Environmental Protection Agency. Washington, DC, Contract No. 68-02-0334. July 1972.

31. Cooper AG: *Carbon Monoxide:* A Bibliography with Abstracts. PHS Publication No. 1503, Washington, DC, US Government Printing Office, 1966.

32. Curphey TJ, Hood LPL, Perkins NM: Carboxyhemoglobin in Relation to Air Pollution and Smoking. Postmortem studies. *Arch Environ Health* 10:179–185, 1965.

33. Dahms TE, Horvath SM: Rapid Accurate Technique for Determination of Carbon Monoxide in Blood. *Clin Chem* 20:533–537, 1974.

34. DeBias DA, Banerjee CM, Birkhead NC, et al: Carbon Monoxide Inhalation Effects Following Myocardial Infarction in Monkeys. *Arch Environ Health* 27:161–167, 1973.

35. deBruin A: Carboxyhemoglobin Levels Due to Traffic Exhaust. *Arch Environ Health* 15:384–389, 1967.

36. Dinman BD, Eaton JW, Brewer GJ: Effects of Carbon Monoxide on DPG Concentrations in the Erythrocyte. *Ann NY Acad Sci* 174:246–251, 1970.

37. Drinkwater BL, Raven PB, Horvath SM, et al: Effect of Carbon Monoxide and Peroxyacetylnitrate on Man's Ability to Work in the Heat. *Arch Environ Health* 28:117–181, 1974.

38. DuBois AB (ed): Effects of Chronic Exposure to Low Levels of Carbon Monoxide on Human Health, Behavior and Performance. National Academy of Sciences, Washington, DC, 1969.

39. Ekblom B, Huot R: Response to Submaximal and Maximal Exercise at Different Levels of Carboxyhemoglobin. *Acta Physiol Scand* 86:474–482, 1972.

40. Estabrook RW, Franklin MR, Hildebrandt AG: Factors Influencing the Inhibitory Effect of Carbon Monoxide on Cytochrome P-450-catalyzed Mixed Function Oxidation Reactions. *Ann NY Acad Sci* 174:218–232, 1970.

41. Forbes WH, Sargent F, Roughton FJW: The Rate of Carbon Monoxide Uptake by Normal Men. *Am J Physiol* 143:594–608, 1945.

42. Friberg L: Proteinuria in Chronic Cadmium Poisoning After Comparatively Short Exposure to Cadmium Dust. (AMA) *Arch Ind Health* 16:30–31, 1957.

43. Garćia-Palmieri MR, Costas R Jr, Cruz-Vidal M: Risk Factors and Prevalence of Coronary Heart Disease in Puerto Rico. *Circulation* 42:541–549, 1970.

44. Goldberg HE, Cohen AA, Finklea JF: Frequency and Severity of Cardiopulmonary Symptoms in Adult Panels: 1970–1971 New York Studies. Health Consequences of Sulfur Oxides: A Report from CHESS, 1970–1971. US Environmental Protection Agency, EPA-650/1-74-004, May 1974.

45. Goldsmith JR, Landaw SA: Carbon Monoxide and Human Health. *Science* 162:1352–1359, 1968.

46. Goldsmith JR, Terzaghi J, Hackney JD: Evaluation of Fluctuating Carbon Monoxide Exposures. *Arch Environ Health* 7:647–663, 1963.

47. Gregg DE, Fisher LC: Blood Supply to the Heart: p. 1547. In: *Handbook of Physiology*. WF Hamilton, P Dow (eds) Section 2, Circulation. Washington, DC American Physiological Society.

48. Hackney JD, Kaufman GA, Lashier H, et al: Rebreathing Estimate of Carbon Monoxide Hemoglobin. *Arch Environ Health* 5:30–37, 1962.

49. Hammer DI, Finklea JF, Creason JP, et al: Cadmium Exposure and Human Health Effects. Trace substances in environmental health. V. A Symposium. DD Hemphill (ed), University of Missouri, Columbia, 1971, pp 269–283.

50. Hammer DI, Collucci AV, Hasselblad V, et al: Cadmium and Lead in Autopsy Tissues. *J Occup Med* 15:956–963, 1973.

51. Horvath SM, Raven PB, Dahms TE, et al: The Determination of Maximal Aerobic Power Consequent to Carbon Monoxide Exposure. *J Appl Physiol* (submitted).

52. Jenkins CE: The Hemoglobin Concentration of Workers Connected With Internal Combustion Engines. *J Hyg* 32:406–408, 1932.

53. Jones RH, Ellicott MP, Cadigan JB, et al: Relationship Between Alveolar and Blood Carbon Monoxide Concentrations During Breathholding. *J Lab Clin Med* 51:553–564, 1958.

54. Kahn A, Rutledge RB, David DL, et al: Carboxyhemoglobin Sources in the Metropolitan St. Louis Population. *Arch Environ Health* 29:127–135, 1974.

55. Kannel W, Gordon T: *The Framingham Study,* Section 26, 1970.

56. *Coronary Heart Disease in Seven Countries.* A Keys (ed). American Heart Association Monograph Number 29. *Circulation* 41 Suppl I:I–1–I–211, 1970.

57. Klausen K, Rasmussen B, Jellerod HG, et al: Circulation, Metabolism and Ventilation During Prolonged Exposure to Carbon Monoxide and to High Altitude. *Scand J Clin Lab Invest* Suppl. 103:26–38, 1968.

58. Kuller L, Radford E, Swift D, et al: The Relationship Between Ambient Carbon Monoxide Levels, Postmortem Carboxyhemoglobin, Sudden Death and Myocardial Infarction. (in press)

59. Mulhausen RO, Astrup P, Mellemgaard K: Oxygen Affinity and Acid-base Status of Human Blood During Exposure to Hypoxia and Carbon Monoxide. *Scand J Clin Lab Invest* 22 Suppl. 103:9-15, 1968.

60. Musselman NP, Groff WA, Yevich PP, et al: Continuous Exposure of Laboratory Animals to Low Concentration of Carbon Monoxide. *Aerosp Med* 30:524–529, 1959.

61. Neilsen B: Thermoregulation During Work in Carbon Monoxide Poisoning. *Acta Physiol Scand* 82:98-106, 1971.

62. Pirnay F, Dujardin J Jr, Deroanne R, et al: Muscular Exercise During Intoxication by Carbon Monoxide. *J Appl Physiol* 34:573–575, 1971.

63. Raven PB, Drinkwater BL, Horvath SM, et al: Effect of Carbon Monoxide and Peroxyacetylnitrate on Man's Maximal Aerobic Capacity. *J Appl Physiol* 36:288–293, 1974.

64. Raven PB, Drinkwater BL, Horvath SM, et al: Age, Smoking Habits, Heat Stress and Their Interactive Effects with Carbon Monoxide and Peroxyacetylnitrate on Man's Aerobic Power. *Int J Biometerol,* 1974.

65. Russell MA, Cole PV, Brown E: Absorption by Non-smokers of Carbon Monoxide from Room Air Polluted by Tobacco Smoke. *Lancet* 1:576–579, 1973.

66. Schroeder HA, Buckman J: Cadmium Hypertension. *Arch Environ Health* 14:693–697, 1962.

67. Sjostrand T: Endogenous Formation of Carbon Monoxide in Man Under Normal and Pathological Conditions. *Scand J Clin Lab Invest* 1:201–214, 1949.

68. Sjostrand T: Method for the Determination of Carboxyhemoglobin Concentrations by Analysis of Alveolar Air. *Acta Physiol Scand* 16:201–210, 1948.

69. Smokler PE, MacAlpin RN, Alvaro A, et al: Reproducibility of a Multistage Near Maximal Treadmill Test for Exercise Tolerance in Angina Pectoris. *Circulation* 48:346–351, 1973.

70. Stewart RD, Baretta ED, Platte LR, et al: Carboxyhemoglobin Levels in American Blood Donors. *JAMA* 229:1187–1195, 1974.

71. Stewart RD, Baretta ED, Platte LR, et al: "Normal" Carboxyhemoglobin Levels of Blood Donors in the United States. (Report No.ENVIR MED MCW CRC-COHb-73–1) Milwaukee: Med. Clg. of Wisc. 1973.

72. Stewart RD, Fischer TN, Hosco MJ, et al: Experimental Human Exposure to Methylene Chloride. *Arch Environ Health* 25:342–348, 1972.

73. Syvertsen GR, Harris JA: Erythroprotein Production in Dogs Exposed to High Altitude and Carbon Monoxide. *Am J Physiol* 225:293–299, 1973.

74. US Department of Health, Education, and Welfare, Public Health Service, Environmental Health Service, National Air Pollution Control Administration. *Air Quality Criteria for Carbon Monoxide.* NAPCA Publication No. AP-62. Washington, DC, US Government Printing Office, March 1970.

75. Weinblatt E, Frank CW, Shapiro S, et al: Prognostic Factors in Angina Pectoris — A prospective study. *J Chronic Dis* 21:231–245, 1968.

76. Weinblatt E, Shapiro S, Frank CW, et al: Prognosis of Men After First Myocardial Infarction: Mortality and First Recurrence in Relation to Selected Parameters. *Am J Public Health* 58:1329–1347, 1968.

77. Vogel JA, Gleser MA, Wheeler RC, et al: Carbon Monoxide and Physical Work Capacity. *Arch Environ Health* 24:198–203, 1972.

78. Vogel JA, Gleser MA: Effect of Carbon Monoxide on Oxygen-transport During Exercise. *J Appl Physiol* 32:234–239, 1972.

79. Wohlrab H, Ogunmola GB: Carbon Monoxide Binding Studies of Cytochrome $a_3$ hemes in Intact Liver Mitochondria. *Biochemistry* 10(7):1103–1106, 1971.

80. Zorn H: The Partial Oxygen Pressure in the Brain and Liver at Subtoxic Concentrations of Carbon Monoxide. *Staub-Reinhalt Luft* 32:24–29, 1972.

# 8 Effect of CO on Angina, Intermittent Claudication, and Hemodynamics

Wilbert S. Aronow, M.D.

## EFFECT OF CO ON EXERCISE-INDUCED ANGINA

We previously demonstrated that patients with angina pectoris due to documented coronary artery disease showed a decrease in exercise performance until the onset of angina following less myocardial work after CO exposure from smoking non-nicotine cigarettes or from heavy freeway traffic.[1,2] These data led us to design a double-blind, randomized study that evaluated the effect of breathing 50 ppm of CO for two hours versus breathing compressed, purified air for two hours on exercise-induced angina in 10 patients with angina pectoris due to angiographically documented significant coronary artery disease.[3] The results from these tests performed on four successive study mornings are presented here.

The mean venous carboxyhemoglobin level significantly increased from 1.03% to 2.68% after the breathing of 50 ppm of CO for two hours ($P<0.001$) and significantly decreased from 1.07% to 0.77% after the breathing of compressed, purified air ($P<0.001$). All of the hemoglobin values were within normal limits.

The mean exercise time until the onset of angina was 226.7 seconds in the control period and 223.0 seconds after compressed, purified air was breathed (P not significant). However, the mean exercise time until the onset of angina significantly decreased from 224.3 seconds in the control period to 187.6 seconds after the breathing of 50 ppm of CO for two hours ($P < 0.001$).

No significant differences were found between the resting mean systolic blood pressure, diastolic blood pressure, heart rate, and product of systolic blood pressure times heart rate values during the eight different study periods.

The mean systolic blood pressure at the onset of angina significantly decreased from 153.5 to 149.4 mm Hg after the breathing of CO ($P < 0.001$) but was not significantly different from the control period after patients breathed compressed, purified air.

The mean heart rate at the onset of angina significantly decreased from 129.2 to 120 beats per minute after the breathing of CO ($P < 0.001$) but was not significantly different from the control period after patients breathed compressed, purified air.

The mean product of systolic blood pressure times heart rate/100 at the onset of angina significantly decreased from 199.2 to 180 after breathing CO ($P < 0.001$) but was not significantly different from the control period after patients breathed compressed, purified air.

Ischemic S-T segment depression was not observed in the Holter electrocardiographic recordings in any of the study periods while the patients were breathing either CO or compressed, purified air. The maximal amount of ischemic S-T segment depression observed after exercise-induced angina pectoris was 1.35 mm in the control period compared with 1.33 mm after the breathing of compressed, purified air and 1.30 mm in the control period compared with 1.45 mm after the breathing of CO. These differences are not statistically significant.

The data from this double-blind study show that the increase in carboxyhemoglobin level after the breathing of 50 ppm of CO for two hours is of sufficient magnitude to cause angina pectoris to develop sooner in patients with angina and after less work. Since our angina patients could not adequately increase their coronary blood flow while exercising, and since their elevated carboxyhemoglobin level made less oxygen deliverable to the myocardium, their myocardial oxygen demand exceeded their myocardial oxygen supply, and angina pectoris was induced sooner, after less myocardial work.

Anderson and associates demonstrated in a double-blind study that patients with angina pectoris who breathed 50 ppm of CO for four hours increased their mean venous carboxyhemoglobin level from 1.3% to 2.9%.[4] Their patients with angina had a significant decrease in exercise performance until angina after exposure to CO when compared to their breathing compressed, purified air ($P < 0.005$). Ischemic S-T segment depression after angina in general appeared earlier and was deeper after 50 ppm of CO was breathed when compared to the breathing of compressed, purified air.

We found an increase in maximal ischemic S-T segment depression from 1.30 mm to 1.45 mm after exercise-induced angina after the breathing of 50 ppm of CO for two hours compared to the control period. This increase in maximal ischemic S-T segment depression after angina was not statistically significant. However, ischemic S-T segment depression after exercise-induced angina occurred earlier and after less myocardial work in our patients after they breathed CO compared to the control period.

The data reported by Ayres and associates,[5,6] our previously rereported data showing a decrease in exercise performance until angina following less myocardial work after CO exposure due to smoking non-nicotine cigarettes[1] or from heavy freeway traffic,[2] the data reported by Anderson and associates in their double-blind study,[4] and the data from our double-blind study[3] led us to conclude,[7] as did Goldsmith,[8] that CO exposure aggravates angina pectoris. Therefore, patients with angina pectoris should avoid heavy atmospheric pollution and smoking.

## EFFECT OF CO ON INTERMITTENT CLAUDICATION

These data led us to wonder what effect CO exposure would have on intermittent claudication due to occlusive arterial disease. We therefore performed a double-blind, randomized study evaluating the effect of breathing 50 ppm of CO for two hours versus compressed, purified air for two hours on intermittent claudication in 10 patients with classic intermittent claudication of the calf or thigh due to angiographically documented occlusive ilio-femoral arterial disease with patent distal vessels.[9] The results from this double-blind study performed on four successive study mornings are discussed here.

The mean venous carboxyhemoglobin level significantly increased from 1.08% to 2.77% after 50 ppm of CO was breathed for two hours ($P < 0.001$). The mean venous carboxyhemoglobin level decreased from 1.12 to 0.90% after compressed, purified air was breathed (P not significant). All of the hemoglobin values were within normal limits.

The mean exercise time until the onset of intermittent claudication was 168.8 seconds in the control period and 173.0 seconds after the breathing of compressed, purified air (P not significant). However, the mean exercise time until the onset of intermittent claudication significantly decreased from 174.3 seconds in the control period to 144.0 seconds after the breathing of CO for two hours ($P < 0.001$).

Since our patients with documented ilio-femoral occlusive arterial disease could not adequately increase the blood flow to their thigh and calf muscles while exercising, and since the elevated carboxyhemoglobin levels caused by carbon monoxide inhalation made less oxygen deliverable to the thigh and calf muscles, the oxygen demand exceeded the oxygen supply to these muscles, so that intermittent claudication was induced sooner after equivalent exercise. That the intermittent claudication-inducing effect was related to carbon monoxide acquisition is supported by the absence of this effect when compressed, purified air was breathed by our patients during their double-blind study.

## EFFECT OF CO ON CARDIOVASCULAR HEMODYNAMICS

We also reported data showing the effect of cigarette smoking[10,11] and breathing carbon monoxide[11] on cardiovascular hemodynamics in patients with angina pectoris. In one of these studies, eight cigarette smokers with severe angina pectoris who needed cardiac catheterization and coronary angiography had two cardiac catheterizations performed one week apart.

Hemodynamic measurements were made during the first cardiac catheterization before and after the patients smoked three high-nicotine cigarettes. The subject began to smoke his second cigarette 5 minutes after he finished smoking his first cigarette, and he began to smoke his third cigarette 30 minutes after he finished smoking his second cigarette. Hemodynamic measurements were made during the second cardiac catheterization before and after the patient breathed 150 ppm of CO from a tank through a mask until his increase in coronary sinus CO level was similar to that experienced after he smoked his third cigarette.

A significant increase in mean coronary sinus, venous, and arterial carboxyhemoglobin levels occurred after the breathing of CO ($P < 0.001$). The mean coronary sinus carboxyhemoglobin level increased 1.82%, from 2.04% to 3.86%. The mean venous carboxyhemoglobin level increased 1.80%, from 2.20% to 4.00%. The mean arterial carboxyhemoglobin level increased 2.16%, from 2.05% to 4.21%.

Significant decreases in mean coronary sinus, venous, and arterial $PO_2$ levels occurred after the breathing of CO ($P < 0.001$). Coronary sinus $PO_2$ changed from 21.3 to 18.1 mm Hg, venous $PO_2$ decreased from 38.5 to 34.3 mm Hg, and arterial $PO_2$ shifted from 85.0 to 81.8 mm Hg.

The systolic arterial pressure, diastolic arterial pressure, heart rate, and aortic systolic ejection period did not significantly change after the patients breathed CO. The mean left ventricular dp/dt significantly decreased from 1,203 to 1,020 mm Hg/sec after CO ($P < 0.001$), and the mean left ventricular end-diastolic pressure significantly increased from 9.8 to 10.8 mm Hg after CO ($P < 0.01$). The mean cardiac index significantly decreased from 3.11 to 2.57 liters/minute/m$^2$ after CO ($P < 0.001$) and the mean stroke index significantly decreased from 43.6 to 35.4 ml/beat/m$^2$ after CO ($P < 0.001$).

These data show that breathing 150 ppm of CO to produce an increase in mean coronary sinus carboxyhemoglobin level of 1.82% caused a negative inotropic effect on the myocardium in patients with angina pectoris with a significant reduction in left ventricular dp/dt, stroke index, and cardiac index and a significant rise in left ventricular end-diastolic pressure.

## REFERENCES

1. Aronow WS, Rokaw SN: Carboxyhemoglobin caused by smoking nonnicotine cigarettes. Effects in angina pectoris. *Circulation* 44:782–788, 1971.

2. Aronow WS, et al: Effect of freeway travel on angina pectoris. *Ann Intern Med* 77:669–676, 1972.

3. Aronow WS, Isbell MW: Carbon monoxide effect on exercise-induced angina pectoris. *Ann Intern Med* 79:392–395, 1973.

4. Anderson EW, et al: Effect of low-level carbon monoxide exposure on onset and duration of angina pectoris. A study in ten patients with ischemic heart disease. *Ann Intern Med* 79:46-50, 1973.

5. Ayres SM, et al: Systemic and myocardial hemodynamic responses to relatively small concentrations of carboxyhemoglobin (COHB). *Arch Environ Health 18:699*–709, 1969.

6. Ayres SM, Gianelli S Jr, Mueller HS: Effect of low concentrations of carbon monoxide. Myocardial and systemic responses to carboxyhemoglobin. *Ann NY Acad Sci* 147:268–293, 1970.

7. Aronow WS: Smoking, carbon monoxide, and coronary heart disease. Circulation 48:1169–1172, 1973 (Editorial).

8. Goldsmith JR: Carbon monoxide and coronary heart disease: compelling evidence in angina pectoris. *Ann Intern Med* 77:808–810, 1972 (Editorial).

9. Aronow WS, Stemmer EA, Isbell MW: Effect of carbon monoxide exposure on intermittent claudication. *Circulation* 49:415–417, 1974.

10. Aronow WS, et al: Effect of smoking cigarettes on cardiovascular hemodynamics. *Arch Environ Health* 28:330–332, 1974.

11. Aronow WS, et al: Effect of cigarette smoking and breathing carbon monoxide on cardiovascular hemodynamics in angina patients. *Circulation* 50:340–347, 1974.

# 9 CO and Cardiac Performance

Edward Haak, M.D.

This presentation is in the form of a progress report on two separate studies recently completed in our laboratory. Both studies are part of a research program to determine the cardiac effects of low-level CO exposure. The first study deals with the effect of low-level CO exposure on the exercise performance of young healthy men. The second study is concerned with the effect of low-level CO exposure on the ischemia associated with acute myocardial infarction in dogs.

Earlier studies in our laboratory demonstrated that carboxyhemoglobin levels of 5% to 7% produced a significant reduction in exercise time to target heart rate in both young and middle-aged men. These studies were without the benefit of a double-blind protocol and sophisticated exercise testing equipment. For these reasons we have re-evaluated the effect of low-level CO exposure on young healthy subjects with the use of more sophisticated equipment and a more elaborate protocol that is similar to that used with the angina patients. In addition to studying the effects of carbon monoxide on exercise performance, we evaluated the effect of CO on cholesterol and triglycerides, blood clotting factors, platelets, and platelet agglutination. This investigation was done in light of reports of changes in these parameters in smokers and the association between smoking and carboxyhemoglobin levels.

The subjects were young healthy men between the ages of 20 and 40, both smokers and non-smokers. Screening procedures included a routine history and physical examination and a battery of laboratory tests, which included chest x-ray, ECG, cholesterol, triglycerides, clotting studies, hematocrit, and platelet count. As in the study of patients with angina pectoris, low levels of carbon monoxide, either 50 or 100 ppm in compressed air, were administered by a closed face mask system for a four-hour period. On control days, compressed air was administered in the same manner so that both the subject and the exercise technician were unaware of the treatment of the day.

The exposure protocol for five consecutive days was as follows: Monday, Wednesday, and Friday were the clean air days that tested for a training effect in order to be sure that the subject did not improve just with repeated exercise testing. The CO exposure was on Tuesday and Thursday, and it was totally randomized as to whether it was a 50- or 100-ppm day. This arrangement also enabled us to test the effect of venesection since we required a great deal of blood from these subjects in order to test for these other parameters.

The daily protocol, which was used every day of the week, was as follows: The subject came to the laboratory fasting, and smokers were asked not to smoke between 2 A.M. and 8 A.M., although they could smoke between 2 P.M. and 8 P.M. on the day preceding a test day. Blood samples were drawn for a baseline, again before exercise four hours later, and post-exercise. Comparison of baseline and pre-exercise values allowed us to look for a resting effect of CO; a comparison of pre-exercise and post-exercise values allowed us to see whether or not CO had any effect on the changes associated with these parameters in exercise. For the four-hour exposure the patient usually sat in a chair at rest. The exercise was performed on a multi-staged Quinton type treadmill with an automatic program. Each subject was exercised on the same program and each subject started at his own point every day. He was exercised until he obtained 85% of his maximum heart rate, and we recorded as the exercise performance the time from the starting of the exercise to the achievement of the 85% level of maximum heart rate. The heart rate was monitored by a Quinton cardio-tachometer that was calibrated daily and that provided a very sharp end point since it displays heart rate continually throughout exercise and post-exercise. Blood pressure was measured at two-minute intervals. The ECG was recorded continuously during exercise and post-exercise, three leads at a time.

The 50-ppm CO exposure raised the COHb level for the group from 1% to 3%, and the 100-ppm exposure resulted in about a 5% increase. There was really not much difference between the step-wise increase of smokers versus non-smokers, but there was some difference in baseline, especially since the smokers did not totally abstain from smoking. The increase, though, from the baseline to the 50-ppm exposure measurement is roughly the same, with no statistical difference in performance. The overall mean exercise time for the group of 22 subjects on air is 507 seconds compared with 476 on 100 ppm and 499 after 50 ppm exposure. This difference of 31 seconds is roughly a 6% decrease in the exercise performance of these young healthy men. Analysis of variance revealed that there was a significant difference between air and CO. Furthermore, air versus 100 ppm is significant at the 0.01 level, while the decreased performance on 50 ppm is not significant. Another important finding is that there was no significant difference between smokers and non-smokers. Since 100 ppm corresponds to 5% COHb and 50 ppm to 3% COHb, there must be a threshold for this effect somewhere between 3% and 5%. There was no significant effect of CO on the screening test of cholesterol, triglycerides, clotting functions, or platelets.

In conclusion, this study demonstrates that in young healthy men carboxyhemoglobin levels of 5% significantly reduce exercise performance in terms of the time to target heart rate and that the threshold of this effect probably lies between 3% and 5%. The burden of 5% carboxyhemoglobin in these young men appears to require a higher level of cardiac work to achieve the same level of exercise performance. The reduced capacity for cardiac work is characteristically seen in people with clinically evident coronary artery disease. This condition makes them more susceptible to the adverse cardiac effects of carbon monoxide, a finding noted in the previous study. This increased sensitivity has been demonstrated in our laboratory in the study of patients with angina pectoris at levels as low as 2% to 3% and confirmed by Doctor Aronow. Doctor Aronow has reported that 5% carboxyhemoglobin has produced a shift to more anaerobic cardiac metabolism in patients with coronary artery disease at the time of catheterization.

These studies, as well as others, provide us with important information on some of the cardiac effects of CO. However, these studies do not allow us to predict or to estimate the effect of low levels of carboxyhemoglobin on an acute myocardial infarction. We studied the effect of carbon monoxide on ischemia associated with experimental myocardial infarction in dogs. This model was developed by Becker et al. and employs a cross correlation of epicardial ECG mapping as a measure of ischemia with measurements of myocardial blood flow by the radioactive microsphere technique.

Twelve male mongrel dogs were used in this study. They were anesthetized and placed on a respirator. The chest was entered through the left thoracotomy, the pericardium was incised, and the left inferior descending coronary artery and its superficial free wall branches were exposed. The aortic root pressure was measured with a catheter placed in the aorta, and the catheter placed in the left atrium was also used for the injection of the radioactive microspheres into the circumflex artery, the left anterior descending coronary artery, and the branches that would be involved in producing an infarct of these small branches on the superficial left ventricular wall. We created a coronary occlusion by ligating the small branches; we then were able to tell whether there was an infarction by watching for ST segment elevation on the ECG. We kept ligating vessels until we did produce an infarct.

After the infarct was stabilized for 60 minutes, we then began a series of CO interventions. The purpose of this whole study was to see if, given a stable infarct and stable levels of ST elevation, we could enhance that ST elevation through raising the carboxyhemoglobin level of these animals. Five CO interventions took place that consisted of respiratory exposures of 10 liters each at 5,000 ppm that lasted about three minutes. The carboxyhemoglobin levels went to 4%, to roughly 8%, and then up to 17%. Radioactive microspheres were injected at three points: at the baseline, after the second intervention, and after the fifth intervention. The radioactive microspheres, 7$\mu$ to 10 $\mu$ in diameter, were injected in solution into the left atrium. The spheres are designed to lodge in the pre-capillary arterial bed, and they do not pass the capillary bed. They are not in sufficient number to change the hemodynamics or the flow characteristics, but they do mark flow. By using three different gamma levels, we are able to go back retrospectively and count the gamma level present in that cube of muscle and determine the three different flows present at the time of injection of each of those microspheres. This is a complicated technique that I am describing only briefly.

After the final intervention, we sacrificed the dog, removed the heart, and dissected the left ventricular free wall, which was now radioactive. We laid it flat, made a map of all the ligated vessels, and then we diced this heart into small cubes for counting. Using a reference area on the posterior wall well removed from the distribution of the left anterior descending artery, we could determine the non-ischemic flow. We then defined ischemic flow as 50% of non-ischemic flow. We then divided the heart up into two areas on the basis of ischemic flow and non-ischemic flow, and we found good agreement between epicardial mapping data delineating the area of infarct and the flow data defining the ischemic muscle. Every time we did epicardial mapping to see what the ST elevation was at each site, we would sum all the sites for a given dog. After each intervention, the mean for the nine dogs increased dramatically fron 34.4 to 55.9. The effect of increasing the levels of carboxyhemoglobin, then, was to increase ischemia and ST elevation for these dogs. There was no significant change in all those parameters that would indicate increased myocardial oxygen demand, such as heart rate, left atrial pressure, or cardiac output.

The left ventricular radioactive microsphere flow data were expressed in terms of cc per gram of part per minute. The two areas, previously defined, consisted of the ischemic area, which corresponded roughly to the infarct area, and a non-ischemic area. The small cubes of heart were then cut into an inner half and an outer half so that we could look at sub-epicardial and sub-endocardial flow. For the ischemic area there was essentially no change from baseline to 8.4% COHb to 17.3% COHb , but dogs have a fantastic collateral circulation. Non-ischemic flow did increase somewhat, and we think that the effect of the carboxyhemoglobin on the non-ischemic tissue is to increase the blood flow.

In conclusion, we have demonstrated that carboxyhemoglobin levels as low as 4.8% are associated with increased myocardial ischemia in terms of ST elevation at the time of experimental acute myocardial infarction in dogs. Based on this study, we are able to demonstrate a dose-response relationship between the level of carboxyhemoglobin and its effect of increasing myocardial ischemia in a setting of an acute myocardial infarction. It is known that dogs have a better collateral network of coronary blood vessels than man; and for this reason, they are more tolerant of coronary occlusion. It could be expected, then, that man would at least be sensitive to the presence of carboxyhemoglobin in terms of its effect of increasing ischemia at the time of acute myocardial infarction.

# 10 Effect of Carbon Monoxide Inhalation On the Vulnerability of the Heart to Induced Fibrillation

D.A. BeBias, Ph.D.

## INTRODUCTION

Ventricular fibrillation (VF) is an irregular, asynchronous, and incoordinate contractile activity of the ventricular myocardium, and is usually a terminal cardiac event. VF is reversible, and its termination can be spontaneous or can be induced by artificial means such as passing a strong electrical countershock through the heart which momentarily depresses all cardiac activity.

VF can be induced artifically by several agents; the most common technique used experimentally to initiate VF is application of an electrical stimulus to the myocardium during the final stage of ventricular repolarization, which is the most vulnerable period for fibrillation.

Organic heart disease predisposes VF, and the probability of VF increases with myocardial infarction. The vulnerability of the myocardium is also enhanced by severe myocardial ischemia resulting from a variety of causes that tend to reduce the availability of oxygen to the myocardium. Of special interest in this investigation is the reduced oxygen-carrying capacity of the hemoglobin following carbon monoxide inhalation.

It was previously reported[1] that continuous inhalation of carbon monoxide (100 ppm) resulted in myocardial effects in normal monkeys and monkeys with myocardial infarction. ECG P-wave amplitudes were elevated in both groups, and inhalation of carbon monoxide resulted in a higher incidence of T-wave inversion in infarcted animals than in noninfarcted animals, a finding suggesting a greater degree of myocardial ischemia in infarcted animals.

Since inhalation of 100 ppm carbon monoxide resulted in demonstrable ECG effects on the myocardium, the present study was designed to determine whether carbon monoxide inhalation also influenced the susceptibility of the ventricles to induced fibrillation.

## EXPERIMENTAL PROCEDURES

**Experimental animal** Cynomolgus monkeys *(Macaca irus irus)* weighing approximately 3.5 kg were used. The animals were stabilized (clinically good health) by the vendor (Primate Imports Corporation) and guaranteed on arrival to be free from TB and other infections and communicable diseases; subsequent checks were made on their general health periodically by the house veterinarian. The animals were fed a standard commercial monkey chow supplemented twice weekly with fresh apples and/or kale. Water was available *ad libitum*.

Twenty animals were randomly selected from the stock and assigned to an air group or a carbon monoxide group. Myocardial infarction was induced in nine of these animals. Air-breathing animals and carbon monoxide-breathing animals were housed in their respective chambers for one week prior to initiation of the carbon monoxide exposure procedures so they could acclimatize to their surroundings. Precautions were taken to maintain uniform surroundings (i.e., noise, lighting, temperature, humidity) and to minimize distractions.

**Exposure chamber** The carbon monoxide exposure chamber was supplied with fresh air (8.4 CMM) from an air-conditioning system which was thermostatistically regulated. The carbon monoxide (100%) was injected into an air inlet elbow located 3 m from the chamber. The inlet duct served as a mixing box to achieve a homogenous mixture of carbon monoxide and air entering the chamber. The carbon monoxide level in the chamber was continuously monitored by a Model 200 Lira Infrared Analyzer (MSA), and the inflow was regulated through a solenoid valve arrangement electrically coupled to the monitor; the chamber carbon monoxide ranged between 96 and 102 ppm. Based on the total volume of the chamber, the CMM inflow, and CMM of the exhaust fans, the exchange rate of the chamber gas mixture was calculated to be 21 times per hour (i.e., the chamber gas mixture was completely renewed 21 times per hour, a procedure that maintained adequate ventilation of the chamber as well as elimination of expired $CO_2$).

**Induced myocardial infarction** The procedure for MI induction has been previously described.[1] The animals were tranquilized with 10 mg phencyclidine hydrochloride (Sernylan, Bio-centic Laboratories) injected i.m. An opaque preshaped catheter (1.4 mm I.D., 1.57 mm O.D., Becton-Dickenson) was guided via the carotid artery into the ascending aorta; the catheter was maneuvered selectively into the ostium of the left coronary artery with the aid of image intensifying fluoroscopy. Approximately 0.05 ml of latex microspheres (styrene divinyl-benzene copolymer; Dow Chemical Co.) measuring $300\mu$ in diameter was suspended in 0.5 ml Renograffin-60 (Squibb) and delivered through the catheter into the left coronary artery. The catheter was then withdrawn and hemostasis achieved. Successful induction of infarction was determined by electrocardiographic changes consistent with myocardial infarction. The hearts were examined grossly and microscopically at autopsy, and infarction confirmed.

**Electrocardiogram** The ECG was recorded on the Hewlett-Packard ECG Model 1511 A. Standard limb leads were used routinely with the animal lying on his back.

**Blood carboxyhemoglobin content** Carboxyhemoglobin (COHb) content of the blood was determined by use of the CO-oximeter (Model 182, I.L. Inc.). The CO-oximeter was calibrated against three standard blood samples: one sample equilibrated with 100% CO gas, one sample equilibrated with a mixture of 95% $O_2$ and 5% $CO_2$ gas, and one sample free of both $O_2$ and $CO_2$ (Sodium hydro-sulfite treated).

**Body temperature** Rectal temperatures were monitored by use of a telethermometer (Model 47, Yellow Spring Instruments, Inc.). Ordinarily, body temperature is maintained quite uniformly in the normal monkey. In this study, however, the animals were "sernylanized" for the duration of the experimental period; since sernyland tends to induce hypothermia, covering the animals with an electric heating pad reduced heat loss and body temperature was maintained within acceptable limits.

**Ventricular fibrillation induction** It was mentioned above that VF can be induced by applying an electrical current to the heart, and the electrical stimulus required to induce fibrillation can be quantified. Ventricular fibrillation was induced by delivering 60 Hertz alternating current (AC) across the heart by placing paddle electrodes on the chest wall. A range of 0 to 110 volts was available through a variable transformer.

The equipment arrangement to induce VF is shown diagrammatically in Figure 1. The heart rate and ECG of the animal were monitored on the oscilloscope (Electronics for Medicine, Model 1R-1). The trace signal was amplified and fed into a stimulator (Grass, Model S48), which was set to deliver 150 milliseconds of AC when triggered by the R-wave of the ECG monitored by the oscilloscope. The AC train was fed into a stimulus isolation unit (Grass, Model SIU-I) which removes artifacts from opening and closing the circuit. The train is then fed through the AC defibrillator (Dallon), which is connected to the external paddle electrodes via a variable transformer. It was determined experimentally that the 150 millisecond train, which was initiated by the R-wave, was adequate to cover the vulnerable period of the ventricle at most encountered heart rates.

The lowest current applied through this system that induced VF for 10 or more seconds was determined for each animal and was established by several inductions at the same voltage; this was possible after defibrillating the heart by delivering single 200-volt shocks via the paddles (by-passing the variable transformer). Successful induction of VF was confirmed by electrocardiographic monitoring and the absence of a detectable heart sound.

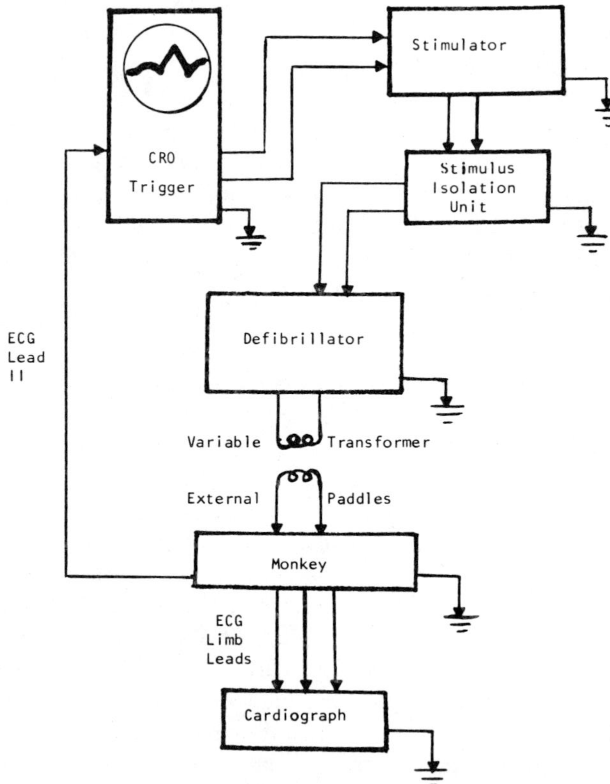


**Figure 1.** The block diagram shows the equipment arrangement for determining VFT and for ECG recording. ECG Lead II is monitored by an Electronics for Medicine ratemeter (CRO); the R-wave remotely triggers the stimulator to deliver a train of impulses over a period of 150 m sec. The pulses are fed into a stimulus isolation unit and through the defibrillator. The intensity of the impulses emitted from the defibrillator and through the chest paddles is regulated by variable transformer. Induction of ventricular fibrillation is confirmed by the ECG tracing.

**Blood determinations** Arterial (carotid) blood $P_{O_2}$, $P_{CO_2}$ and pH were determined with a physiologic gas analyzer (Beckman, Model 160). Blood was drawn into a heparinized syringe (air excluded), stored in ice, and analyzed within 15 minutes from the time it was drawn. Hb and $HbO_2$ blood levels (CO-oximeter) were monitored to establish that the monkeys were not anemic.

**Experimental design** The animals were divided into four groups: Group I consisted of normal, air breathing animals; Group II consisted of infarcted, air breathing animals; Group III consisted of normal animals exposed to carbon monoxide; Group IV consisted of infarcted animals exposed to carbon monoxide.

Immediately after infarction the carbon monoxide animals were placed in the exposure chamber for six hours and without assisted breathing. Air breathing animals remained in ambient laboratory air for the six-hour period. The susceptibility of the heart to induced fibrillation was determined for each animal immediately after the exposure period.

Blood samples were drawn via the carotid artery catheter immediately before infarction, after infarction was induced, and at end of the six-hour exposure period (pre-fibrillation).

**Pathology** At autopsy the monkey hearts were examined in the fresh state, placed in separate plastic bags with cold, fresh saline, and presented within two hours after sacrifice to the pathologist who was uninformed as to the pre-sacrifice status of the animal. The right atrial appendage was cut, and a cut was made through the lateral border of the right atrium and right ventricle. The left atrial appendage was cut and a similar lateral slit was made in the left atrial and ventricular borders. After the right and left borders of the heart were opened, the endocardial surface was examined. Serial sections in ''breaf-loaf'' fashion were made from the apex to, and including, the mitral and tricuspid rings. The tissues were fixed in 10% buffered formalin, embedded in wax, and stained with haematoxylin and eosin. The sections were examined for evidence of infarction consistent with the time of production, for extension of the infarction, and for evidence of more recent changes.

## RESULTS

The blood COHb levels of monkeys exposed to CO for six hours (9.3 ± 1.8%) were significantly higher than those of air breathing animals (1.1 ± 0.4%) (Table 1). Although the COHb levels appear to be slightly lower in non-infarcted animals than in the infarcted animals, they were not statistically different; the slightly higher COHb levels may be attributable to the respiratory pattern of the infarcted animals, which were probably still in cardiogenic shock when placed in the CO chamber. In a previous study[1] it was established that the blood COHb at equilibrium for animals breathing 100 ppm CO was 12.4% after 24 hours of continuous CO inhalation.

Blood gases ($P_{CO_2}$, $P_{O_2}$) and pH were monitored as an index of respiratory efficiency; also it is well known that wide variations in the $P_{CO_2}$ and pH can affect the susceptibility of the heart to fibrillation. The values for these parameters are presented in Table 2. As would be expected *a priori*, the blood $P_{O_2}$ of CO breathing animals was significantly lower than air breathing animals. Small fluctuations in blood $P_{CO_2}$ and pH throughout the experimental period were not statistically significant and therefore were not considered to be of consequence.

Hb concentrations did not change during the experimental period, whereas $HbO_2$ levels were significantly lower in the CO-inhaling animals, as would be expected.

Mean VF induction levels for each of the groups are shown in Table 3. Heart rates and body temperatures are included as indicators of the general condition of the animals. VF was induced in non-infarcted air-breathing animals at 78.5 volts. Animals that were infarcted and maintained on ambient air had a significantly lower VF level of 50.6 volts, an indication that susceptibility to VF was enhanced by MI. While normal animals inhaling CO fibrillated at 45 volts, those that were infarcted and inhaled CO fibrillated at 30 volts. It is, therefore, apparent that MI ($P(F) > 0.025$) as well as CO inhalation ($P(F) < 0.01$) are significant factors in enhancing VF although there does not appear to be interaction ($P(F) > 0.05$) between the two.

Since it is known that heart rate, especially very high rates, can increase the susceptibility of the ventricular fibrillation, an analysis was made to determine if the respective heart rates of each group of animals may have influenced the susceptibility to VF. An analysis of variance and

**Table 1**
**Mean Blood COHb of Monkeys Following 100 ppm Carbon Monoxide Inhalation for Six Hours and of Monkeys Breathing Ambient Air**

| Group Status | Number of Animals | % COHb ± S.D. | P(t) |
|---|---|---|---|
| Non-infarct, ambient air | 6 | 1.2 ± 0.4 | |
| Infarct, ambient air | 4 | 1.1 ± 0.4 | N.S. |
| Non-infarct, CO | 5 | 8.4 ± 1.6 | |
| Infarct, CO | 5 | 10.2 ± 1.8 | N.S. |
| All air-breathing | 10 | 1.1 ± 0.4 | |
| All CO-breathing | 10 | 9.3 ± 1.8 | 0.001 |

Note: Values are means ± S.D.

**Table 2**
**Mean Arterial Blood Parameters of Monkeys Breathing Ambient Air or 100 ppm Carbon Monoxide for Six Hours**

| Animal Group | Number of Animals | Sample Time* | $P_{O_2}$ mm Hg | P | $P_{CO_2}$ mm Hg | P | pH | P | $HbO_2$ % S | P | Hb g% | P |
|---|---|---|---|---|---|---|---|---|---|---|---|---|
| Non-infarct, air | 6 | Initial | 96.7 ± 1.2 | NS | 43.3 ± 2.6 | NS | 7.4 ± 0.1 | NS | 97.0 ± 2.0 | NS | 13.0 ± 1.3 | NS |
| | | Pre-VF | 96.7 ± 1.6 | | 42.7 ± 3.0 | | 7.4 ± 0.1 | | 98.0 ± 1.3 | | 12.8 ± 0.4 | |
| Infarcted, air | 4 | Initial | 96.0 ± 3.2 | NS | 41.0 ± 6.1 | NS | 7.4 ± 0.1 | NS | 96.7 ± 2.9 | NS | 12.1 ± 1.5 | NS |
| | | Pre-VF | 96.3 ± 6.3 | | 40.3 ± 10.0 | | 7.4 ± 0.1 | | 98.3 ± 2.2 | | 10.8 ± 1.1 | |
| Non-infarct, carbon monoxide | 5 | Initial | 94.0 ± 3.2 | <0.01 | 44.8 ± 3.9 | NS | 7.5 ± 0.1 | NS | 96.2 ± 1.4 | <0.01 | 12.8 ± 1.3 | NS |
| | | Pre-VF | 85.6 ± 5.1 | | 46.2 ± 16.0 | | 7.4 ± 0.7 | | 84.2 ± 3.5 | | 12.8 ± 1.4 | |
| Infarcted, carbon monoxide | 5 | Initial | 92.2 ± 2.3 | <0.02 | 46.2 ± 12.5 | NS | 7.5 ± 0.1 | NS | 97.2 ± 2.8 | <0.01 | 11.8 ± 0.8 | NS |
| | | Pre-VF | 84.0 ± 5.4 | | 41.4 ± 8.0 | | 7.5 ± 0.1 | | 81.4 ± 4.4 | | 10.6 ± 1.2 | |

*Initial samples were taken and immediately prior to induced fibrillation (after 6 hours CO inhalation the second sample was taken) (Pre-VF).
Note: Values given are means ± S.D. P (T) values > 0.05 were considered non-significant (NS).

the regression co-efficient indicate that heart rates recorded immediately before VF induction did not contribute to changes in the susceptibility that were attributed to MI and to Co inhalation.

In a previous study[1] utilizing the same technique for inducing MI as that used in the present study, gross and microscopic pathology readily confirmed the presence of MI; however, in previous studies the experimental periods ranged from three to six months, and the hearts with MI were therefore presented to pathologist three to six months post-infarction. Since the experimental period in the present study was only six hours, pathological confirmation of MI was more difficult, especially since in all animals, experimental and control, the coronary artery was catheterized and flushed with Renograffin in an attempt to simulate identical surgical treatment to all animals. The focal inflammatory infiltration observed in non-infarcted controls at six hours post-Renograffin was sufficient to prevent definite identification of the animals that also received the microspheres.

It should be noted that in a previous study serum enzyme changes were also employed to confirm successful induction of infarction. Since in the previous study the enzyme changes did not peak until 24 hours post-infarction, this parameter was not utilized in the present study because of the short experimental period.

Confirmation of MI was therefore predominately based on ECG changes consistent with MI as observed in animals in the previous study. The electrocardiograms of representative animals of each group are presented in Figures 2 to 5 which MI and VF are demonstrated.

## DISCUSSION

VF determination represents a reliable technique of assessing the vulnerability of the ventricles to fibrillation and can be useful in ascertaining the effects of various interventions on this parameter.[4–13] By use of this technique in the present study it was possible to demonstrate the effects of CO inhalation on the VF of normal monkeys and monkeys with induced MI.

**Table 3**
**Induced VF, Heart Rate, and Body Temperature of Monkeys Breathing Ambient Air or 100 ppm Carbon Monoxide for Six Hours**

| Group | VF (Volts, AC) | Pre-Fibrillation | |
|---|---|---|---|
| | | Heart Rate (Beats/min) | Body Temperature (°F) |
| Non-infarct, air breathing | 79 ± 14 | 158 ± 30 | 96.8 ± 2.9 |
| Infarcted, air breathing | 51 ± 39 | 158 ± 39 | 97.7 ± 1.6 |
| Non-infarct, CO-breathing | 45 ± 20 | 148 ± 34 | 96.8 ± 3.4 |
| Infarcted, CO-breathing | 30 ± 16 | 174 ± 35 | 96.8 ± 5.4 |

Note: Values are means ± S.D.

Oxygen availability to the myocardium is dependent on the extraction of oxygen from the coronary blood and on the coronary blood flow. Since oxygen extraction from the coronary blood is very high at rest, the delivery of sufficient oxygen to the myocardium during stress or exertion becomes very critical and additional oxygen must be provided by an increase in coronary blood flow.

It has been demonstrated by Erickson[12] that coronary blood flow increased 21% in the normal dog heart after inhalation of 100 ppm CO for five hours resulted in COHb of 8.3%. Since there was no increase in heart rate during the exposure to CO, the primary mechanism for increase in coronary blood flow was through dilation of coronary arteries; that the increased blood flow through the coronary arteries was mediated by coronary artery dilation is further substantiated by a corresponding significant decrease in the coronary artery resistance.

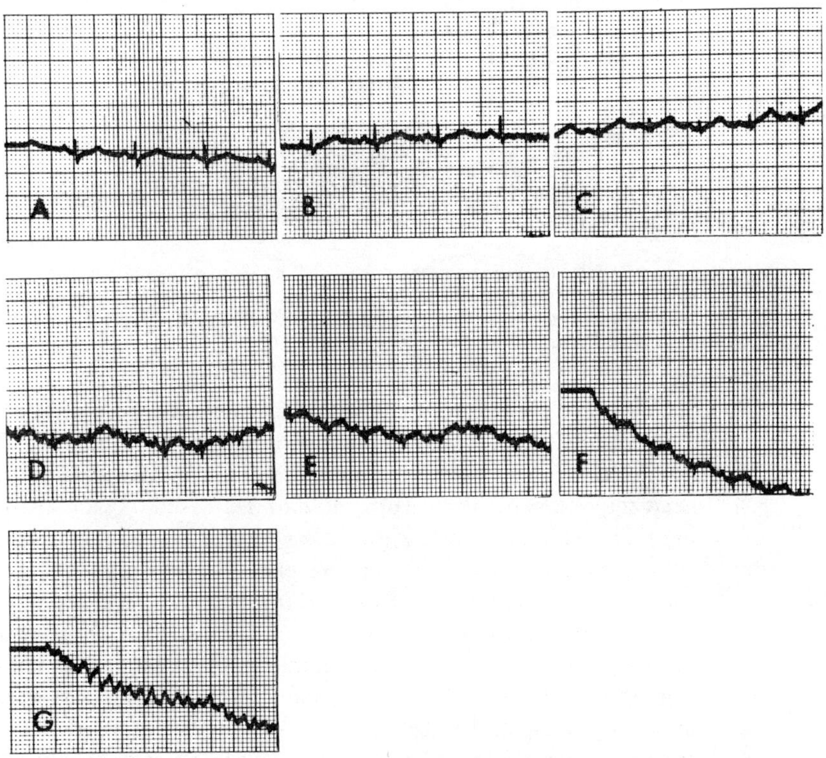


**Figure 2.** Electrocardiograms (Lead II; chart speed 25 mm/sec) of an air breathing, noninfarcted monkey (#129) taken during the experimental period: (A) control, (B) post-catheterization (Sham), (C) pre-fibrillation; unsuccessful fibrillation at: (D) 20 V, (E) 40 V, and (F) 60 V; fibrillation induced at 70 V (G).

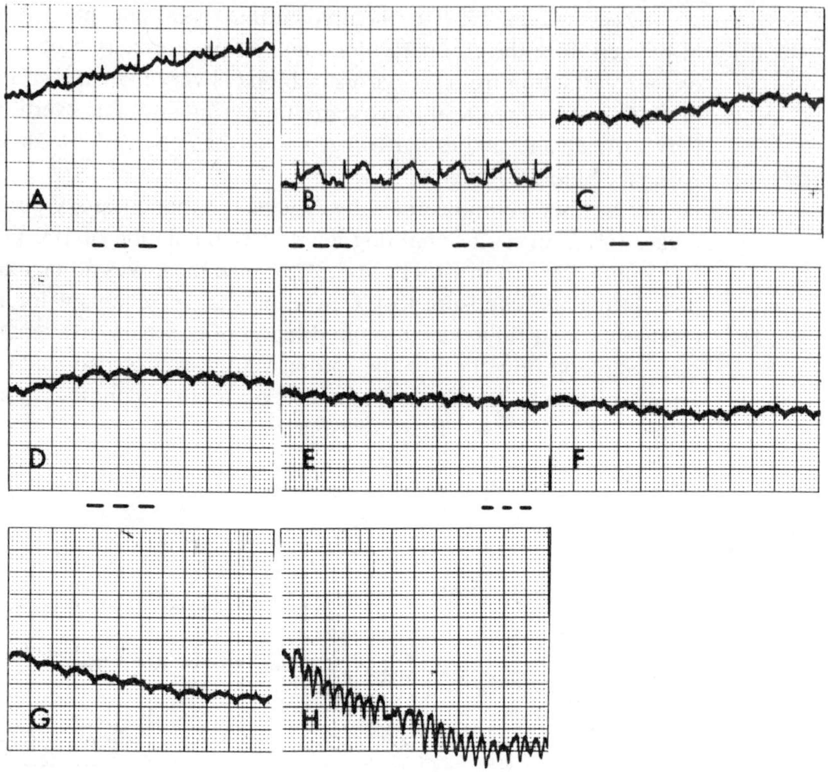


**Figure 3.** Electrocardiograms (chart speed 25 mm/sec) of an air breathing, infarcted monkey (#151) taken during the experimental period: (A) control, (B) post-infarction, (C) pre-fibrillation; unsuccessful fibrillation at (D) 10 V, (E) 20 V, (F) 30 V, (G) 35 V; fibrillation induced at 40 V (H).

While exposure to this level of CO may not be deleterious to a normal, healthy heart, which can compensate for impending myocardial oxygen deficiencies by delivering more blood to the myocardium, the infarcted heart may be particularly vulnerable to decreased oxygen availability during CO inhalation because the reserve coronary oxygen delivery may be already compromised. And, the presence of even small amounts of CO in the blood will hinder further extraction of oxygen from the blood.[13,14] We also[1] demonstrated previously that the incidence of T-wave (ECG) inversion was higher in infarcted monkeys exposed to 100 ppm CO than in normal monkeys similarly exposed. This finding was interpreted as indicating a greater degree of ischemia in the infarcted hearts from an already compromised coronary circulation that was unable to further dilate to compensate for the reduced oxygen availability induced by the circulating COHb.

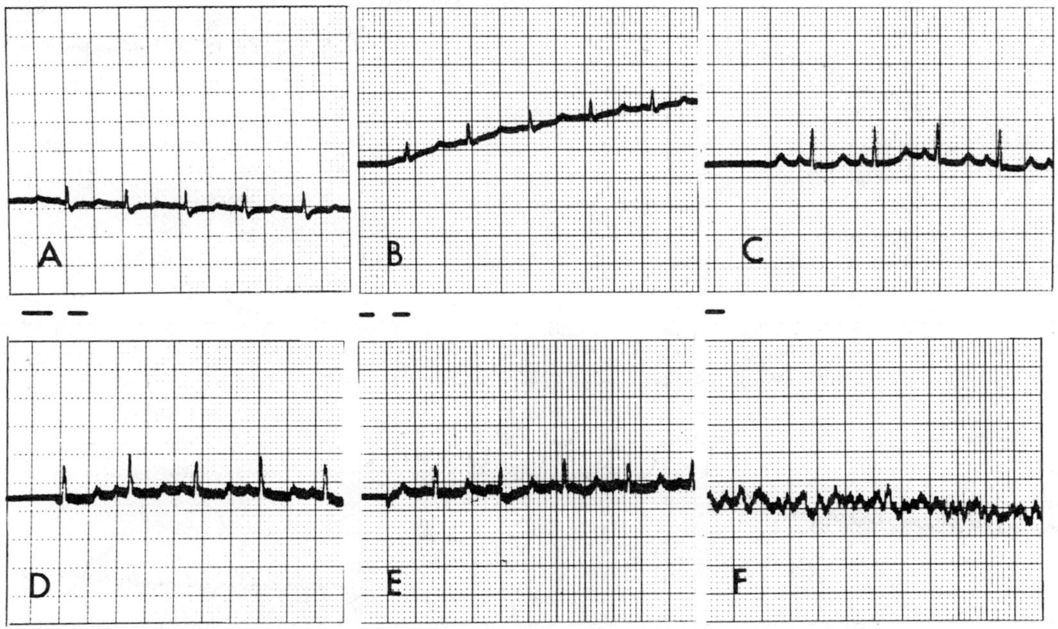


**Figure 4.** Electrocardiograms (Lead II; chart speed 25 mm/sec) of a CO-exposed, non-infarcted monkey (#133) taken during the experimental period: (A) control, (B) post-catheterization (Sham), (C) post-CO exposure (pre-fibrillation); unsuccessful fibrillation at (D) 10 V, and (E) 20 V; induced fibrillation at 30 V (F).

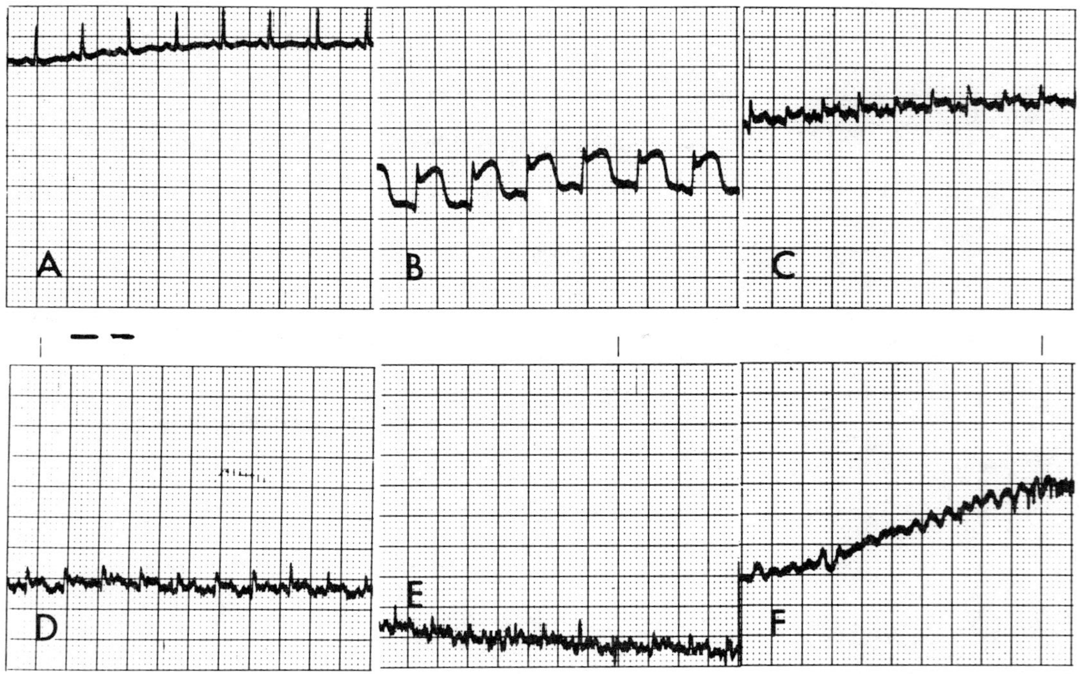


**Figure 6.** Electrocardiograms (Lead II; chart speed 25 mm/sec) of a CO-exposed, infarcted monkey (#148) taken during the experimental period: (A) control, (B) post-infarction, (C) post-CO exposure (Pre-fibrillation); unsuccessful fibrillation at (D) 10 V, (E) 20 V; induced fibrillation at 30 V (F).

It is well known that the probability of VF increases with acute myocardial infarction, and that the vulnerability of the myocardium is further enhanced by myocardial ischemia. It is therefore noteworthy that in the present study the susceptibility to VF was significantly increased in the acutely infarcted animals exposed to 100 ppm CO for six hours; this finding indicates that a blood COHb level of approximately 9% was sufficient to produce a degree of myocardial ischemia that enhanced the vulnerability of the myocardium to induced fibrillation.

Bellet et al.[10] have reported that the susceptibility to VF was significantly increased in normal dogs and dogs with acute myocardial infarction after exposure to cigarette smoke (three cigarettes in 10 minutes via smoking machine). They attributed their findings to the effects of nicotine on adrenergic activity as well as decreased conductibility and increased automaticity of the myocardial Purkinje fibers. Unfortunately, blood COHb levels were not determined in their study, and in light of our findings it is tempting to speculate that the inhaled CO in the cigarette smoke may have been the primary factor that contributed to the decrease in VF "threshold" of the dogs used in Bellet's study. Such a speculation is of interest in view of the increased incidence of sudden death in coronary patients who are heavy smokers.

## REFERENCES


1. DeBias DA, Banerjee CM, Birkhead NC, et al: Carbon Monoxide Inhalation Effects Following Myocardial Infarction in Monkeys. *Arch Environ Health* 27:161–167, 1973.
2. Han J: Ventricular Vulnerability During Acute Coronary Occlusion. *Am J Cardiol* 24:857–864, 1969.
3. Wiggers CJ, Wegria R, Pinerax P: The Effects of Myocardial Ischemia on the Fibrillation Threshold — the Mechanism of Spontaneous Ventricular Fibrillation Following Coronary Occlusion. *Am J Physiol* 131:309–316, 1940–41.
4. Shrimway NE, Johnson JA, Stich RJ: The Study of Ventricular Fibrillation by Threshold Determination. *J Thor Surg* 34:643–653, 1957.
5. Phibbs CM, Levy LM, MacLean LD: The Infuence of Temperature and Coronary Occlusion on the Ventricular Fibrillation Threshold. *Surg Gynecol Obstet* 109:457-461, 1959.
6. Van Tyn RA, MacLean LD: Ventricular Fibrillation Threshold. *Am J Physiol* 201:457–461, 1961.
7. Turnhill AD, MacLean LD, Dobell AR, et al: The Influence of Hyperbaric Oxygen and Hypoxia on the Ventricular Fibrillation Threshold. *J Thorac Cardiovas Surg* 50:842–848, 1965.

8. Gerst PH, Fleming WH, Malm JR: An Evaluation of the Effects of Acidosis and Alkalosis upon Ventricular Fibrillation Threshold. *Surgery* 59:1050–1060, 1966.

9. Hans J, DeJalon PG, Moe GK: Adrenergic Effects of Ventricular Vulnerability. *Circ Res* 14:516–524, 1964.

10. Bellet S, DeGuzman NT, Kostis JB, et al: Effect of Inhalation of Cigarette Smoke on Ventricular Fibrillation Threshold in Normal Dogs and Dogs With Acute Myocardial Infarction. *Am Heart J* 83:67–76, 1972.

11. Spear JF, Moore EN, Gerstenblith G: Effect of Lidocaine on the Ventricular Fibrillation Threshold in the Dog During Acute Ischemia and Premature Ventricular Contractions. *Cir* 46:65–73, 1972.

12. Erickson HH, Buckhold DK: Coronary Blood Flow and Myocardial Function During Exposure to 100 ppm Carbon Monoxide. *Physiologist* 15:128, 1972.

13. Douglas CK, Haldane JS, Haldane JBS: The Laws of Combination Hemoglobin With CO and $O_2$. *J Physiol* 44:275–304, 1912.

14. Haldane JBS: The Dissociation of Oxyhemoglobin in Human Blood During Partial CO Poisoning. *J Physiol* 45:XXII, 1912–13.

# 11 Clinical Implications of Air Pollution Research for Cardiovascular Disease

Joseph F. Boyle, M.D.

The significance of the information obtained in a program such as this takes two forms for me as a clinician. The first is with respect to the kind of health education that I can make available to my patients, plus the effect that it may have on my day-to-day care of patients in private practice. The second, and I think equally as important, is the information that will assist me in a role as physician advisor to the community. I have such an advisory role not only to people who ask my opinion as a physician, but to the general public and their elected officials; and in that role, I am also an advocate working toward attaining a healthier environment in which to live.

I might make one other point: those of you who know me well know me to be an inveterate, chronic, heavy cigarette smoker. Up until the recent past it has been possible for me to be relatively comfortable with the thought that if I am addicted and ruining my health, that is my problem. But now I am receiving increasing amounts of information that would indicate that I may also be seriously impairing the health and future health of others, a matter that becomes a totally different responsibility on my part.*

*Since the Conference Dr. Boyle has quit smoking—*Editors*.

The levels of carbon monoxide that were described by Dr. Hackney in an excellent summary of current knowledge are levels to which people in our cities are regularly exposed. They are levels that produce significant increases in carboxyhemoglobin in many individuals. As has been noted, people, when surveyed in large numbers, are found to have significantly increased levels of carboxyhemoglobin even in areas not considered to have excessively high CO in the atmosphere. The data that were presented by Dr. Aronow support some epidemiologic studies done previously in California that now appear to be increasingly supported by other evidence of harmful effects on cardiovascular function. They indicate a harmful effect on patients with arteriosclerotic heart disease and particularly those who have had or are experiencing a coronary occlusion and myocardial infarction. The significance of these data, although their methodology may be challenged by some, is sufficiently real to me, and, inasmuch as we are concerned with real morbidity and real mortality, I must accept them for what they are and act upon them accordingly. It would be unwise of me as a clinician to say, "No, I need to wait until we have some more data before I take some personal action in advising my patients." This means, then, that I have even more reasons, and better reasons, to encourage patients with arteriosclerotic heart disease to discontinue cigarette smoking. It would also appear to be advisable to suggest that they decrease physical exertion when there are increased levels of CO in the atmosphere and that they might seriously consider discontinuing driving on the freeways at times of peak travel, since there have been measurements of CO alongside the freeways of Los Angeles that approach the levels that are used in the experimental model in the laboratory.

I would certainly also have to suggest to myself and to others that there is evidence that when a cigarette smoker is in a closed environment with non-cigarette smokers, there are significant increases in COHb in the blood of all these persons. These changes might be in the blood of the passengers in an automobile, for example, and one would have to consider the potential impact of the smoker on a passenger in an automobile who is a non-smoker but may or may not have arteriosclerotic heart disease.

It would also appear to me that we have adequate information to advise our legislators and the public that the current ambient air quality standards for carbon monoxide are adequate and would appear to be safe, but that they certainly should not be tampered with at this time.

I was most seriously impressed by other data presented by Dr. Aronow that demonstrated a decrease in the cardiac index in patients exposed to carbon monoxide. Papers by Dr. Haak and Dr. BeBias add further evidence in support of this conclusion. In my view, Dr. DeBias' paper was especially thoughtfully written and would suggest that perhaps we need to do two things together: explain to patients who have had a myocardial infarction and who may have occasional premature ventricular extrasystoles that this condition places them at risk of instant death and that they might also on occasion be treated aggressively with antiarrhythmic drugs if they do work or travel in areas of high CO pollution in the atmosphere. Again, I believe that the information that we have is real enough to provide me, as a clinician, with a basis for action.

# DISCUSSION: PART TWO

**Doctor Hackney:** I hope that we will be able to keep the problem of long-term effects and pathogenesis separate from short-term effects and aggravation.

**Doctor Paul Wehrle:** As a member of a committee involved with human investigation and with an ethics subcommittee, I can say that nothing that has come before us has troubled us quite as much as volunteer studies involving potentially noxious agents or hazardous procedures, such as smoking, in patients with the kinds of difficulties that you describe.

I am sure that this has troubled each of the participants in this conference. How you resolve this dilemma would be very helpful to some of us. This problem is particularly pertinent to investigations involving children, because unless you can find the ideal epidemiological situation, many of the kinds of volunteer challenge studies that have been done on adults are simply not possible at the present time on children.

**Doctor Aronow:** This is a very important question. I think that first of all, before undertaking a volunteer challenge research study, one must consider the risks involved to the participants versus the benefits anticipated from the information gained from the study. For example, within the past two years in our cardiac catheterization laboratory, we have not had any mortality or major complication caused by cardiac catheterization. Under these circumstances, I think that one can do volunteer challenge cardiac catheterization research provided one obtains properly informed consent. The individuals who participated in our research studies knew that the information to be gained was important and that this information would be of value to many patients with cardiovascular disease. These individuals were willing to participate in our studies for humanitarian purposes. They were not paid at all. They participated voluntarily, willingly, and with full understanding and realization of the risks involved. Under these circumstances, I feel volunteer challenge research is justified.

**Doctor Paul Wehrle:** Is there any possibility of benefit to the individual subject? What about some possibly useful information for the patient?

**Doctor Aronow:** The only benefit to the subjects who participated in our volunteer challenge research studies involving carbon monoxide and smoking is that we have demonstrated to them that certain levels of carbon monoxide or smoking are harmful and that they should avoid heavy atmospheric pollution and smoking. But one must have good data before one can take intelligent action. Dr. Crocker described to us an excellent experimental model today that used fetuses for investigating the effects of pollutants. However, there is strong opposition to using fetuses for research. I would make a plea that fetal research be permitted under these circumstances because, by using Dr. Crocker's experimental model, it is possible to determine the effects of various pollutants and their combinations on human health.

Dr. Wehrle brings up the problem of drugs. Unless drug research in children is performed by ethical investigators, we will not be able to obtain the data necessary for treating on a rational basis children who need these drugs. The problem is that without good drug research — performed by honest, ethical investigators who are interested in people as human beings and who are interested in properly getting data that will benefit mankind by use of experimental protocols approved by human rights committees — there will be no knowledge to treat people with these drugs on a rational basis. This is an important problem because, unless you can perform drug research on children, I don't know how you are going to be able to treat them with any new drugs.

**Doctor Wehrle:** I think that our big problem, particularly in pediatrics, is that we have a number of useful drugs for which the package insert clearly says "Pediatric dosage not available" or "Inadequate information available for use in infants." Yet these drugs are used quite regularly during infancy and childhood. I think the increasing appearance of the package insert in a magnified form on the wall of the courtroom during some medical-legal litigation gives many of us very cold chills. I think this problem applies not only to drug usage but also to the kinds of situations that we have here, the search for various environmental hazards. I think that one of the things that we have to turn to is the kind of natural experiments that are going on, such as the increased level of COHb in smokers or the kinds of things that go on in schools or in particularly dense atmospheres. I am afraid of investigations in children at this point.

**Doctor Haak:** I feel very much like Dr. Aronow. I think we make a great effort to be sure that we have informed consent. I think that people in general, when approached properly and given an understanding, do feel that they are contributing to knowledge in a sufficient manner to want to participate. Most of our subjects participate voluntarily, certainly those persons with coronary artery disease. Other subjects, such as the young healthy men, were paid for their participation but, in our opinion, they were also taking far lower risks and were reimbursed because some of them left their jobs to be subjects. We certainly try to make the point, especially with our smoking subjects, that it could be of some benefit to be able to tell them just how much decrement in performance on the treadmill or how much loss of ventricular function in the cathlab was associated with the certain level of CO. While this is meaningful to the subjects, I don't really see that as the main reason for seeking informed consent. I also think that we need to get both smokers and non-smokers as subjects in our studies in order to be able to be certain of the effects that we are measuring. Therefore, I approach these people and request that they volunteer to be subjects willingly.

**Doctor Wehrle:** I think another dimension just slipped in here: the question of payment for volunteers. One of the things that we have tried to adhere to very carefully is that no payment at all be provided for individuals. When Dr. Haak indicates payment for volunteers, I assume that this is in the nature of lunch or car fare, rather than $25 or $50, or $100 because some of the patients, at least in our hospital, would regard $25 or $50 as a windfall of rather substantial dimensions.

**Doctor Haak:** We considered this problem very carefully. You will remember that we take about five hours per day five days running, and over the course of this time period we draw approximately 300 cc of blood for various analyses, a procedure that lowered the hematacrit about 2 points of these healthy subjects, from 46 to 44. Donors normally receive $20 or $25 for a unit of blood — at least in our hospital at the University of North Carolina, this is possible. We would not be able to get subjects very easily who would be able to donate a week of their time. We felt it not unreasonable to pay these persons at the rate of $3 or $4 per hour, including the blood, so that for the entire week they would receive roughly $100, and we did not feel that this was a significant enticement to have people who did not want to participate but would be willing to be a participant only for the money. We felt that the fact that a number of people did not want to participate in the study indicated that we were not offering a huge financial incentive to our subjects.

**Doctor Aronow:** We go one step further. We have never paid anybody for doing anything. We don't even give our subjects lunch unless they are hospitalized at the Veteran's Hospital and would get lunch anyway. Our subjects who have exercised in a research study until the onset of anginal pain have done so on a volunteer basis without any payment. They volunteer for these research studies without any payment because "they feel they are doing something for humanity."

**Doctor Richard J. Lescoe, Los Angeles:** Dr. Aronow, I would like to know if the time of the onset of the angina is significant? In your 1973 study 11 of 20 persons had angina sooner after exposure to purified air than before they were exposed to the purified air.

**Doctor Aronow:** First of all, when patients are exercised until the onset of angina pectoris, there are many factors that determine the exact onset of anginal pain. What we are interested in is the overall exercise performance of each individual, the mean exercise performance until the onset of anginal pain. Carbon monoxide is only one of a number of factors that affect the time until one develops angina pectoris. Nevertheless, in our double-blind study evaluating the effect of carbon monoxide on exercise-induced angina pectoris published in 1973, the mean exercise time until the onset of angina pectoris was significantly reduced ($P < 0.001$) after patients breathed 50 ppm of carbon monoxide for two hours compared to the control period or after they breathed compressed, purified air. All 10 of our 10 patients with angina pectoris, or 100%, developed angina pectoris sooner after exercise after breathing carbon monoxide compared to the control period or after breathing compressed, purified air.

**Doctor Lescoe:** If that is the end point, how do you square it with Anderson's and Neilson's study that indicated that the time of onset of angina was not significant either with 50 ppm CO or 100 ppm, but it was the duration of the pain that was important?

**Doctor Aronow:** I would have to disagree with you because Dr. Anderson specifically found that the onset of anginal pain after patients breathed either 50 ppm or 100 ppm of carbon monoxide was significantly sooner than after they breathed compressed, purified air. As a matter of fact, the P value was less than 0.005 for patients developing exercise-induced angina pectoris sooner after breathing carbon monoxide than after breathing compressed, purified air.

**Doctor Haak:** I am very familiar with the study since it was done in our laboratory; Dr. Aronow is correct. The primary finding was that there was an earlier onset of chest pains with 100 ppm exposure for four hours in patients with angina pectoris and less emphasis was placed on the duration of pain.

**Doctor Lescoe:** In Dr. Anderson's study the duration of the angina was said to be no worse on 100 ppm than on 50 ppm and yet 4 and 8 subjects developed angina sooner on air than on 50 ppm, and subject number 2 developed it at the same time. This means that only 7 out of the 10 were correlated with his findings of the development of the pain. Then, for the duration on 100 ppm CO, the duration was actually shorter for test subjects number 2, 6, 9, and there were no data on number 5 so that we have only 6 subjects having pain lasting longer at 100 ppm of carbon monoxide out of 10 and I just wondered whether this 6 out of 10 is statistically significant.

**Doctor Haak:** These data were examined by the analysis variance with each person serving as his own control. Those who did change, changed enough so that for the group there was a significant effect on the provocation of angina by the carbon monoxide.

**Doctor Hammer:** Selection bias may be introduced into a study when subjects are selected by the volunteer method. For example, if you don't pay people, you obviously select highly motivated volunteers — for several reasons. I realize that you cannot fool the electrocardiograph, but do you think that your data may either overestimate or underestimate the true effect due to selection bias?

**Doctor Aronow:** A double-blind, randomized study would eliminate any bias caused by using highly motivated volunteers. I don't think there would be any difference between the results obtained with highly motivated volunteers or people who are paid.

**Doctor Speizer:** Dr. Aronow, I am not quite sure what you mean by double-blind. What are you doing to protect yourself from an untoward tragic accident? Does Dr. Aronow know what the individual is receiving; if he does, then it is not double-blind?

**Doctor Aronow:** I don't know what the individual is receiving. However, during these studies, the patients are being monitored by telemetry, and if there is marked S-T segment depression or significant arrhythmia, one can stop the particular study to protect the patient. It is extremely important that none of the investigators, including myself, nor the subjects know what intervention is being performed in order to avoid biased data. It is also extremely important to monitor the subjects for safety. If one needs another physician to monitor the patient's safety during the study, this physician must monitor the patient as an independent observer outside the study.

**Doctor Burrows:** I gather that carbon monoxide represents an additional stress or, if you will, a provocative agent in terms of angina. Two questions: (1) Is anything more than hypoxemia induced? and (2) Why do you conclude so definitely that this is bad for people? It seems to me that you regularly prescribe mild provocation all the time in the form of exercise; if a little exercise is good for people, why isn't a little carbon monoxide good for them?

**Doctor Haak:** In most of the studies that Dr. Aronow and we have embarked upon, we have individuals who have a very compromised blood supply, either to calf muscle in the case of claudication or in terms of coronary artery disease involving some area of the heart. Perhaps this cardiac area is stimulated by ischemia but certainly could be lost with myocardial infarction, whereas in the case of claudication one gets pain and cramping, and has to stop. If someone has a limited capacity for blood flow to some region, either of skeletal muscle of cardiac muscle, exposure to CO does nothing but further limit his activity rather than encourage collateral blood flow.

**Doctor Aronow:** First, carbon monoxide not only displaces oxygen from hemoglobin but it also shifts the oxyhemoglobin dissociation curve to the left, a change causing tighter binding of oxygen to hemoglobin and thereby reducing the availability of oxygen to the tissues. Second, hypoxia and exercise are not the same thing. Hypoxia will reduce exercise performance until angina pectoris. Gradual progressive physical exercise will significantly improve exercise performance until angina pectoris by reducing the heart rate and arterial pressure response to exercise. This is a physical training effect. Exercise also causes a positive inotropic effect on the left ventricle. The improvement in left ventricular contractility is beneficial. Carbon monoxide causes hypoxia and depression of myocardial contractility.

**Doctor Burrows:** I am not sure my question was understood. I do understand that there are differences between hypoxemia and carbon monoxide. What I was asking is: Has the difference actually been shown to be of practical importance in experimentally induced angina? Has anyone shown that the effect is any different from what you would get with the amount of reduction in oxygen content in the blood? As for the second point, I would have thought on theoretical grounds that exercise was a terrible thing for people with coronary insufficiency. Although you now have a lot of empiric data to say that exercise is probably good, I have not seen any evidence that repeated exposures to small amounts of carbon monoxide might not have exactly the same effect. Indeed, I don't know that tolerance can't be built up. Is there real evidence that it cannot be, as the panel implies?

**Doctor Aronow:** First of all, the data from exercise programs in coronary heart disease show that after physical training, patients with this disease will be able to perform the same amount of exercise with less of an increase in product of systolic blood pressure times heart rate. In other words, the trained person increases his mechanical efficiency. The lower heart rate and blood pressure response to exercise means that the trained patient can do more exercise at a lower oxygen cost to his myocardium. This is important. In addition, a number of studies have shown that physical training significantly increases the amount of exercise that patients with angina pectoris can perform before developing angina. Moreover, Redwood and associates have demonstrated that patients with angina pectoris in a physical training program will develop angina at a significantly higher product of systolic blood pressure times heart rate. Although changes in ventricular size and inotropic state must be considered in interpreting their observations, the data suggest that physical training may enhance oxygen delivery to the myocardium. Therefore, physical training improves the arterial pressure and heart rate response to exercise, improves the inotropic state of the myocardium, and may improve myocardial oxygen delivery. For these various reasons, people in physical training programs may significantly improve their exercise performance. Moreover, there are data that show that a physical training program will reduce the likelihood of recurrence and death from myocardial infarction in a selected group of patients who have survived an acute myocardial infarction.

On the other hand, carbon monoxide does not help the mechanical efficiency of the heart in any way. Carbon monoxide in concentrations inhaled during heavy atmospheric pollution or after cigarette smoking does not affect the resting heart rate or blood pressure but causes exercise-induced angina pectoris to develop at a significantly lower product of systolic blood pressure times heart rate. This means that carbon monoxide has caused less oxygen to be supplied to the heart. This is a very important point.

In summary, there is a difference between the effect of carbon monoxide exposure and the effect of exercise on the patient with coronary heart disease. Exercise training may be helpful to the patient with coronary heart disease. Carbon monoxide is not helpful and may be harmful to the patient with coronary heart disease both by impairing oxygen delivery to the myocardium and by depressing myocardial function.

**Doctor Hackney:** I understand that Dr. Goldsmith would be willing to share some information on commuter carbon monoxide levels studies in Los Angeles.

**Doctor Goldsmith:** There is need for determining whether the kinds of effects that Dr. DeBias has demonstrated in monkeys and that Dr. Aronow demonstrated in humans with angina pectoris are also observed at progressively lower levels. We do not have enough information about the nature of the dose-response relationship to know whether the present objectives and present levels of regulation are as protective as they should be for these populations. One interesting set of points on the dose-response curve is the measured level of carboxyhemoglobin in a study by Margaret Deane and Norman Perkins supported by the California Air Resources Board. This was a study in which commuters employed by the county public utility, Los Angeles Light and Power, and a group from the Los Angeles Police Department had exhaled carbon monoxide values measured before and after both morning and evening commuting. The commuting durations ranged from 30 to 90 minutes. The times of the year of the study were chosen so that some of them occurred when CO is expected to be at a relatively low level (for Los Angeles) and some when it might have been high (that is, late December, January, and February). This is a well-designed and well-carried out study. It involves replicate determinations of what the carbon monoxide exposure was during commuting, and what the increase in the body burden of carbon monoxide was, as estimated by expired air carbon monoxide measurements. These were validated by carboxyhemoglobin determinations in a sample of the population.

Very simply, the average morning trip for these commuters is sufficient to increase the expired air CO by between 3.3 and 5.8 ppm. Now, just for purposes of reference, this amount is approximately equivalent to an increase of about 1% carboxyhemoglobin during the average commute under average circumstances of the average person working in downtown Los Angeles. Virtually the same effect was found in smokers, but, because each smoker has so much intrinsic variability, the stability of the estimates is not quite as good. The carbon monoxide samples in the car during these ordinary commuting periods was approximately 18 ppm. This means essentially that during commuting approximately a quarter of the ppm of measured CO is in fact absorbed and becomes part of the body burden. This is based on measurements in comparable units, namely 4.3 ppm increase in expired air and 18 ppm exposure. An increase in homeward commuting was not observed, probably because, during the day, the exposure increment led to a background body burden level that was not increased by the homeward journey. The air monitoring data show a somewhat greater contrast for the morning CO values, which are generally high just during the periods of commuting. So we know, I think, a good deal about what the burden is. If we add the probable variation to such increases, some exposed subjects are just about at the point at which patients with coronary disease would have an aggravation of their condition.

I would like to end, if I may, by thanking the speakers who have done a brilliant job of following some leads that epidemiologic work produced. Dr. DeBias perhaps was too generous concerning our contribution. We did not show anything about the sensitivity of patients with myocardial infarction. We did say that patients who are hospitalized with myocardial infarction had a higher case fatality rate during the periods of the year and at places in the air basin at which the highest CO value were occurring. This does not necessarily prove that the carbon monoxide exposure was doing this but it does suggest it. Certainly the clinical studies and experiments reported here today provide ample evidence that our concerns were justified.

**PART THREE**

# IMMUNOLOGY, HYPERSENSITIVITY, AND HOST DEFENSE

# 12 Air Pollution, Host Immune Defenses, and Asthma; A Review

Gareth M. Green, M.D.

## INTRODUCTION

Asthma, defined as intermittent, reversible airways obstruction, is a syndrome that represents, at the clinical level, a response of the airways to noninfectious immunogenic particulates, infectious agents with immunogenic potential, and organic and inorganic irritant gases and particles. At the pathophysiologic level of cellular and tissue responses, asthma represents the effects of the interaction between the inhalation of such agents and of specific and nonspecific bronchopulmonary defense mechanisms of the host. Since these interactions are initiated by the inhalation of foreign material in gaseous or particulate form, the clinical syndrome may be thought of as a collection of air pollution diseases caused by specific airborne pollutants such as pollens, infectious agents, and gaseous and particulate chemicals. Thus, the literature on asthma is found in discussions of allergies to pollens, house dust mites, and fungal spores; discussions of complications of infectious diseases, reactions to occupational, household, and environmental agents; and an analysis of the clinical effects of air pollution exposures. In addition, the contributions of immunology have added significantly to the understanding of specific cellular and humoral immune mechanisms and to the role of mediators in the pathogenesis of specific immunologically mediated asthma.

Developing understanding of the integrated systems of specific and nonspecific host defense mechanisms in the lung and growing awareness that resident nonspecific defense mechanisms are intimately related to local and systemic specific immune systems suggests a rationale for unifying pathogenic concepts of asthma by considering the clinical syndrome as a common response of the organ — i.e., bronchospasm and hypersecretion — to a varied collection of airborne air pollutants which affect, in very specific ways, cells participating in the host defensive response. This review attempts to integrate concepts of asthma by considering the effects of air pollutants of diverse nature on these defensive target cells; it also considers how these cellular reactions and injuries can alter the host defensive responses to infectious and other agents that may be present concurrently in the lung. Also discussed are the ways in which the management of asthma impinges on concepts of host-defense and how treatment itself may either improve or injure further these host defense mechanisms. Since air pollutants, host defense mechanisms, and immunologic responses and mediators are well reviewed in the literature, this review emphasizes the integration of such knowledge into a concept of asthma as an air pollution-host defense disorder.

## AERODYNAMIC FACTORS

Air pollutants are presented to the lungs as mixtures of particulates of heterogeneous nature and particle size plus vapors and gases, both toxic and nontoxic. Both gases and particles are influenced by vapor pressure and humidity and by the oxidant and reductant state of the mixture. The physical and chemical interactions among the particles, gases and vapors are incredibly complex, constantly changing and poorly understood. The sciences of aerobiology, chemistry, and physics have only recently come of age, so that knowledge of any given exposure to air pollutant mixtures is largely inadequate.

The architecture of the respiratory tract is so designed that inhaled particulates are sequentially filtered out of the inhaled airstream according to aerodynamic properties of particle size, shape, and density that give to each particle a characteristic falling speed dependent on those characteristics and on the velocity of airflow. Although the penetration of particulates down the tracheobronchial tree is generally inversely proportional to particle size or density, this simplistic concept is complicated by the shape of particles and alterations in particle size and density produced by physical interaction of particles and by particle growth or shrinkage under the influences of humidification of the airstream in the upper respiratory tract.

Gases and vapors, by contrast, are distributed along the respiratory tract in proportion to their solubility in the respiratory tract fluids lining the airspaces. Soluble gases, such as $SO_2$, tend to be totally absorbed in nasal passages, while insoluble gases, such as ozone and nitrogen dioxide, penetrate along with the respiratory gases to the respiratory bronchioles and alveoli. This simplistic concept is, likewise, complicated by the interaction of soluble gases with accompanying particulates, present as both solid particles and liquid droplets, and again by chemical interaction of component gaseous materials. To complicate matters even further, air pollutant components, such as $SO_2$, undergo transformation in the environment, including both oxidation, such as to sulfates, by liquefaction to micronic particulates, by solubilization in water and other vapors, and by adsorption on solid particles. Such transformation processes greatly alter the characteristics of the airborne exposure and the ability of a soluble gas, such as sulfur dioxide, to penetrate to the depths of the lung as a sulfate particulate or as a solubilized gas or as a vapor adsorbed on a particle. Thus, measurement of a soluble gas, such as sulfur dioxide, as an index of air pollution may reflect poorly the actual hazard of exposure to the air pollution mixture.

Because of the dependence of deposition on effective falling speed, the penetration of a particle into the lung is influenced by air velocities and breathing patterns. Studies with radiotracer particulates have demonstrated that central deposition of particles is favored by rapid and shallow breathing, whereas peripheral deposition is favored by deep inspiration, slow respiratory rate, breath-holding, and airways obstruction. It thus becomes apparent that the risk of exposure to an airborne pollutant will depend, in part, on the respiratory pattern of the exposed individuals and on the level of physical activity, a factor that has great influence on respiratory breathing patterns. The first point to be made, therefore, in regard to the effects of air pollution on host defense mechanisms, is that the risk of any given exposure varies with both the physical-chemical characteristics of the air polluting agent and the above-mentioned state of the respiratory system of the host.

These observations have as obvious an application to the therapeutic use of aerosols and inhalation agents as they do to the environmental exposure. Thus, in evaluating the effectiveness of aerosol therapy to thin bronchial secretions, the physician must consider the particle size and vapor compositions of a given inhalation device.

In view of the fact that only micronic particulates penetrate to the depths of the lung, only small total volumes of liquid aerosols can be expected to reach the distal portions of the tracheobronchial tree. This scientific realization has made the medical community aware of the futility of attempting to liquefy tracheobronchial secretions through administration of vapor particulates by the inhalation route. This latter statement represents the *first* clinical conclusion to be drawn from consideration of the aerodynamic defense system of the tracheobronchial tree.

## SECRETORY BARRIERS

The secretory system of the respiratory tract is probably the most appreciated but least understood defense mechanism. The secretory system has four major components: respiratory tract fluid, surfactant, mucus, and secretory immunoglobulins. Respiratory tract fluid is the major product of the normal lung and is a mixture produced by secretory cells in the alveoli and bronchioles, epithelial secretions of the bronchi, serous secretions of the bronchial glands and water introduced in the process of secretion and by transudation across the epithelial lining. Respiratory tract fluid exists on the bronchial-epithelial surface as a biphasic fluid, lining layer, consisting of a sol phase in which the cilia exert their activity and a gel phase on the lumenal surface, which is more viscid, receives the propellant force of the ciliary tips, and acts to carry particles and debris out of the tracheobronchial tree.

Under normal conditions, approximately 100 ml of respiratory tract fluid are produced per day, but this amount is insufficient to be appreciated or produced as phlegm. The sol phase is probably in a state of active equilibrium with the underlying epithelium regarding water content. These secretions contain proteins and mucopolysaccharides, as well as small amounts of mucus. The physical state of this material is most important for mucociliary function.

Mucus is the secretory product of the goblet cells and the mucus glands at the bronchial wall. Mucus is a complex mucopolysaccharide whose maximum water content is determined at the time of synthesis. Once secreted onto the bronchial surface, mucus is capable of dehydration but cannot be rehydrated. Contrary to popular belief, mucus is not a prominent part of the normal respiratory tract and is not present as a continuous lining on the bronchial epithelium. Rather, mucus tends to occur as ropy strands of secretion exuded from the mural bronchial mucus glands.

Surfactant is the name given to the lipid secretions that lie in the alveoli and are secreted by the type 2 alveolar epithelial cells. This material has the physical properties of low surface tension and, among other functions, serves to maintain alveolar stability. Although the origin of surfactant is from the type 2 alveolar epithelial cells, it is not clear how it passes through or along the alveolar surfaces or whether it enters as part of the respiratory tract fluid at the bronchioles. In disease states, it does appear in the sputum, and thus, presumably, this is a normal course of exit out of the lung.

Secretory immunoglobulins form an important component of respiratory tract fluid. Immunoglobulin A is secreted from submucosal plasma cells as a dimer joined by a small assembly piece. Immunoglobulin G, or IgG, is normally present in low concentrations in respiratory tract fluid, but may rise with inflammation. Immunoglobulin E (IgE) is present in varying amounts and is particularly important in considerations of asthma as the reagenic antibody.

The functions of the secretory barrier of the tracheobronchial tree are physical, chemical, immunological, and biological. As described earlier, the bronchial surface and its lining fluids, as well as the alveolar surface and its surface layer of surfactant, provide a broad absorptive surface for the dissolution of inhaled gases and liquid and solid particles. The presence of an aqueous phase in the bronchial tree and a lipid phase in the alveoli suggests that both aqueous and lipid-soluble materials are absorbed from the air onto the surface and into the fluids of the surface lining layer. In the case of aqueous, soluble gases, such as $SO_2$, this vast absorptive surface is capable of completely removing significant quantities of air pollutant vapors before they reach the alveoli. The presence of a lipid phase in the alveoli is probably important in terms of hydrocarbon and other insoluble gaseous exposures.

In addition to the dissolution of gases, the respiratory tract fluid provides the vehicle for humidifying inspired air and maintaining 100% humidity below the glottis. Maintenance of humidity is vital for ciliary function, and interference with this humidification mechanism results in deterioration of the mucociliary and transport function of the lung.

In the alveoli, the surface lining fluid has important opsonic and surface active functions. Particles deposited in the alveoli are immediately subjected to coating by surface active lipid materials, which may lyse particles that are subject to surface activity, or may alter the surfaces in such a way that they are rendered more susceptible to particle uptake by phagocytic cells. The specific activity of this function is poorly understood.

Surface lining fluids, particularly in the bronchial tree, also have important neutralizing actions, both from a chemical and an immunological point of view. The abundant polyionic organic molecules of the proteins and mucopolysaccharides in the respiratory tract fluid provide a mechanism for buffering acid or alkaline aerosols; these materials also act as a sulfhydryl source for neutralizing oxidant gases and particles. It is, perhaps in response to an increased demand for such neutralizing functions that respiratory tracts chronically exposed to acid or oxidant aerosols demonstrate a hyperplasia and hypertrophy due to an increased volume of secretion of all phases of respiratory tract fluid, most noticeably of mucus.

Neutralization, in an immunologic sense, is the function of the secretory immunoglobulins. Bronchial secretory immunoglobulins have an important role in limiting the antigen load in the lung. These secretory immunoglobulins are normally found as far down as the terminal bronchioles, but it is not clear whether this function exists at the alveolar level of the lung. Secretory IgA is capable of directly killing or neutralizing viruses, but the effects on bacteria are less clear. Type specific IgA antibodies against organisms such as *Streptococcus pneumonii* and *Hemophilus in-*

*fluenzae* have been demonstrated in the sputum of patients with chronic bronchitis, although these antibodies do not appear to be bactericidal. This inability is perhaps related to failure of the antibodies to activate complement by the classical pathway, although complement activation by the alternate properdin pathway has been shown to occur. One would think IgA would process bacteria for phagocytosis, but this has not been shown to occur, since IgA generally is not opsonic.

The major function of secretory IgA may be to limit the penetration of inhaled antigen to underlying respiratory epithelial tissue. IgA appears to limit antigen absorption at mucosal surfaces and binds potentially antigenic material, rendering it nonantigenic. Secretory IgE is produced in some individuals, and it is obviously of significance for the pathogenesis of asthma because of its attachment to mast cells and its role in the antigen-mediated release of pharmacologically active broncho-constrictor agents.

Finally, the transport functions of the secretory defense mechanisms are obviously of vital importance. Alterations in transport activity of respiratory tract fluid have been demonstrated in smokers and in chronic diseases such as bronchitis and cystic fibrosis. However, studies done with radiotracer-tagged particles indicate that as long as the cough is intact, net tracer transport out of the lung along the mucociliary stream is maintained, despite impairment of respiratory tract fluid flow. Clinical abnormalities of secretory defenses take the form of hypersecretion due to hyperplasia and hypertrophy of the secretory apparatus, dehydration of respiratory tract secretions, oxidation of proteins and mucopolysaccharides, and deficiencies of immunoglobulins or abnormalities of secretory composition, such as occur in hypoglobulin states or in cystic fibrosis. Selective IgA deficiency leads to a greatly increased incidence of mild to moderate upper respiratory tract infections, an increased incidence of atopy, and an unexplained high incidence of auto-immune disorders. Abnormalities of protein content of respiratory tract fluid, such as occur in cystic fibrosis, are not well understood. Perhaps they lead to abnormalities in ciliary function and decreased transport rate of respiratory tract fluid. Or they might alter the opsonic functions that interfere with phagocytic or other cellular functions in the lung.

Hypersecretion is clearly the most evident of respiratory tract fluid abnormalities, being a common response to a variety of infectious, allergic, or chemical insults to the tracheobronchial tree. Hypersecretion is not an acute response to the lung, but occurs after chronic exposure to irritant materials; such exposures result in the hyperplasia of goblet cells and hypertrophy of mural mucosecretory glands of the bronchi. The hyperplasia of the goblet cells is characterized by an extension to the bronchiolar levels of the lung where they are not ordinarily found. This extension of muco-secreting cells to these small airways results in an alteration of their surface active properties and in a tendency to constrict the lumen. This alteration in surface active characteristics of the small airways is perhaps what accounts for a functional airways obstruction and is associated with air pollutant or cigarette smoke exposure. Hypersecretion due to hyperplasia and hypertrophy of the mucus-secreting apparatus is also a complication of bronchial asthma induced by infectious or allergic agents and is one of the problems that gives rise to most clinical management difficulties.

Air pollutants commonly exert their tissue-damaging action through their oxidant properties. Acid sulfates, nitrogen dioxide, ozone, and certain oxidant hydrocarbons react with the respiratory tract fluids, with the proteins and mucopolysaccharides in the respiratory tract fluids, and mucus to oxidize sulfhydryl groups and increase the proportion of disulfide bonds. This chemical change results in an increased viscosity of the mucus and is probably one of the principal forms of injury that leads to hyperviscous sputum in patients with chronic bronchitis, smokers, and individuals subjected to chronic air pollution exposure.

The action of a drug such as acetyl cysteine is to break the disulfide bonds and to restore the sulfhydryl state of the oxidized mucus polysaccharides. It is through this action that acetyl cysteine diminishes sputum viscosity and aids clearance of mucus from the bronchial tract. Unfortunately, to achieve this action, acetyl cysteine and its sulfhydryl groups have to be present in considerable quantity; and, consequently, significant volumes of material must be instilled into the tracheobronchial tree. Such volumes cannot be introduced by aerosol because of the factors discussed under the above section that prevent particles of large volume from penetrating to any significant depth into the tracheobronchial tree. For that reason, aerosolized acetyl cysteine has no beneficial effect on the state of mucus in the tracheobronchial tree in the clinical setting. This is the *second* clinical therapeutic point to be made.

Dehydration is the most common causative condition for inspissated secretions and the most likely to be undermanaged clinically. Dehydration of respiratory tract fluid and of mucus secretions rapidly leads to an increase in viscosity and an impairment of transport. As stated previously, the maximal water content of mucus is attained at the time of secretion, and it is impossible to rehydrate a mucus globule once it has become dehydrated. The simple experiment of placing a globule of mucus into a glass of water will demonstrate that even after hours or overnight there has been no substantial hydration of the mucus. This observation again demonstrates the futility of attempting to manage respiratory tract secretion problems by aerosol inhalation therapy, and it is the *third* clinical point to be made in this paper.

The *fourth* clinical point follows from the above observation; that since the water content of mucus is maximal at the time of synthesis, it becomes necessary to assure that adequate free water is available to the bronchial secretory cells at the time of mucus synthesis and secretion. Since there is no adequate way to determine when free water is available from studies of bronchial secretions, the clinician is best guided by evidence of a water diuresis as indicated by observation of the urine specific gravity. Thus, hydration therapy is inadequate until the urine specific gravity drops below approximately 1.010; the astute clinician will observe that as the urine specific gravity begins to drop below 1.010, respiratory tract secretions begin to flow and airways improve in function. Often large volumes of water are required to achieve this effect since water depletion may be at a massive level. Particularly in asthmatics, in hot weather, under humid conditions where perspiration may be profuse, and where hyperventilation is massive in the acutely ill patient, a check of urine specific gravity at the time of admission of an acute asthmatic will invariably show a state of severe water retention with high urinary specific gravity. Several liters of fluid therapy yielding free water is often necessary before the urine specific gravity begins to drop and the mucus and respiratory tract fluid begins to flow adequately again. In an out-patient setting, the patient is instructed how to use a urinometer or at the very least is instructed to assure enough fluid intake so that the urine is always colorless in the bowl 24 hours a day, 7 days a week.

## Cellular Responses

Cellular defense mechanisms of the lung may be classified as nonspecific and specific defenses. In the past ten years, extensive progress has been made in understanding both types of cellular response in the lung; and for detailed consideration of specific aspects of cellular defenses, the reader is referred to appropriate reference sources.

Nonspecific cellular defense mechanisms in the lung rest principally on epithelial cell elements and several types of phagocytic cells. Epithelial cell defenses include the ciliated epithelial cells, goblet cells, and other secretory epithelial cells of the tracheobronchial tree. In the alveolar spaces, the squamous type 1 epithelium provides a continuous lining to the alveolar surface and the large secretory type 2 cell is responsible for the secretion of the surfactant. These epithelial elements arise embryologically from foregut endoderm in the process of bronchial tubular growth. In embryological development, the tracheo-bronchial tree buds from the foregut into the mesenchymal capillary meshwork that subsequently differentiates to become the vascular and supporting structure of the mature lung. The endodermal bronchial structures terminate at the terminal bronchiole, but the bronchial epithelium is thought to continue to divide and spread over the developing respiratory bronchioles and the several orders of alveolar structure. The existence of a continuous epithelial covering of the alveoli was debated for decades in the anatomic literature and final proof of the existence of the continuous epithelium was developed only 20 years ago through electron microscopic studies.

The origin of the type 2 epithelial cell, the secretory cell of the alveolus, is not well known, but it is likewise felt to be an endodermal derivative. The type 2 epithelial cell occurs singly as a large corner or niche cell with characteristic secretory inclusions or myelin bodies, which constitute the pre-secretory form of surfactant.

The tracheobronchial epithelium is not a continuous ciliated coating as is often imagined. Scanning electron microscopic studies have demonstrated clearly that there are large numbers of non-ciliated epithelial cells interspersed as islands among the ciliated cells. Although the cilia are not continuous, a continuous coating to the bronchi is provided by the overlying respiratory tract fluid layer.

Ciliary action is a primitive defense mechanism that can be traced way back in phylogenetic development. It has long fascinated investigators and clinicians alike, and serves a transport function in the tracheobronchial tree much as it does for lower animal forms. Mucociliary transport is a fascinating process to observe, and recent studies of ciliary activity with fiberoptic bronchoscopy have provided new insights into the way in which these cells function — for example, transport is not uniform at different levels of the tracheobronchial tree or even at the same level of the tracheobronchial tree. In the trachea and large bronchi, some longitudinal segments of the lining fluid travel at greater rates than others; the mucociliary stream is like an expressway where particles are moved in lanes at different speeds.

Phagocytic cells function chiefly in particulate uptake, isolation of particles from the respiratory membrane, and detoxification of potential irritant or viable particles. The principal resident phagocyte of the lung is the alveolar macrophage, a large mononuclear cell with an eccentric nucleus and a cytoplasm rich with organelles, endoplasmic reticulum, lysosomes, and phagocytic vacuoles. The origin of the alveolar macrophage is both from the bone marrow with migration to the lung and locally from a source of phagocytes resident in the lung. Whether the macrophage divides in the lung once it has differentiated at that site is as yet undetermined. The aveolar macrophage pool reflects the general exposure level to inhaled particles so that individuals exposed to large amounts of particulate material, such as smokers or those working in dusty trades or living in polluted atmospheres, show larger numbers of phagocytes than individuals with low exposures to particulate material. Under high loads, the alveolar macrophage system recruits new cells, either from the bone marrow or local lung pools.

The alveolar macrophage is the only phagocyte of significance in the normal lung; however, when particle and irritant burdens to the lung are high or when the alveolar macrophage is unable to handle a load, either through excessive burden or injury to its phagocytic capacity, polymorphonuclear leukocytes are recruited from the bloodstream in variable numbers and appear either as inflammatory cells in the sputum or as dense exudates as in pneumonia. Polymorphonuclear leukocytes are, therefore, secondary, nonspecific cellular defense mechanisms and function in the lung very much as polymorphonuclear leukocytes at any site in the body.

A third cell type that appears in a variety of clinical conditions, often involving antigen antibody reactions and immunologic events, is the eosinophil. The function of the eosinophil has yet to be adequately explained, but its appearance in a variety of diseases involving immunologic events suggests it has an important function in regard to host defense responses to antigen-antibody interactions. Its presence in allergic asthma is well known and the presence of the eosinophil in large numbers in the sputum is a useful means of differentiating allergic from chemical, irritant, or infectious causes of asthma. Its protective role is suggested by the spectrum of diseases spanning eosinophilic pneumonia through allergic granulomas to the necrotizing vasculitides represented by Wegener's granulomatosis and periarteritis nodosa. Asthma is frequently associated with many of these specific disease entities. The inverse relationship between the presence of large numbers of eosinophils and the chronicity, severity, and necrotizing potential of these different diseases suggests that the eosinophil is, in some way, involved in the successful host defense in these diseases and that the absence of eosinophilia may portend a more virulent and necrotizing course.

A fourth cell type that appears to be of nonspecific function is the histiocyte-epithelioid cell/giant cell sequence. These cells are part of a specific immune response to antigenic particles, but the occurrence of the sequence as a response to inorganic dust particles suggests that the sequence is a nonspecific event that augments an unsuccessful phagocytic defense of the lung. These cells, found usually in granulomas, serve to isolate an antigenic or inorganic particulate material that cannot otherwise be destroyed in or removed from the lung.

The alveolar macrophage has received a lot of attention in the past few years, and a number of reviews exist in the literature. Progress in understanding this important cell type continues to be sufficiently rapid so that these reviews do not present all of the latest information, but extensive details are beyond the scope of this review. The importance of the cell in host defenses, however, requires some further disucssion here. Although the alveolar macrophage is derived largely from the bone marrow, it undergoes extensive morphologic, metabolic, and biochemical differentiation after its arrival in the lung. Scanning electron microscopy has shown the cell to have an elaborate membrane with a multitude of veil-like foldings over its membranous surface that serve as the phagocytic arms when a particle is encountered. These cells do not take the form of the extremely distorted and ameboid tissue histiocyte, but remain somewhat rounded or slightly flattened and extended on the alveolar surface. The mature cell contains an abundant array of intracellular granules and vacuoles of varying size and shape. After phagocytosis, a phagosome is formed that is enlarged as lysosomal bodies approach the phagosome and empty their contents into what then becomes a phago-lysosome. Metabolic studies have shown that this cell is highly aerobic in its energy metabolism with a high resting oxygen consumption that increases further during phagocytosis and with an increase in aerobic glucose metabolism via the hexose monophosphate shunt during the process of particle digestion. This highly aerobic character of the macrophage metabolism renders it peculiarly susceptible to certain types of air pollutant injury, as is described below.

Evidence of the central role of the alveolar macrophage in the cellular response of the lung to inhaled particles arises from many sources. Macrophages laden with inhaled dust particles have been described in the lung in connection with occupational dust exposure, and electron microscopic studies have shown that these particles do end up within the phagolysosomes. Bacteria inhaled as an aerosol have been demonstrated very soon after inhalation within the alveolar macrophage cells by techniques using fluorescein conjugated antibody. Studies using radiotracer-labelled aerosolized bacteria, which have shown rapid exponential decay of bacterial viability in the lungs of experimental animals while the tracer remained behind, indicate the central role of the macrophage in the in situ detoxification of an infectious challenge. This bactericidal activity precedes by hours to days the transport and elimination of the infectious material from the lung.

This fundamental concept bears emphasis as it leads to the conclusion that antigenic material may persist within the phagocytic cells in the lungs for long periods of time after an infectious challenge, even when no evidence of viability of the organism can be found. The persistence of antigenic material in the lung, that may subsequently be released and react repeatedly with the immune mechanisms of the lung, is an important concept underlying the etiology and pathogenesis of asthma associated with infection. According to this concept, the total amount of infectious antigen resident in the lung determines the potential of antigenic challenge to the lung at any time the macrophage system might be injured by viral infection or air pollutant exposure. With subsequent release of antigenic material from damaged macrophages, immunologic reactions with release of mediators, and genesis of asthmatic pathophysiology and symptomatology would follow. This concept points to the advisability of minimizing bacterial antigenic load within the lung by adequate control of infectious diseases over the lifetime of the asthma-prone individual and leads to the *fifth* important clinical conclusion that infections in asthmatics, particularly those who have a history of infectious asthma, should be controlled promptly and adequately with antibiotic therapy.

Phagocytosis begins with chance contact between inhaled particle and cell surface. Adherence of the particle to the cell membrane is the first step in the process, and this is enhanced by specific membrane binding sites and/or antibodies or opsonins coating the particle. After adherence, the particle is enveloped by cell membrane, in part by the contractile action of cytoplasmic myofibrils. That the processes of adherence and particle ingestion are two separate processes, dependent on different cellular functions, can be demonstrated by drug-induced paralysis of the contractile fibrils and the observation that the particles continue to adhere to the cell but are not ingested.

The process of particle uptake by the alveolar macrophage is dependent on energy produced by anaerobic glycolysis and ATP produced at or near the membrane surface. Once particles are ingested, the external membrane becomes sealed off within the cell as a phagosome. Lysosomes rapidly move into contact with the phyagocytic vacuole and empty their digestive enzymes into the phago-lysosome. The alveolar macrophage is particularly rich in the quantity and variety of proteolytic enzymes that are sequestered in the lysosomal membrane bound packets. In addition to the bactericidal capabilities of the lysosomal enzymes, potent intracellular bactericidal action may be exerted through the myeloperoxidase-hydrogen peroxide-halide bactericidal system. This system is dependent on the presence of active myeloperoxidase, abundant hydrogen peroxide, rather high local concentration of halogen ions (chloride or iodide), and generates aldehyde from amino acids on the bacterial cell wall surfaces. Though active in PMN leukocytes, its role in alveolar macrophage is uncertain.

The complex physical and biochemical processes involved in particle adherence, particle digestion, and intracellular processing make this system particularly vulnerable to a multiplicity of genetic deficits and to disturbance or suppression of environmental agents such as air pollutants. It is quite likely that the individual variation in tissue responses to environmental agents and air pollutants is determined by subclinical genetic deficiencies that are expressed when the resident host system is stressed by exposure to the varied particulate and gaseous chemical substances incorporated in air pollutants. The system appears to be particularly vulnerable to oxidant air pollutants, such as ozone, nitrogen dioxide, and acid sulfate aerosols. The potential complexity of etiologic and interactive factors is considerable and this whole area of lung cell physiology requires extensive additional research for adequate understanding and application to clinical medicine.

The *specific* cell defense system of the lung is a local expression in the lung of the principles of systemic, humoral, and cellular immunology involving the macrophage, lymphocyte, and plasma cell sequence and interaction. Recent evidence suggests that, in addition to participating in systemic immune events, the lung possesses its own local immune responsive system. Although evidence is not yet clear, this system probably focuses on the peribronchiolar lymphoid nodules that are found at bifurcations of the bronchi at all levels of the bronchial tree from the major bronchi to the terminal bronchioles. These collections of lymphoid cells are barely noticeable in the young or antigenically unstimulated individual but can be shown to become enormously hyperplastic when chronic antigenic stimulation is present in the lung as in chronic bronchitis and other forms of chronic infectious disease in the lung.

The respiratory tract fluid in relatively undiseased lungs contains largely macrophages with only a small proportion of lymphocytes. In certain disease processes in the lung, however, the lymphocyte population increases markedly. The resting lymphocyte population has been reported largely as the B series of lymphocytes, whereas under antigenic stimulation the T lymphocyte series enters into the lymphocyte pool to an increasing degree. The presence of both B and T lymphocytes in respiratory tract fluid points to the ability of the lung to respond locally to antigenic stimulus by both the production of humoral antibodies in the form of IgA, IgE, IgG (IgM not having been described) and also to respond to antigens through the development of cellular immunity, the elaboration of lyphokines (macrophage migration inhibition factor), and lymphoblast proliferation in the face of antigenic challenge.

The presence of these immunologic systems resident in the lung explains the vigorous participation of the lung in a wide spectrum of immunologic events and permits the interpretation of hypersensitivity disorders in response to airborne antigenic challenge. The concept of local immunity also suggests that the immunologic apparatus of the lung is available as a target system for suppression or injury by or for excessive stimulation and reactivity to the wide variety of organic and chemical materials that are present in air pollution exposures. This concept also explains why the total antigenic burden of the lung, particularly that which may be stored within macrophages, is an important determinant in the individual variation of responses to air pollutant exposures. Therefore, minimizing total life-long exposure to antigenic, organic and inorganic particulates is a sound clinical goal and is the *sixth* critical conclusion to be drawn.

The diversity of cellular defense systems in the lung and their central role in the response of the lung to environmental exposures suggests that these systems are of key significance in the mediation of pathologic responses to air pollution exposure. A multitude of biologic effects of general and specific gaseous and particulate chemicals contained in air pollutants has already been described, but a great deal more information is required before a satisfactory level of understanding of the interaction of air pollutants, cellular defense systems, and asthmatic conditions in the lung is attained. Air pollutants both initiate and aggravate a variety of respiratory diseases, including asthma. Patients with existing asthma show increased symptoms when exposed to air pollution. Similarly, patients with other bronchial illnesses, such as ordinary respiratory viral infections, show increased symptoms when exposed to even customary levels of air pollution.

It is well known that patients with chronic bronchitis show exacerbation with elevated air pollution levels, and that even children, without the confusing variables of occupation and smoking history, show an increased morbidity of respiratory diseases and impaired expiratory air flow, this latter measurement, of course, being the subclinical corollary of what is recognized at the clinical level as airways obstruction or asthma. These air pollutant effects can be distinguished from effects of smoking, occupation, age, or social class.

Animal experiments have shown that even "ordinary" levels of oxidant air pollutants, such as ozone and nitrogen dioxide, impair phagocytic activity in the lung and at concentrations below that which will impede ciliary activity. Similarly, exposure to oxidant pollutants or even high levels of $O_2$ result in a proliferative response of epithelial cells, including the goblet cells of the tracheobronchial tree and the type 2 epithelial cell of the alveolus. Chronic exposure to air pollutants, particularly cigarette smoke, results in hypersecretion of these secretory cells, loss of cilia in bronchial epithelium, and metaplasia of the columnar epithelium of the bronchial tree to a squamous epithelium.

Oxidant injury is a particularly important form of cellular injury in the tracheobronchial tree and results in oxidation of sulfhydryl groups in mucus to produce disulfide bonds and consequent increased visocity, oxidation of important membrane enzymes involved in glucose metabolism, specifically glyceraldehyde phosphate dehydrogenase, and oxidation of reduced pyridine nucleotides (NADPH). Similarly, individuals exposed to pollutants such as cigarette smoke show suppressed immunologic responses, both in the form of reduced humoral antibody and in the reduction of cell-mediated immunity as expressed by the elaboration of lymphokines and the ability to achieve a lymphoblast proliferative response. It is premature to indicate the clincial and pathogenic significance of such impaired immunologic responses as far as the production of asthma is concerned, but it is likely that such impaired responses complicate the disease process in asthma at the very least.

The cytotoxic injury of air pollutants, especially particulate pollutants, is also mediated through injury to the lysosomal digestive apparatus of the alveolar macrophage. Certain particulates such as silica are known to injure the phago-lysosomal structures leading to leakage of digestive enzymes in the cytoplasm and into the surrounding medium. The effects of such lysosomal injuries may be mediated through the resultant inflammatory responses leading to collagen production, through injury to supportive collagen and elastin fibers leading to emphysema, or through irritation of bronchial musculature with subsequent broncho-constrictor responses. Although studies to date have focused principally on fibrogenic inorganic particulates, it is not unlikely that the components in certain forms of air pollution might also operate to induce bronchospastic responses by this mechanism. These air pollutant effects may be accentuated where genetic deficiencies exist, either in components of membrane structures or in enzyme systems involved in cellular metabolism and in the elaboration of detoxifying enzymes.

The lungs of patients with clinical asthma are obviously the site of immunologic events that, by their existence, might alter cellular defensive mechanisms leading either to an aggravation of the asthmatic response or to additional forms of immunologic injury. The occurrence of IgE mediated immune responses has not been shown to have an effect on macrophage function in the lung. Indeed, the inflammatory mediators released by IgE antigen interaction appear to be focused principally on vascular permeability and smooth muscle contraction. Type 3 hypersensitivity reactions, however, which are mediated by IgG-antigen interactions, are stimulated by antigenic material reaching the alveoli (such as the microspores of thermophilic actinomycetes) and are amplified into biological effects largely through the complement cascade. Although activated complement components may enhance the uptake of particles to which they are bound and may alter the behavior of smooth muscle and other cells, there is no available evidence to suggest a direct effect of complement fragments on macrophage cell function.

The type 4 or cellular immune responses of the lung have a direct effect on defense mechanisms by activation of macrophages. Manifestation of this response is the elaboration of larger and more numerous intracellular organelles, accentuation of metabolic activity, increased respiratory rate, enhancement of hydrolytic enzymes, increased phagocytic ability, and an increased microbicidal and cytopeptic activity. Although this type 4 hypersensitivity response is antigen specific and due to the presence of committed T lymphocytes, there are nonspecific effects of the presence of activated macrophages with an enhancement of antibacterial responses. Similarly, macrophages activated to one antigen as part of the cell mediated immune response may facilitate the production of antibodies to a second and separate antigen. In this way, the existence of a hypersensitive state may alter macrophage function and alter the intrapulmonary responsiveness to unrelated particles and antigens, possibly rendering them more immunogenic. This may be one mechanism by which an individual allergic to one antigen progressively expands the family of antigens to which he becomes allergic. This points to clinical conclusion *seven* on the advisability of keeping antigen loads to a minimum in individuals who have already existing allergic or hypersensitivity disease in the lungs.

Finally, the role of atopy or type 1 IgE mediated hypersensitivity in predisposing to the later development of type 3 and possibly type 4 immunologic lung disease is illustrated in allergic bronchopulmonary aspergillosis. This disease is uncommon except among patients with an antecedent history of atopic respiratory disease. In this disease, both type 1 and type 3 hypersensitivity can be demonstrated and circulating IgG antibodies against aspergillus are present. In fact, type 1 response appears to be necessary for the expression of the allergic alveolitis since suppression of the type 1 response aborts the later development of the manifestations of type 3 hypersensitivity alveolitis. Another peculiarity of this disease is that in contrast to the normal situation in which aspergillus does not grow in the lungs, in these patients aspergillus can often be found growing in the bronchial lumen, an observation indicating some other alteration of inherent host defense mechanisms that permits the establishment of the microcolony in the bronchial secretions.

## INTERACTIONS OF AIR POLLUTANTS, DEFENSE MECHANISMS AND INFECTIOUS AGENTS IN ASTHMA

Clinical experience and epidemiologic data both hold that attacks of asthma in children and adults are associated with acute episodes of air pollution exposure on the one hand and with respiratory tract infection on the other. Viral infections are associated with, and implicated as a cause of, up to 50% of asthmatic attacks in children. Influenza group viruses are well known for predisposing to secondary bacterial infections. These data suggest that simultaneous impairment of pulmonary defenses by several insults multiplies the clinical effect, so it is not surprising to find that studies of influenza epidemics have shown that allergic patients with influenza suffer a twofold greater incidence of secondary bacterial pneumonia than nonallergic individuals similarly affected. Virus infection, particularly with influenza viruses, is known in experimental models to suppress the bactericidal activity of the alveolar macrophage system. The effect of asthma on these same cells is not well understood. The increased frequency of well-defined viral upper respiratory infections in asthmatics as compared to normal siblings indicates that asthmatics have an increased susceptibility to viral illnesses. The increased incidence of bacterial infection in asthmatics, therefore, may be as well related to differences in susceptibility to the virus infection as to the effects of asthma on the antibacterial defense mechanisms of the lung.

Air pollution exposure likewise adversely affects antibacterial defense mechanisms of the lung. Since the mechanisms of suppression of phagocytosis by alveolar macrophages produced by virus infection are probably different from that produced by air pollutants, one would expect that a combined exposure to viral infection and air pollutants would enhance the risk of secondary bacterial infection. Similarly, the suppressive effect of air pollutant exposure on the immune response in the lung may contribute not only to the increased susceptibility to bacterial infection in the lung but may impair the development of effective immunity to viral infection as well.

Exposure to air pollutants that cause cellular proliferation and that induce a hypersecretory response in the tracheobronchial tree would understandably aggravate attacks of asthma by increasing the secretory response to the asthmatic stimulus. This fact would make asthma more difficult to control in individuals with a hypertrophied tracheobronchial secretory apparatus. The clinician should appreciate that the lung chronically exposed to air pollutants, be they of immune origin or induced in the smoking habit, is an abnormal lung in respect to its complement of defense mechanisms and that the superposition of other clinical conditions such as asthma would be modified and probably worsened in such individuals. What is less well understood is the extent to which the effects of air pollutants on the cellular and humoral immune responses of the lung alter the allergic aspects of the disease and the ability of the lung to deal with the complications of the illness. Much additional research will be required to clarify these important relationships and to turn speculation into clinically useful data. It is clear that air pollutants exert wide ranging and damaging effects on the multiple defense mechanisms of the respiratory tract. In order to reduce the morbidity of asthma, significant changes in air pollution will be required. The specific details of the mechanisms and interactions of such effects await further research for clarification.

# 13 Clinical and Epidemiological Studies on Childhood Asthma in Air Polluted Areas in Japan

Ryo Yoshida, M.D.
Ken Motomiya, M.D.
Hiroyasu Saito, M.D.
Shigeru Funabashi, M.D.

## INTRODUCTION

The prevalence rate of bronchial asthma in school children has been on the increase in the air polluted areas of Japan in recent years. According to our epidemiological surveys made on schoolchildren at all primary schools in Chiba City, the prevalence rate was 0.56% in 1964, 1.17% in 1968, and 2.02% in 1972 — that is, it increased with each survey. Surveys for the prevalence rates of chronic bronchitis and other respiratory diseases in adults have been conducted by use of the questionaire card provided by the British Medical Research Council (BMRC) for respiratory symptoms, but there is no similar method used worldwide for investigation of the prevalence rate of bronchial asthma in children.

Ever since 1968, surveys for the prevalence rate of bronchial asthma in school children in Fuji City in Shizuoka Prefecture, Chiba Prefecture, and Tokyo (Figure 1) have been made by a screening method devised in our laboratory. The results of the surveys are presented first.[1]

In an attempt to investigate whether or not the clinical findings of bronchial asthma in schoolchildren in the air polluted areas would differ from those in the unpolluted areas, surveys were made by the questionnaire system on schoolchildren with bronchial asthma who were redressed by the Japanese Government in respect of medical fees, etc., as victims of air pollution, and on school children with bronchial asthma in Saku City in Nagano Prefecture, a city that can be considered to be free of air pollution. The results of these comparison surveys will be described next.

Finally, results will be presented of clinical immunological tests made on schoolchildren with bronchial asthma in Fuji City and the role of air pollution in the development and onset of bronchial asthma will be discussed.

## SURVEYS OF THE PREVALENCE RATE OF BRONCHIAL ASTHMA IN SCHOOL CHILDREN

Surveys by the questionnaire method were made by distributing survey cards of a given form to the homes of all schoolchildren in three cities so as to have their parents enter answers therein. The answers were classified by criteria that included past history of asthma as well as combinations of wheezing, dyspnea, and their recurrences. Physicians interviewed the parents of such schoolchildren who were suspected of having bronchial asthma and questioned them for history of allergy, intensities of wheezing,

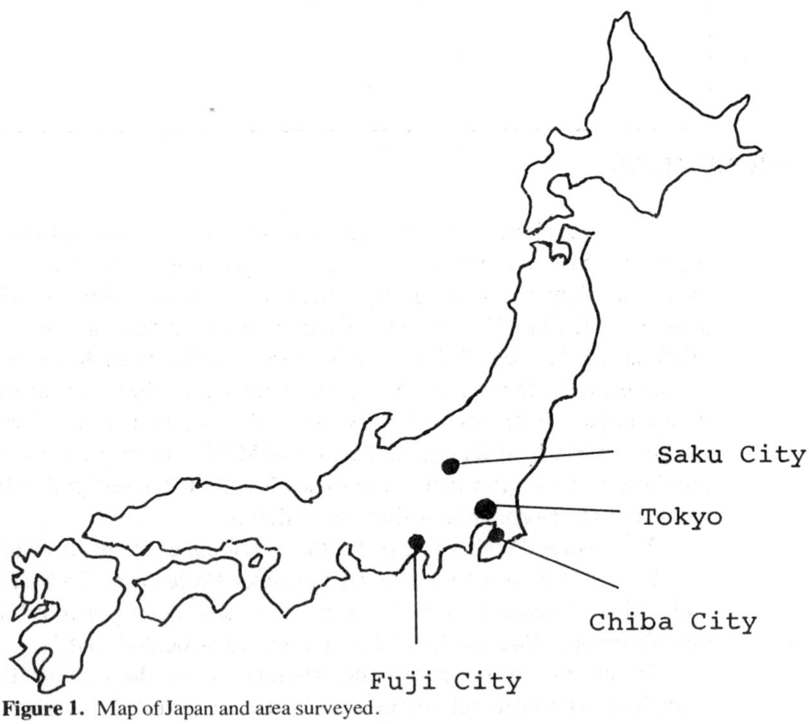


**Figure 1.** Map of Japan and area surveyed.

and dypsnea and their recurrences. A judgment was made by this two-step screening as to whether the given schoolchildren were suffering from bronchial asthma or not. In the surveys, the definition presented by the Childhood Clinical Allergy Research Group — namely, a disease characterized by repeated attacks of expiratory dyspnea accompanied by wheezing — was followed as the definition of bronchial asthma.

The results of the surveys are given in Table 1. The air in Fuji City is severely polluted by pulp and paper mills; there are iron mills, electrical power plants, and heavy chemical industry combinations such as petroleum refineries in Chiba Prefecture; and severe air pollution by urban type public nuisance due to an intense concentration of population exists in Tokyo. The prevalence rates of bronchial asthma at primary schools in the air polluted areas (Fuji City, Chiba Prefecture, and Tokyo) were not less than twice as high as the rate in the unpolluted area.

Figure 2 give the prevalence rates of bronchial asthma in schoolboys at 11 primary schools near the air pollution monitoring sites in Chiba City and their correlation with the concentrations of sulfur dioxides. The correlation coefficient of + 0.705 is statistically significant ($P < 0.05$).

## BRONCHIAL ASTHMA IN SCHOOLCHILDREN IN AIR POLLUTED AREAS AND THOSE IN UNPOLLUTED AREAS

As described above, the air in Fuji City is markedly polluted by pulp and paper mills. Under a governmental ruling relating to severe damages to health due to air pollution, patients with chronic obstructive pulmonary diseases, such as chronic bronchitis and bronchial asthma, in this city as well as those in Yokkaichi City are compensated by the Japanese Government for medical fees. On the other hand, Saku City in Nagano Prefecture is one of those areas with the least air pollution in Japan. In order to compare the clinical findings of bronchial asthma in schoolchildren in the air polluted areas with those in the unpolluted areas, surveys by the questionnaire system were made on children recognized as patients with diseases due to public nuisance in Fuji City and children treated as outpatients at the Department of Pediatrics, Saku General Hospital.

These surveys were made on 134 primary schoolchildren in Fuji City and 49 similar children in Saku City. As shown in Table 2, males constituted a larger part of the subjects in both cities, and 85.8% of the subjects in Fuji City and 91.8% in Saku City had past medical histories of allergic diseases. Allergic diseases such as asthma, urticaria, and eczema were found in family histories of 50.8% of the subjects in Fuji City and in 65.3% in Saku City.

Alimentations given in the first three months after birth are shown in Table 3. In both cities, breast feeding alimentation was most frequent, followed by mixed feeding and cow's milk. There was no difference in the

**Table 1**
**Prevalence of Bronchial Asthma in Primary School Children**

| Prefecture | Air Polluted Area | | | | | Non-Polluted Area | | | | |
|---|---|---|---|---|---|---|---|---|---|---|
| | Number of Schools | Sex | Number of Pupils | Asthmatics | | Number of Schools | Sex | Number of Pupils | Asthmatics | |
| | | | | Number | Percent | | | | Number | Percent |
| Fuji City | | Male | 1,411 | 35 | 2.48* | | Male | 1,470 | 19 | 1.29 |
| (Oct. 1970) | 2 | Female | 1,371 | 26 | 1.90† | 3 | Female | 1,403 | 8 | 0.57 |
| | | Total | 2,782 | 61 | 2.19† | | Total | 2,873 | 27 | 0.94 |
| Chiba | | Male | 2,566 | 92 | 3.59† | | Male | 2,014 | 32 | 1.57 |
| Prefecture | 5 | Female | 2,448 | 57 | 2.33† | 7 | Female | 1,866 | 20 | 1.07 |
| (Feb. 1972) | | Total | 5,014 | 149 | 2.97† | | Total | 3,880 | 52 | 1.34 |
| Tokyo | | Male | 2,635 | 87 | 3.30* | | Male | 651 | 9 | 1.38 |
| (Oct. 1971) | 7 | Female | 2,478 | 53 | 2.14* | 1 | Female | 581 | 5 | 0.86 |
| (Nov. 1972) | | Total | 5,113 | 140 | 2.74† | | Total | 1,234 | 14 | 1.14 |

*$P < 0.05$
†$P < 0.01$

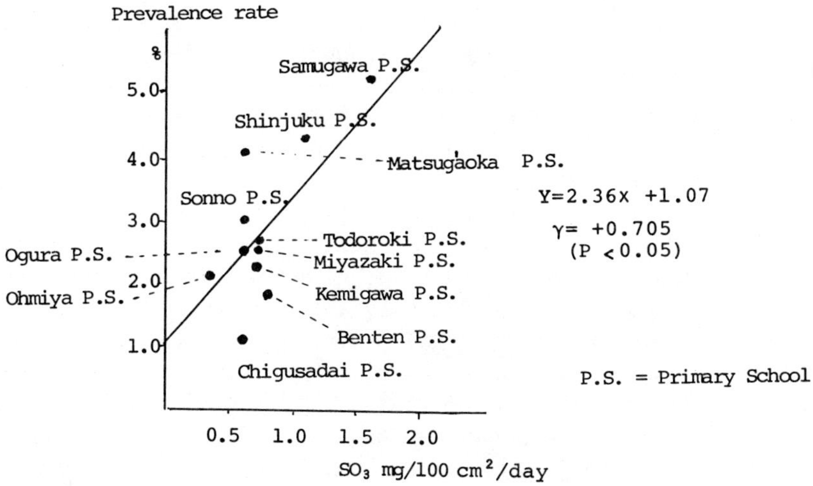


**Figure 2.** Correlations between the prevalence rate in primary school boys and the sulfur oxides concentrations (by $PbO_2$ method; 1972) in Chiba, Japan.

**Table 2**
**Sex Distribution and Allergic History**

| | Patients in Fuji City (Air-Polluted Area) | Patients in Saku City (Non-Polluted Area) |
|---|---|---|
| Number of Cases— | 134 | 49 |
| Sex Distribution (Male : Female) | 1.6 | 2.8 |
| With allergic diseases in past medical history | 115 (85.8%) | 45 (91.8%) |
| Eczema in babyhood | 66 (49.3%) | 29 (34.7%) |
| Complaint of itchiness due to rough skin in winter | 65 (48.5%) | 17 (34.7%) |
| Strophulus, urticaria | 53 (39.6%) | 15 (30.6%) |
| Allergic rhinitis | 54 (40.3%) | 23 (46.9%) |
| With allergic diseases in family history | 68 (50.8%) | 32 (65.3%) |
| Asthma | 24 (35.3%) | 15 (46.9%) |
| Urticaria | 23 (33.8%) | 14 (43.8%) |
| Eczema | 17 (25.0%) | 6 (18.8%) |
| Allergic rhinitis | 12 (17.7%) | 10 (31.3%) |
| Others | 30 (44.1%) | 16 (50.0%) |

frequency of any one of these alimentations between the two cities. The tendency of being attacked by common cold in the previous year is given in Table 4: 49.3% of the subjects in Fuji City had frequent cold attacks seven times or more in the last one-year period; this ratio was higher than that in Saku City, namely, 18.4%, and this difference proved statistically significant ($P < 0.01$).

Comparisons between attacks of asthma in the two cities are shown in Table 5. No fewer than three attacks of asthma had occurred in 69.4% of the subjects in Fuji City and 65.3% in Saku City in the preceding one year. Thus, there was no difference in the frequency of attacks during this period between the two cities. Examinations of seasons with high incidences indicated that the highest incidence occurred during the period between two seasons in both cities, while 18.7% of the subjects in Fuji City and 16.3% in Saku City were patients with perennial attacks. In the evaluation of conditions that could be considered liable to cause an attack or be an inducing cause, it was of particular interest that many patients in Fuji City cited conditions that could be considered associated with air pollution, such as days with strong odors and windy days, while many patients in Saku City cited physical conditions such as fatigue.

Variations in severity of the asthma in the preceding one-year period are presented in Table 6. Exacerbations had occurred in 11.2% of the subjects in Fuji City, the rate being slightly higher than 6.1% in Saku City, while remission had been achieved in 71.4% of the subjects in Saku City, the rate being slightly higher than 59.7% in Fuji City. Both of these differences, however, proved to be of no statistical significance.

School attendance is given in Table 7. It was noted that 32.7% of the subjects in Saku City had seldom been absent from school, this rate being

**Table 3**
**Alimentations in Babyhood**

| | Patients in Fuji City | Patients in Saku City |
|---|---|---|
| Breast feeding | 69 (51.5 %) | 23 (46.9 %) |
| Mixed feeding | 37 (27.6 %) | 19 (38.8 %) |
| Cow's milk feeding | 28 (20.9 %) | 7 (14.3 %) |

**Table 4**
**Tendency to Being Attacked by Common Cold in the Past One Year**

| | Patients in Fuji City | Patients in Saku City |
|---|---|---|
| 7 times or more | 66 (49.3 %)* | 9 (18.4 %) |
| Less than 6 times | 68 (50.8 %) | 40 (81.6 %)* |

*$P < 0.01$

**Table 5**
**Attacks of Asthma**

| | Patients in Fuji City | Patients in Saku City |
|---|---|---|
| More than 3 Attacks of Asthma in the Last One Year | 93 (69.4%) | 32 (65.3%) |
| Seasons with high incidence of asthmatic attacks | | |
| Spring | 27 (20.2%) | 8 (16.3%) |
| Summer | 16 (11.9%) | 7 (14.3%) |
| Autumn | 41 (30.6%) | 19 (38.8%) |
| Winter | 24 (17.9%) | 9 (18.4%) |
| Changing stage between seasons | 77 (57.5%) | 24 (49.0%) |
| Rainy season | 32 (23.9%) | 13 (26.5%) |
| Unsettled | 27 (20.2%) | 4 ( 8.2%) |
| Perennial | 25 (18.7%) | 8 (16.3%) |
| Inducing causes | | |
| Very smoky days | 17 (12.7%) | 2 ( 4.1%) |
| Days with strong odors | 27 (20.2%) | 1 ( 2.0%) |
| Rainy days | 31 (23.1%) | 5 (10.2%) |
| Cold days | 14 (10.5%) | 7 (14.3%) |
| Windy days | 34 (25.4%) | 5 (10.2%) |
| Fatigue | 50 (37.3%) | 27 (55.1%) |
| Unsettled | 48 (35.8%) | 17 (34.7%) |

**Table 6**
**Variations in Condition in the Last One Year**

| | Patients in Fuji City | Patients in Saku City |
|---|---|---|
| Exacerbation | 15 (11.2%) | 3 ( 6.1%) |
| No change | 39 (29.1%) | 11 (22.5%) |
| Remission | 80 (59.7%) | 35 (71.4%) |

**Table 7**
**School Attendance Records**

| | Patients in Fuji City | Patients in Saku City |
|---|---|---|
| Frequently absent | 24 (17.9%) | 9 (18.4%) |
| Occasionally absent | 79 (59.0%) | 24 (49.0%) |
| Seldom absent | 31 (23.1%) | 16 (32.7%) |

**Table 8**
**History of Antigen Tests**

| | Patients in Fuji City | Patients in Saku City |
|---|---|---|
| No | 49 (36.6%) | 2 ( 4.1%) |
| Yes | 85 (63.4%) | 47 (95.9%) |
| Under hyposensitization therapy | 64 (44.8%) | 45 (91.8%) |

**Table 9**
**Eosinophilia in Peripheral Blood**

| | Patients in Fuji City | Patients in Saku City |
|---|---|---|
| Number of patients tested | 134 | 26 |
| Range | 0 ~ 28% | 0 ~ 22% |
| Mean | 7.4% | 6.9% |
| Percentage of eosinophilia (≧ 5%) | 67.9% | 65.4% |

slightly higher than the rate in Fuji City, but there was no statistically significant difference in the attendance rates between the two cities.

Shown in Table 8 are the histories of antigen tests in these children. Many schoolchildren in Fuji City had never received antigen tests, and as a result there were few children receiving hyposensitization therapy.

The results of peripheral blood tests are shown in Table 9. Hypereosinophilia with more than 5% eosinophilia was found in 67.9% of the subjects in Fuji City and 65.4% in Saku City.

The results of surveys made by the questionnaire system on children recognized as patients with diseases due to public nuisance in Fuji City and children with bronchial asthma in Saku City revealed that there were some differences in the frequency of attacks by common cold and so forth between the two cities, but the clinical findings of bronchial asthma in the air polluted areas did not fundamentally differ from those in the unpolluted area, and it was inferred that this disease in both areas had originated in allergic reactions.

## CLINICAL IMMUNOLOGICAL TESTS ON SCHOOLCHILDREN WITH BRONCHIAL ASTHMA IN FUJI CITY

A total of 163 schoolchildren with bronchial asthma in Fuji City were examined, including the aforementioned children recognized as patients with diseases due to public nuisance; at the same time, some clinical immunological tests were made on them. The results of the tests will be described in terms of degree of air pollution. The areas inside Fuji City were classified into the highly polluted area with the average annual $SO_2$ concentration in the range of 0.07 to 0.05 ppm, the moderately polluted area with the average annual concentration of 0.05 to 0.03 ppm, and the slightly polluted area with the average annual concentration of less than 0.03 ppm. The results are presented along with those of the same tests made on non-asthmatic children.[2]

Presented in Table 10 are the results of the decrease in the forced expiratory volume in the first second ($FEV_{1.0}$) after inhalation of methacholine compared with the ($FEV_{1.0}$) before the inhalation, which was done to check bronchial hypersensitivity. Since the decrease was in the range of 35% to 38%, irrespective of intensity of pollution, it was obvious that the majority of the patients had bronchial hypersensitivity.

The rates of positive intracutaneous reactions to inhalant antigens are given in Table 11. The rates of positive reactions to pollens of ragweed, pine, and cypress and to house dust were low. The rate of positive reactions to the house dust commercially available was in the range of 20.0% to 37.5%, which was, indeed, higher than the rate of 9.1% in normal control children but lower than the positive reaction rate of 53.6% in generally asthmatic children, such as children with asthma in Saku City. Intracutaneous reaction tests were also made with wood chip extracts and pulp materials, but the positive reaction rate in asthmatic children was similar to that in nonasthmatic children. The rates of positive reactions to mold antigens — namely, fungi of the genera *Alternaria* and *Penicillium* — were higher than those in non-asthmatic children.

Assay for immunoglobulins was made on a small number of children. As shown in Table 12, $\gamma$-A globulin was absent in one child with bronchial asthma, but there was no significant difference in the average value between children with bronchial asthma and non-asthmatic children. There was no significant difference either in the level of $\gamma$-M globulin between asthmatic children and non-asthmatic ones. $\gamma$-G globulin was decreased in children with bronchial asthma, and that in the highly polluted area was less with statistical significance that that in non-asthmatic children. The level of $\gamma$-E globulin in children with bronchial asthma was two to four times as high as that in non-asthmatic children.

**Table 10**
**Methacholine Inhalation Test**

| Degree of Air Pollution | Number of Cases Tested | Mean Decrease Percent in $FEV_{1.0}$ | Percent of Positive Cases* |
|---|---|---|---|
| High | 44 | 36.3 % | 84.2 |
| Moderate | 30 | 37.9 % | 93.3 |
| Low | 10 | 36.1 % | 80.0 |

*We defined as "Positive" a decrease in $FEV_{1.0}$ of more than 20%.

It was clear from these results of clinical immunological tests that bronchial asthma in children in Fuji City had originated in allergic reactions, and bronchial sensitivity of these children was enhanced. It was, however, noteworthy that the rate of positive intracutaneous reaction to the house dust commercially available was low in these children.

**Table 11**
**Intradermal Reactions (Inhalant Antigens)—Percent of Positive Cases**

| Degree of Air Pollution | Commercial House Dust | Pollens | | |
|---|---|---|---|---|
| | | Ragweed | Pine | Cedar |
| High | 20.0 (85) | 17.4 (69) | 8.1 (37) | 0 (35) |
| Moderate | 37.5 (40) | 8.8 (34) | 3.2 (31) | 3.2 (31) |
| Low | 25.0 (24) | 15.4 (13) | | |
| Normal controls | 9.1 (635) | 14.9 (415) | 0.48 (415) | 0.96 (415) |

| Degree of Air Pollution | TATAMI | Extracts of Chips | | | |
|---|---|---|---|---|---|
| | | Gum | Mangrove | Hemlock | Douglas Fir |
| High | 6.7 (30) | 30.0 (40) | 40.0 (37) | 39.0 (39) | 47.1 (66) |
| Moderate | 0 (30) | | | | |
| Low | | | | | |
| Normal controls | | 30.5 (36) | 30.5 (36) | 33.3 (36) | 33.3 (36) |

| Degree of Air Pollution | Molds | | | | |
|---|---|---|---|---|---|
| | Candida | Alternaria | Penicillium | Cladosporium | Aspergillus |
| High | 47.1 (50) | 16.3 (68) | 10.5 (58) | 8.8 (45) | 6.7 (45) |
| Moderate | 35.5 (34) | 17.6 (34) | 11.7 (34) | 0 (34) | 0 (34) |
| Low | 33.4 (21) | 15.4 (13) | 30.0 (10) | | |
| Normal controls | 28.0 (635) | 3.3 (635) | 4.5 (220) | | |

Note: Numbers in parentheses—cases tested.

**Table 12**
**Immunoglobulins**

| Immuno-globulin | Degree of Air pollution | Number of Cases Tested | Range | Mean ± SD | >700 (%) |
|---|---|---|---|---|---|
| Gamma-A (mg/dl) | High | 22 | 95 – 380 | 194 ± 68 | |
| | Low | 16 | 0 – 370 | 212 ± 84 | |
| | Normal controls | 35 | 38 – 480 | 189 ± 85 | |
| Gamma-M (mg/dl) | High | 22 | 55 – 165 | 111 ± 32 | |
| | Low | 16 | 50 – 350 | 148 ± 76 | |
| | Normal controls | 35 | 30 – 300 | 129 ± 64 | |
| Gamma-G (mg/dl) | High | 22 | 530 – 1400 | 906 ± 236 | |
| | Low | 16 | 620 – 1800 | 1198 ± 309 | |
| | Normal controls | 35 | 550 – 3800 | 1429 ± 725 | |
| Gamma-E ($\mu$/mg) | High | 22 | 150 ~ 3500 | 971 ± 995 | 40.9 |
| | Low | 15 | <150 ~ 1750 | 1754 ± 494 | 53.3 |
| | Normal controls | 22 | <150 ~ 700 | 372 ± 205 | 0 |

## CONCLUSION

In recent years, the prevalence rate of bronchial asthma in schoolchildren has been on the increase in Japan, and the rates in air polluted areas are higher than the rates in unpolluted areas. It has also been shown that the clinical findings in the air polluted areas are not fundamentally different from those in the unpolluted area, that bronchial asthma in the air polluted areas originates in allergic reactions, and that many children with bronchial asthma show high $\gamma$-E globulin levels.

Induction of hypersecretion from the secretory glands of the bronchial mucosa and bronchial spasticity may be considered to be the effects of air-polluting substances on bronchial asthma, and in our experiments on animals it has also been proved that $NO_2$ and $O_3$ possess an action that releases histamine in the lung tissues, an adjuvant effect that facilitates sensitization with other antigens.

For symptomatic treatment of children with bronchial asthma in the air polluted areas, it is necessary to quarantine them from air-polluting substances by use of a dust-free room and so forth, but we believe that it is fundamental to institute hyposensitization therapy by a positive study on antigens.

We are grateful to Prof. Seiji Kubo, at the Department of Pediatrics, Chiba University School of Medicine, Chiba; Dr. Shun'ichi Wakatsuki, Director of the Saku General Hospital, Saku, Nagano Prefecture; Dr. Jo

Kanbe, at the Department of Pediatrics, of the same hospital; and Dr. Heisuke Kato, at the Department of Allergy, of the same hospital, for cooperation throughout the present study.

## REFERENCES

1. Yoshida R: Effects of Environmental Pollution on Children. Japan Medical Association, Igaku Koza, 1974.

2. Funabashi S, Yoshida R: Air Pollution and Childhood Asthma. *Shonika Shinryo* (JAPAN), 35:(4), 460–464, 1972.

# 14 Environmental Respiratory Disease (Tokyo-Yokohama Asthma): The Case for Allergy

George W. Meyer, Lt. Colonel, U.S.A.F., M.C.*

In the fall of 1946, some American servicemen stationed in the Yokohama area, 30 miles from Tokyo, developed nocturnal cough, shortness of breath, and wheezing. By the 1950s this syndrome had been recognized in American servicemen and their dependents stationed throughout the Tokyo-Yokohama area. Huber[1] et al. in 1954 and, later, Phelps[2] in the early 1960s developed rigid diagnostic criteria for their studies:

1. The ages of the persons affected were almost always between 18 and 40 years.
2. In Phelps' study, few of the patients had a personal or family history of allergy.
3. 95% of the patients were cigarette smokers.
4. Symptoms almost always occurred at night or in the early morning hours.
5. Phelps[8] documented the fall-winter prevalence of this disease by tabulating the monthly incidence of new diagnoses.

*The views expressed herein are those of the author and do not necessarily reflect those of the Department of the Air Force.

6. Most persons developed their symptoms within the first 15 months after arrival in Japan, but almost 10% of these patients developed initial symptoms in a second or third tour of duty in this area.

7. Patients usually experienced dramatic symptomatic improvement when they left the area.

The clinical features as originally described usually include a subtle onset, with the illness presenting initially as a nocturnal cough that progresses to wheezing, nocturnal dyspnea, and, later, frank asthma — often resistant to bronchodilators. On physical examination there is wheezing and evidence of an increased expiratory phase. The white blood count may be normal or slightly elevated and peripheral eosinophilia may be present.[3] The sputum is sterile but may contain eosinophiles.[1,7] The chest x-ray is usually normal and pulmonary function tests are variable depending upon the severity of the disease.

Although Phelps originally felt that most patients' symptoms cleared after leaving the area, he showed that permanent changes of chronic obstructive airway disease could develop in patients who remained in the area for extended periods after their first attack.[2] He also found that, although patients obtained symptomatic relief immediately after leaving the area, pulmonary function studies did not necessarily improve within the two week period during which he studied them at bases in northern Japan.[3]

Spotnitz[4] in 1965 felt that 50% of his patients had allergic backgrounds. He found that, although pulmonary function testing usually showed obstruction, the response to bronchodilators was significant early in the disease. He also believed that there were atopic patients who were re-afflicted with asthma when they came to the Tokyo-Yokohama area. He suggested, however, that there was no such specific entity as what was then known as Tokyo-Yokohama asthma. Japanese investigators felt that our diagnostic criteria had been too rigid and so had been unable to recognize the entity in their patients.[5]

For the following reasons, I believe there is strong evidence to suggest that this disease is an allergic reaction to something in the environment, probably in the air:

1. There is evidence that environmental dust could be causing an allergic response. In 1952 it was reported that "Clinical testing of pollen spores found during the period of high incidence of Yokohama asthma did not yield any positive results."[6] Later, the 406th Army Medical General Laboratory reported that both ether-soluble aerosols and dust levels correlated well with the increased attack rate during the October through March period but levels of sulfur dioxide, ozone, and nitrogen oxide did not.

In 1955, preliminary work suggested that a causative agent may have been identified. Many patients suffering with so-called Tokyo-Yokohama asthma had an immediate skin test reaction to an extract of coarse dust. In 1956, the Laboratory prepared more extracts "for more definitive clinical testing." Due to technical problems, testing was not performed and, "in view of the personnel and budget cuts," all their work on the air pollution problem was stopped in early November 1957.

2. Numerous observers have documented significant eosinophilia in up to 84% of affected patients during the fall and winter season. Phelps et al.[1] reported occasional sputum eosinphilia. Tremonti reported that 58% of his patients had sputum eosinophilia.[7]

3. Many patients diagnosed as having so-called Tokyo-Yokohama asthma in previous tours have been immediately affected on return to the Tokyo area. Two severe examples of this problem are summarized:

> One patient, a 40-year-old white male, returned to the Tokyo area on 30 November 1972. So-called Tokyo-Yokohama asthma had been diagnosed during each of three previous tours in the Tokyo area. He denied cough, dyspnea, or wheezing when stationed outside that geographic area. Within two weeks of his arrival, he developed marked dyspnea and wheezing and his pulmonary functions were markedly abnormal. He had no peripheral eosinophilia at the time he was seen. He was so severely affected he was immediately transferred to the United States.
>
> Another patient had developed asthma in a previous tour in the Tokyo area which did not resolve completely after he returned to the United States. Within three months of his return to the Tokyo area, he developed status asthmaticus. He too returned to the United States after almost losing his life.

4. In spite of the fact that the Japanese have changed from coal base fuels in the 1940s and 1950s to higher grade petroleum based fuels used at the present time, the disease continues to attack Americans stationed in the Tokyo-Yokohama area.

5. In spite of the fact that air pollution was said to be worse in the winter in the past,[8] but is now much worse in the summer, the asthma season occurred in the same months in 1972 as it did in the 1950s and 1960s.

Data I have collected for the past two years is also very suggestive of allergy:

1. Alpha-1 antitrypsin levels and phenotypes performed by crossed immunophoresis[9] on affected patients did not differ significantly from that in the population at large. Doctor Robert Fallat, in San Francisco, found that 19 of 22 patients studied were normal MM phenotypes with normal enzyme levels. Three patients were heterozygotes, one each with an MS, MZ, and SZ pattern. Doctor Fallat predicted that the patient with the SZ phenotype and only 76 mg% of the enzyme present would have underlying obstructive airway disease. This is not the case, but studies revealed an obstructive pattern during two different winters (January 1973; February 1974) with a normal pattern in between (May 1973).

2. In 1964, it had been suggested that frequency of asthmatic attacks could be related to the presence of low level temperature inversions.[3] Review of present data by members of the 20th Weather Squadron, stationed in Japan, does not confirm that theory.

3. My data show that 8 of 13 patients studied in the winter of 1972–1973 had peripheral eosinophilia. By May 1973, 6 of the 8 patients had normal levels and a seventh had halved his eosinophilic count.

4. IgE levels were measured by the method of Arbesman, Ito et al.[10] by Doctor Ito at Tokyo University Hospital. The IgE levels seemed to be lowest in May 1973, rising during the following winter. This corresponds to the rise seen in IgE levels during ragweed season as studied by Yunginger and Gleich[11] in ragweed sensitive patients at the Mayo Clinic and to that shown by Berg and Johanssen[12] in children sensitive to timothy and birch pollens.

Air pollution data collected at the Tachikawa train station,[13] just one mile from the base, were incomplete but suggested that dust levels were highest during the September–January period. Data for $SO_x$ also revealed higher levels during the asthma season.

I therefore feel that this environmental respiratory disease is an allergic disease — rare in non-smokers. Those non-smokers who are symptomatic have bronchi that are sensitive to an inhaled agent or agents. Smoking may make otherwise resistant bronchi more responsive to this agent or these agents.

There are several possible mechanisms for this sensitivity:

1. Although it is unlikely that there are any undiscovered pollens that have caused this strange syndrome in Americans, there is a fall pollinating grass in Japan. Japanese allergists must make their own pollen extracts and it is therefore very difficult to obtain this allergen for skin testing purposes.

2. In mid-1973, the Tokyo English language newspaper *The Mainichi Daily News* reported that the air of many of Japan's major cities contained metallic dust in concentrations of up to several hundred times the acceptable amount.[14] Nickel concentrations in parts of Tokyo were as high as 12.5 times the so-called "normal" amount. Platinum has been associated with respiratory allergy[15] and recently, Dr. Fink's group in Milwaukee described asthma caused by nickel sensitivity.[16]

3. Popa et al.[17] have described acute asthma attacks associated with peripheral and sputum eosinophilia elicited by exposure to the simple chemicals ethylenediamine, sulfathiazole, and chloramine with molecular weights greater than 1,000. Other chemicals elicited variable lung responses, usually delayed and not consistently associated with eosinophilia. Reactions to skin tests were delayed and there was little evidence to implicate type I allergic reactions to these chemicals.

4. Hitchcock and co-workers[18] found that methyl piperonylate, extracted from cotton bracts, caused histamine release from chopped human autopsy lung. Although this is not necessarily a true allergic reaction, it demonstrates again that simple chemicals may be responsible for similar reactions.

5. Cowan and co-workers[19] studied allergic asthma at the Minneapolis campus of the University of Minnesota. They found that grain dust in the environment caused a significant increase in attack rates when weather conditions permitted increased exposure of sensitized students on that campus.

6. Sulfate levels seem to be increased in the Tokyo-Yokohama area during the fall-winter period. It is conceivable that suspended sulfates may be related to the symptomatology in these patients.

The data strongly support the idea that Environment Respiratory Disease in Americans stationed in Japan is an allergic disease. The investigations of the 406th Army General Laboratory up to 1957 and present information strongly implicate dust as the offending agent. The dust itself may be the offender, possibly as a rice dust, or it could act as a carrier introducing simple chemicals or metallic particles to persons with sensitive lungs.

## REFERENCES


1. Huber TE, Joseph SW, Knoblock E, et al: New Environmental Respiratory Disease (Yokohama Asthma). AMA *Arch Indus Hygiene* 10:399–408, 1954.
2. Phelps HW, Sobel GW, Fisher NE: Air Pollution Asthma Among Military Personnel in Japan. *JAMA* 175:990–993, 1961.
3. Smith RBW, Kolb EJ, Phelps HW, et al: Tokyo-Yokohama Asthma. *Arch Env Health* 8:805–817, 1964.
4. Spotnitz M: The Significance of Yokohama Asthma. *Am Rev Resp Dis* 92:371–375, 1965.
5. Oshima Y, Ishizaki T, Miyamoto T, et al: Air Pollution and Respiratory Diseases in the Tokyo-Yokohama area. *Am Rev Resp Dis* 90:572–581, 1964.
6. Annual Reports of the 406th Army Medical General Laboratory 1949–1957.
7. Tremonti LP: Tokyo-Yokohama Asthma. *Ann Allergy* 28:590–595, 1970.
8. Phelps HW, Koike S: Tokyo-Yokohama Asthma. *Am Rev Resp Dis* 86:55–63, 1962.
9. Fagerhol MK: Quantitative Studies in Inherited Variants of Serum -1 Antitrypsin. *Scand J Clin Lab Invest* 23:97-103, 1969.

10. Arbesman CE, Ito K, et al: Measurement of Serum IgE by one-step single radial radiodiffusion method. *J Allergy* 49:72–80, 1972.

11. Yunginger JW, Gleich GJ: Seasonal Changes in IgE Antibodies during Immunotherapy for Ragweed Hay Fever. *J Clin Invest* 52:1268–1275, 1973.

12. Berg T, Johansson SGO: In Vitro Diagnosis of Atopic Allergy; IV. Seasonal Variations of IgE Antibodies in Children Allergic to Pollens. *Int Arch Allergy* 41:452–462, 1971.

13. Kindly supplied by Tachikawa City.

14. *Mainichi Daily News.* Enormous Amount of Metallic Dust in Tokyo Air, 1973.

15. Freedman SO, Krupey J: Respiratory Allergy Caused by Platinum Salts. *J Allergy* 42:233–237, 1968.

16. McConnell LH, Fink JN, et al: Asthma Caused by Nickel Sensitivity. *Annals Int Med* 78:888–890, 1973.

17. Popa V, Techulescu D, et al: Bronchial Asthma and Asthmatic Bronchitis Determined by Simple Chemicals. *Dis Chest* 56:395, 1969.

18. Hitchcock M, Piscitelli D, Bouheys A: Histamine Release from Human Lung by a Component of Cotton Bracts. *Arch Env Health* 26:177–182, 1973.

19. Cowan DW, Thompson HJ, et al: Bronchial Asthma Associated with Air Pollutants from the Grain Industry. *J Air Poll Control Asso* 13:546–552, 1963.

# 15 Clinical Implications of Air Pollution Research for Hypersensitivity Diseases

Robert E. Smith, M.D.

I personally am very pleased to be able to act as a substitute for Dr. John P. McGovern since I have had a continuing interest in the subject and some experience in the area of the papers to be summarized. I was stationed at Tachikawa, Japan in 1969 to 1972 and either performed or supervised the workup on the patient reported on in the paper by Dr. Tremonte and referred to by Dr. Meyer this morning. The McGovern Allergy Clinic has been, and is, involved in research related to the effects of air pollution on our patients. Houston, like most of our major cities, has a significant amount of air pollution; and our patients have very definite reactions to air pollution, some of it emotional as well as physical.

Air pollution can be thought possibly to cause or influence allergic diseases in two basic ways — depending upon one's willingness to accept the view that allergic disease is immunologically based, which, of course, we allergists do. The first basic way that air pollutants could cause allergic disease is directly — that is, to act as an antigen, as a haptene combining with carrier proteins. In the broad sense, ragweed as an air pollutant is a complete antigen. As Dr. Meyer mentioned, ethylenediamine has been implicated as a small molecular weight chemical causing asthma in some patients, and by our current concept it must be a haptene that combines with a carrier protein before it can cause the production of specific antibodies. The second broad way in which air pollutants can cause or influence allergic disease is indirectly. There are several possible mechanisms for this to take place, such as an additive effect, a nonspecific inflammatory effect, and a decrease in the ciliary mechanisms of the respiratory tract that allows antigens to remain in contact with the respiratory tract and cells longer. In some way the mechanism of handling potential allergens by the lungs is changed, and they are made more potent. There may also be direct and indirect effects on the autonomic nervous system. Perhaps we will discover many other ways by which this effect can be influenced as we learn more about this problem.

The commissioned review paper by Dr. Green was most interesting and informative. He aptly demonstrated many of the mechanisms by which air pollution affects the immune mechanism so that allergic disease might develop directly or be influenced indirectly. Certainly, our growing understanding of the basic mechanisms of the protective functions of the respiratory tract and the interrelationship of damage caused by infection, allergy, and air pollution are important factors in our future prevention and control of many of the diseases of the respiratory tract.

The paper presented by Dr. Meyer on "Tokyo-Yokohama Asthma — The Case for Allergy" is very important in that it emphasizes that something in the air besides large airborne biological materials, such as pollen and mold spores, can probably cause true allergic disease. From experience with patients with Tokyo-Yokomama asthma (TY asthma) and from the reported cases in the literature, we, too, are convinced that TY asthma is an allergic disease and not toxic or pharmacologic. As Dr. Meyer points out, the physical and laboratory findings of the TY asthma patients are identical with those of patients with asthma due to known allergens, such as ragweed and house dust. The response of these patients to treatment with sympathomimetics, theophyllin, and corticosteroids is similar if not identical to that of atopic asthmatics; and three of the patients improved markedly, if not completely, when removed from the source of the antigens. Also as Dr. Meyer points out, the three most characteristic features of allergic or immune reactions are present. *Induction period:* to my knowledge, no one with this problem has developed it on the first day of exposure. It takes a minimum of several weeks and usually a few months before a patient will develop symptoms. *Anamnestic response:* once the patient has developed the response, he will quickly develop the symptoms upon subsequent exposure despite a long absence from the area and the unknown allergen. Finally, *specificity:* the antigen seems to be specific for the Kanto Plains area and does not cross-react with any known allergens or air pollution from other countries.

While I was at Tachikawa, our public health people used an air sampler to run a large amount of air through regular 10 inch No. 41 filter paper for 24 hours. These samples were collected for several weeks during the winter of 1970–1971. The filter papers from days when our TY asthma patients were having marked symptoms were chopped up and placed in an Erlenmeyer flask containing 100 ml of buffered saline. After the flasks were agitated for 24 hours, the liquid was squeezed out of the filter paper and passed through a millipore filter. After culture for sterility, the resultant extracts were used to test 10 patients with TY asthma. All tests were negative, although these patients did have significant positive reactions to allergens such as house dust and pollen. Although this crude experiment failed to give positive results, it did seem to indicate that the allergen causing TY asthma is of small molecular weight and probably gaseous, or else the allergen is not water-soluble and needs to be extracted with some other solvent.

Since up to 90% of the patients were cigarette smokers, one wonders if the allergen might not be an aldehyde. I believe that Dr. Vincent Derby at Tulane once reported on some patients who were cigarette smokers and who developed asthma when exposed to aldehyde from cigarettes and to other laboratory aldehydes.

The paper by Dr. Yoshida and his colleagues is most significant in that it indicates that air pollution can not only cause specific allergic disease such as TY asthma but can lower the threshold for other allergies and thus increase the incidence of allergic disease in an exposed population. This underscores one of the points that Dr. Green originally made in his paper that I think is important: the goal of minimizing total life-long exposures to antigens or to organic or inorganic particles is a sound clinical goal. Thus, it seems to us that the papers presented give sound evidence that exposure to air pollution for significant variable periods of time can produce harmful effects in several ways:

1. True immunologic disease as illustrated by Tokyo-Yokohama asthma
2. Increase in susceptibility of an exposed population that develops atopic disease due to the common air biological allergens, dust and molds and pollen
3. Harm the immunologic barrier to infectious diseases caused by viruses, bacteria, and possibly granulomas and fungal diseases

The papers also give sound evidence that air pollution contains variable agents not only from one area to another, but seasonally within the same area. This point is most dramatically demonstrated by TY asthma, which occurs almost exclusively in the winter months about the time that home heating is needed, although air pollution in the Tokyo-Yokohama area is as bad if not worse, in the summer months. With some apologies to our Japanese friends here, I might say that I lived at Tachikawa for about five months from July to November before the air was clear enough for me to discover that the Kanto Plains area was in a basin surrounded on three sides by mountains. I never saw them during the first five months.

I think that the papers have demonstrated that we are now beginning to understand the many harmful effects of these multiple pollutants in the air. The papers presented here this morning give sound evidence that immunologic effects are a significant factor in both morbidity and mortality. I personally know of three patients who died of TY asthma at Tachikawa during the first year I was there. I also underscore our need for further studies if we are to make strides in the prevention and treatment of the chronic problems so produced.

# DISCUSSION: PART THREE

**Question:** I am having some trouble sorting out why pollutants like $NO_x$ and ozone act like irritants while the large molecule pollutants might act as antigens. Therefore, I have two questions. How does Dr. Meyer know that Tokyo-Yokohama asthma is not a mild form of pulmonary edema and was this evaluated in these episodes? Second, are there any animal studies being done to see if exposure to pollutants leads to any formation of immunoglobulin?

**Dr. Meyer:** I can give two answers to the first question. You have commented on ozone, $SO_x$, $NO_x$. There are data from the 406th Medical General Laboratory back in the 1950s, which have not been repeated, that suggest the incidence of this disease is not correlated with an increase in the atmospheric levels of these simple chemicals. In fact, these levels seem to be higher in the summer than in the winter. I am not aware of mild pulmonary edema being associated with eosinophilia or IgE elevations. I had not considered the possibility that these people might have mild pulmonary edema secondary to chemical reaction. And I cannot answer your second question.

**Dr. Green:** Dr. Yoshida, were the immunoglobulin measurements made on blood or respiratory tract secretions in the children?

**Dr. Yoshida:** In blood.

**Dr. Green:** That is an important point because again we are dealing with an area that is not too well known. We do know, however, that the respiratory tract does have its own local immunologic system that does respond to locally administered antigens even when specific antibodies are not found in the serum. Conversely you can speculate that this system may respond to suppressive agents administered locally — that is, by the airborne route, when such a suppressive effect is not observed in the systemic, specific immunologic system. Again, I think that this is an area that is unclear. Until one has these kinds of data on the several species of immunoglobulin levels in respiratory tract fluids in relation to air pollution exposure, the question is not closed.

Further, I think that the question of whether the effect of an exposure is both short- and long-term is a very important point. Dr. Meyer referred to the effects of removal from the Tokyo-Yokohama area on the symptomatology of patients with asthmatic episodes. Is the persistence of impaired flow rates interpreted as a permanent functional defect in these patients, and is there any other evidence of persistent functional alteration in the lungs such as persistent respiratory symptoms or susceptibility to respiratory tract infections or other agents?

**Dr. Meyer:** It has been well documented by Dr. Phelps and his group, when they were in Japan in the 1960s, that if patients who had this diagnosis remained in the Tokyo-Yokohama area and continued to have symptoms they would go on to develop irreversible obstructive pulmonary disease. That is one of the reasons why I made a big point of getting patients out as quickly as I could. Dr. Smith mentioned that in his first year in Japan there were three deaths due to this illness. This was another major reason why when someone looked as if he were getting sick, we got him out on the next plane.

One thing that has made evaluation of this phenomenon so confusing is that there are patients who have developed their asthma while in Japan and who, after they left the area, followed a pattern not unlike that of patients with true Environmental Respiratory Disease. When these patients leave the area, they do not become asymptomatic; they may have their asthma symptoms all year round. They may be sensitized to certain pollens and perhaps have a springtime asthma or fall asthma. This is what prompted Dr. Spotnitz to claim that there was no such thing as Tokyo-Yokohama asthma. Some of the patients that we diagnosed in the past, and some who are currently suffering from this disease, may really have bronchitis. Some of them may have an unmasked inherent tendency to asthma, but I do honestly believe there is a true illness that has always been called Tokyo-Yokohama asthma. The terminology is not important, but I think there is definitely a group of patients in whom this illness recurs if they go back to the area and who are relieved of their symptoms once they leave it.

**Dr. Walter Linaweaver, Riverside:** In this discussion no one has referred to any studies of provocative pulmonary challenge, which would seem to be one of the few ways by which this problem could be studied in more detail. Has anyone challenged people with graded dilutions of grain dust or sulfur dioxide or any other antigens or irritants in the atmosphere in the Tokyo area? We can't possibly answer the question as to whether it is allergic or not unless we can demonstrate the antigen and then elicit the antigen-antibody reaction.

**Dr. Meyer:** I agree completely, it is speculation. The problems are time and money and manpower. Back in 1957 the Army was running what I considered to be a good program. They did, in fact, find a coarse dust extract to which patients had a skin sensitivity. One of the reasons they stopped their work was that they had done their processing of this alleged antigen in plastic tubes, and they found a lot of one of these plastic components in their antigens or their suspected antigens. They felt that if they started skin testing with this material they would sensitize people to what was then, and may be now, a very common product. They felt they might be inducing more disease than they were actually investigating. That was one of the reasons why this part of the project was discontinued. The other reason was that the Army ran out of money.

**Dr. Smith:** One of the major problems is that the analysis of the materials in the air in this area has not been shown to be different from any other place. We know these patients have a specific problem that is related to some specific material from the Kanto Plains area. We know something is there because when we take patients up in aircraft and circle around for an hour or two — as we have done many times — the wheeze will go away, only to recur on landing. We still have not defined the specific cause. It would be a major contribution if we could do this.

**Dr. Green:** I think it would be worth pointing out that we have addressed allergic disease as response to agents. But there are two other aspects of which we should be aware: One is the reflex by which cold, odors, or other irritant materials presented to the upper respiratory tract can produce airway constriction. Another area that perhaps is most important in regard to air pollution exposure is acute inflammatory bronchial constriction induced by chemical inflammation, which may lead to permanent bronchial hyper-reactivity. This has been described in response to exposures to sulfur dioxide, ammonia, ozone, and so forth. Dr. Yoshida, you pointed out the increasing incidence or prevalence of asthma during the 1960s. Are you aware of changing oxidant levels in the two cities that you have been studying over the same period of time as your asthma studies have been occurring?

**Dr. Yoshida:** In Japan, sulfur dioxide and nitrogen oxide were measured in past years; in recent years oxidants have been measured. So, when I studied the prevalence rate of bronchial asthma in Chiba, there were no measurements of oxidant levels.

**Dr. Green:** In other words, in the earlier studies there was no measurement of oxidant levels in those cities. It is very difficult to put all these variables together in some rational way because when one is looking at changes or prevalence rates over time in populations, it is terribly important to know what is going on in the atmosphere.

**Question:** Is there any effect of cromolyn sodium on TY asthma?

**Dr. Meyer:** About a year and a half ago we were able to get some cromolyn sodium from one of the pharmaceutical manufacturers. We had hoped to study its effect because we felt that it would add to the circumstantial evidence we already had that this was, in fact, an allergic mediated disease. One of the problems we had was deciding which patients should get this medication because we have always been taught *primum non nocere* and we did not feel that it would be wise to give it to people we thought were going to become more ill, especially in our military population whom we could very easily air-vac back to the United States. At the Naval Hospital at Yokosuka, 11 Japanese patients who are civilians employed by the United States Government were treated with cromolyn sodium. According to their data, which were given to me verbally, 10 of these patients were very easily controlled with 4 doses a day. The eleventh patient required up to 6 doses a day. Since the best cure rate that I have heard is about 80%, I am skeptical about this 100% cure rate. But if the data are correct, they add further circumstantial evidence that there is a mast cell degranulation disease that is prevalent.

**Question:** Dr. Meyer, what has been done with these people? Have they all come back to the United States?

**Dr. Meyer:** Occasionally we were able to get some of them transferred to other Air Force bases in the Pacific area, but most of them went back to the United States.

**Dr. Green:** What is the possibility of follow-up of that population?

**Dr. Meyer:** Dr. Phelps has apparently followed some of his patients very well. A couple of my patients have told me that for about 10 years, they have had regular pulmonary function studies sent to him. His follow-up is much better than mine.

**Dr. Smith:** I think that we have underscored the fact that there probably is significant allergy-immunologic disease caused by air pollution. I think that we also established that we do not know much about it yet, that we have more questions than we have answers, and that comprehensive studies involving the patients and measurement of air pollutants all done at the same time is an important program that we need to pursue.

## PART FOUR

# CENTRAL NERVOUS SYSTEM AND SENSE ORGANS

# 16 Effects of Vapor Phase Pollutants on Nervous System and Sensory Function

Robert I. Henkin, M.D.

## INTRODUCTION

The relationship of changes in the central nervous system and sense organs to environmental pollutants is a vast and difficult topic. It is clear that the effects of these environmental pollutants are very real, and when concentrations of these pollutants are high, the effects are quite obvious. These effects can be very clearly delineated in animals, and somewhat in man. Difficulties arise in a number of ways: Some of these tests involve psycho-physical measurements that we as physicians have always been reticent to accept. The problem of individual variability has always been difficult and chemical markers have been problematical. In addition, there have been problems in terms of finding specific pathological changes, sometimes because it has been difficult to obtain tissue for examination. So, there are difficulties inherent in this field that may be a little different from those that have been discussed previously. Nevertheless, these problems are very meaningful, very important; they are ones about which people complain, but which we as physicians have some trouble defining specifically. Consequently, the data are somewhat soft. Moreover, the problems are so vast and extend over so many pollutants, each of which has its own characteristics both pathologically and physiologically, that the whole field is difficult to grasp. I shall try to present an overview of the topic, and the subsequent contributors will deal with specific problems.

Vapor phase pollutants are toxic or hazardous chemical agents that arise from natural or from man-made sources and that enter the atmosphere through a variety of pathways. Their impact on man is influenced by their physical properties or by environmental factors such as precipitation, wind, temperature, and inversion. They enter the body by inhalation or they affect the surface of the body by direct contact. These pollutants may be divided into three classes depending upon their physical properties: (1) true gases, (2) vapors of volatile or volatilizable liquids, and (3) aerosols.

*True gases* that are toxic to man and that occur normally in the gaseous state may be generated from either natural or man-made sources; carbon monoxide, chlorine, and sulfur dioxide are examples of these gases. *Vapors* of volatile or volatilizable liquids are usually the vapor phase of organic solvents and other substances that vaporize under normal conditions or upon combustion and are then released into the atmosphere; various hydrocarbons and chlorine gas are examples of this category. *Aerosols* are airborne solid particles and/or small liquid droplets that enter the environment as clouds of solid particulate matter such as smokes, dusts, pollens, fogs, or sprays and may include toxic agents such as metals or pesticides.

The purpose of this paper is to examine the effects of several vapor phase pollutants, largely ignoring their source, and to concentrate entirely upon the effects of these agents on the central nervous system and on sense organs. In addition, changes in sensory function resulting from these toxic agents will be identified as potential diagnostic signs of excessive levels of these agents in the body. The effects on man will be emphasized in all examples.

## TRUE GASES AFFECTING THE NERVOUS SYSTEM

The major environmental gases that affect central nervous system function are aliphatic and aromatic hydrocarbons, halogenated hydrocarbons, alcohols, aldehydes, ketones, esters, and ethers. These enter the body by inhalation, are quickly absorbed across the alveolar capillary membranes and can attain high enough concentrations in the nervous system to produce toxic effects. These gases can also act as sensory irritants to produce tears (lacrimators) or sneezing (sternutators).

**A. Aldehydes** Aldehydes are products of incomplete combustion of hydrocarbons and other organic materials. They are emitted into the atmosphere from exhausts from motor vehicles, from incineration of wastes or from combustion of fuels. They are also formed by photochemical reactions between nitrogen oxides and some hydrocarbons. The principal effects of these vapors in man are irritation of the eyes and skin.[1,2] The olefins or unsaturated aldehydes and the halogenated aldehydes generally produce a more noticeable irritation than do saturated aldehydes. Aromatic and heterocyclic aldehydes generally cause less irritation than saturated aldehydes. Toxicity of aldehydes decreases with increasing molecular weight and generally decreases as chain length increases.[2,3] The addition of a double bond increases its toxicity. Although aldehydes possess anesthetic properties[2,4] these effects are generally obscured by their irritant effects and by their rapid metabolism. The severity of effects are obviously related to duration of exposure as well as to concentration of the pollutant.

Formaldehyde is one of the common aldehydes that produces irritation of the mucous membranes of the eyes, nose and other portions of the upper respiratory tract.[2,4,5] In addition to irritation it stimulates lacrimation and sneezing. Irritant effects have been reported to occur at 1 ppm or below (1,200 $\mu g/m^3$),[6] but other workers reported little discomfort until concentrations reached 2–3 ppm[2] or even 4–5 ppm. Concentrations of 10 ppm can be endured for only a few minutes and cause profuse lacrimation in all people. The odor threshold of formaldehyde has been reported to be as low as 70 $\mu g/m^3$ and a cortical reflex threshold has been observed at 98 $\mu g/m^3$.[7] All reported fatalities from formaldehyde have resulted from ingestion.

Acrolein vapors are also extremely irritating to the eyes and upper respiratory tract[2] and can stimulate lacrimation at concentrations as low as 625 $\mu g/m^3$ (0.25 ppm) for as short a period as five minutes. The odor threshold has been reported to beat 525 $\mu g/m^3$.[8] There has been one fatal case of acrolein inhalation at 150 ppm (375,000 $\mu g/m^3$) for 10 minutes.[9]

**B. Ammonia** The primary sources of ammonia as an air pollutant in cities are the combustion of fuels, the incineration of waste products, and the use of the internal combustion engine. Industrial sources include chemical plants, coke ovens, and refineries. In stock yards ammonia is formed by biological degradation. However, the main atmospheric source of ammonia is from natural processes on land and in the oceans.

Inhaled ammonia is an irritant that affects primarily the upper respiratory tract.[10] Hyperesthesia has been reported at concentrations of 350,000 $\mu g/m^3$.[10] Zinc ammonia sulfate aerosols were suggested to be partly responsible for the irritant effects during the Donora Smog Episode in 1948,[11] and other observers have noted similar effects in other incidents involving the same aerosol. The smaller the particle size the greater the irritative action.[12] If ammonia concentrations are high or exposure prolonged, the eyes may be seriously damaged, with resultant chemosis, infiltration opacities of the cornea, iritis, and cataracts. Sneezing and salivation have also been reported following exposure to ammonia.

**C. Chlorine** The major commercial source of chlorine is the electrolysis of alkali chloride solution. Chlorine is usually liquified before use or storage. If the liquefaction process is not carefully controlled, chlorine may escape as a gas and become an atmospheric pollutant. The major effect of chlorine gas is its irritating and corrosive action on the mucosa of the eyes and upper respiratory tract as a result of its strong oxidizing action on the sensory receptors in the mucous membranes. At levels of 3,000 $\mu g/m^3$ of air (1ppm) or less, some noticeable irritation has been reported[13]; however, many reports indicate that people chronically exposed to chlorine develop some degree of tolerance.[14] Headaches have been reported to be caused by irritation of the nasal sinuses[15] and sneezing, coughing, and loss of voice have been reported to occur. Dizziness, anxiety, syncope, and anorexia have also been reported as well as profuse tearing and photophobia.[16,17] The odor threshold is reported to be below 3,000 $\mu g/m^3$, but some reports of thresholds range from 150 to 940 $\mu g/m^3$.[18,19]

**D. Hydrogen sulfide** The major sources of this pollutant are kraft paper mills, industrial waste disposal ponds, sewage treatment plants, tanneries, and coke oven plants. Hydrogen sulfide can act as a cell and enzyme poison and can cause irreversible changes in nervous tissues.[20,21] At high concentrations (over 1,000,000 $\mu g/m^3$) it frequently produces death by causing paralysis of the respiratory center.[21] However, if the victim is moved quickly to an uncontaminated environment and respiration initiated before heart action stops, rapid recovery can be expected.[22] At lower concentrations it produces conjunctivitis, psychic changes, disturbed equilibrium, nerve paralysis, spasms, and unconsciousness.[21,23] Common symptoms of exposure affecting the nervous system are a metallic taste, fatigue, blurring of vision, intense aching of the eyes, insomnia, and vertigo.[21,24]

The odor threshold for hydrogen sulfide is between 1 and 24 $\mu g/m^3$, much below the toxic level, and it is easily recognized by its characteristic and obnoxious odor of rotten eggs. However, exposure may produce olfactory fatigue or habituation,[22] and some workers have suggested that at a concentration of 320,000 $\mu g/m^3$ its smell is not as pungent as it is at lower concentrations, probably because of olfactory fatigue.[23] Loss of the sense of smell has been reported at concentrations of 150,000 $\mu g/m^3$ after exposure from 2 to 15 minutes.[25] The dulling of olfactory acuity through the effects of hydrogen sulfide on the olfactory system constitutes a major hazard to people exposed to moderate and high concentrations for extended periods of time.[23] At concentrations over 1,120,000 $\mu g/m^3$ there is practically no sensation of odor and death can rapidly occur.[21]

**E. Sulfur dioxide** Exposure to sulfur dioxide is mainly limited to those people who work in areas where this gas is industrially generated, such as where sulfide ores are smelted, coal or fuel oils containing sulfur as an impurity are burned, paper is manufactured, or where petroleum is refined. The rapidity with which sulfur dioxide forms sulfurous acid on contact with moist mucous membranes explains its most prominent biological effect — namely, irritation of the upper respiratory tract. Acute exposure to concentrations above 20 ppm additionally produces choking and sneezing.[26,27] Acute exposure to concentrations of 50 ppm produces irritation of the upper respiratory tract that cannot be tolerated for more than 15 minutes.[27] Mixtures of sulfur dioxide and aerosols often have a greater effect than might be expected from the two components acting independently.[28,29] This enhanced effect is called synergism and is due to the transformation of sulfur dioxide into a variety of products, including sulfuric acid and sulfate salts, which are more highly irritating than sulfur dioxide itself. This synergism occurs with increased frequency in humid air, as sulfur dioxide dissolves in the vapor in the air and forms aerosols of soluble salts and acids. The tiny droplets carry the sulfur dioxide more deeply into the respiratory tract than would occur with sulfur dioxide alone. The droplets also provide an aqueous medium in which the chemical transformations to sulfuric acid and metallic sulfates may take place.

**F. Carbon monoxide** This gas is released from incomplete combustion in heating stoves and furnaces and from gasoline powered motor vehicles. Automobile exhaust gas may contain 7% to 30%, coal gas about 10%, water gas about 30%, blast furnace stack gas about 30%, and high-explosive blasts about 30% to 60%.[30,31] Toxicity to the nervous system varies with the concentration in air, the duration of exposure, and the presence of other pollutants.

No adverse effects have been reported upon exposure to 0.04% carbon monoxide for as long as one hour. Exposure to 0.06% to 0.07% for one hour has produced headache, fatigue, and nausea. The effects are dependent upon the combination of carbon monoxide with hemoglobin to form carboxyhemoglobin, which in turn interferes with the transport of oxygen to tissues. Carbon monoxide would be relatively nontoxic if its reactivity with hemoglobin were the same as that of oxygen; instead, hemoglobin reacts approximately 200 times more readily with carbon monoxide than it does with oxygen. For example, breathing air containing 0.1% carbon monoxide results in a 50% blood oxygen saturation and a 50% carbon monoxide concentration. With a blood saturation of 10% to 20% carbon monoxide, a tightness across the forehead and a slight headache are the only sensory changes observed. At a blood saturation of 20% to 30%, headache and throbbing in the temples have been reported. At concentrations of 30% to 40%, severe headache, weakness, dizziness and nausea have been reported; whereas at 50% to 70% concentration, coma, respiratory paralysis, and death have been reported.

After severe carbon monoxide poisoning numerous neurological changes occur that may persist for weeks or even years.[32,33,34] These changes are due either to prolonged neural anoxia or to anoxemic damage to vascular walls that results in perineural hemorrhage. These changes lead to neuritis, especially of the peripheral nerves of the lower extremities, commonly the sciatic, tibial, and peroneal nerves. Severe pain of the lower extremities has also been reported in relationship to this neuritis. Disturbance of speech, complete or partial loss of vision or hearing, and losses of taste and smell have been commonly reported as residual and chronic effects. Dizziness and headaches have been reported to persist for months after severe, prolonged exposure, and amnesia, with or without loss of the capacity for memorization, and psychoses have also been reported. Disturbances of mental state have taken the form of toxic psychoses of a manic or depressed type. If unconsciousness occurs and persists for too long a period degeneration of the globus pallidus and of the cerebrum may result. Recovery rarely occurs at this stage; but when it does, neurological impairments such as hemianopsia, mental symptoms and other neurological deficits may persist. Death may occur within hours or days.

There is no evidence that carbon monoxide is a cumulative type of poison among occupational groups continuously exposed such as taxi drivers, garage workers, or tunnel guards. On the other hand, some populations are particularly susceptible to carbon monoxide exposure. These include subjects with a respiratory disease that interferes with oxygen uptake, subjects with anemia or others in whom there is a reduced oxygen transport capacity in blood, subjects with any form of atherosclerosis, particularly when blood flow to the brain or myocardium is impaired, subjects with reduced tissue oxygenation because of high altitude, and subjects with peripheral vascular disease.

There are numerous effects of carbon monoxide on behavior. At low concentrations these effects are small and variable; but at a carboxyhemoglobin content of 2% to 5%, there are decrements in vigilance tasks[35,36,37] and in the ability to discriminate differences between auditory signals.[38] The latter studies also implied that exposure to carbon monoxide altered time discrimination and estimation. These decreases in perceptual ability can occur at carboxyhemoglobin levels lower than those reached by chronic smokers. These results have an important impact upon automobile drivers where misperception of such sensory stimuli are of great importance in preserving life. However, studies of the carboxyhemoglobin levels in automobile drivers have produced inconclusive results, most showing little effect even when the carboxyhemoglobin concentration was as great as 15%.[39,40]

The effects of carboxyhemoglobin on vision have been studied in some detail. Carboxyhemoglobin content as low as 4.5% interfered with brightness discrimination and decreased the ability of subjects to distinguish a slight increase in the intensity of a dimly lit field.[41] In later studies exposure to carbon monoxide that produced carboxyhemoglobin concentrations of as little as 3% was associated with significant increases in brightness thresholds[42] and with decreases in visual acuity. Other investigators did not observe these changes.[43,44]

Some studies suggest that exposure to carbon monoxide decreased several aspects of complex intellectual behavior, including ability to do arithmetic and to detect plural nouns.[45] Although this study has been criticized,[46,47] other investigators have confirmed several aspects of this work.[48,49] It is useful to point out that these decrements in intellectual behavior were found in firemen exposed to carbon monoxide at 100 ppm for varying periods of time,[45] and that these men are commonly exposed to high concentrations of carbon monoxide during the course of their daily work.

**G. Carbon dioxide** Carbon dioxide results from complete combustion of carbon-containing compounds and occurs in flue gases, sewers, storage bins of grain elevators, caves, old wells, cisterns, tunnels and mines. At concentrations of 5%, dizziness, headache and confusion occur; at 8% to 10%, severe headache, dimness of vision, and tremor occur. If these levels are maintained for 5 to 10 minutes, consciousness may be lost. Concentrations exceeding 10% carbon dioxide in the air have been associated with ataxia followed by unconsciousness. Concentrations of 25% to 30% in air have been associated with diminished respiration, hypotension, coma, areflexia, and anesthesia. At these high concentrations consciousness is lost almost immediately and death may ensue.[50]

**H. Ozone** Ozone is a highly reactive oxidizing agent that commonly occurs in the environment as a result of the interaction of unsaturated hydrocarbons and nitrogen oxides in the presence of ultraviolet rays in sunlight. It is a constituent of smog in cities where motor vehicle traffic is heavy. It has been alleged that ozone itself does not produce eye irritation, but that it coexists with formaldehyde in automobile emissions which do produce eye irritation. In spite of these findings, other investigators have suggested that ozone does produce eye irritation at concentrations of 200 $\mu g/m^3$.[51] Neural effects of ozone have been reported to follow exposure to 2 ppm ozone for two hours with production of a temporary exhilaration that is followed by depression. These effects have been noted to persist for two days. Higher concentrations (5 to 10 ppm) have produced headache and drowsiness.[52]

**I. Nitrous Oxide** Nitrous oxide pollution of the atmosphere occurs usually from the nitration of cellulose and other organic materials. Irritant gas exposure is characterized by pain and by burning and choking sensations in the upper respiratory tract at concentrations of 50 to 150 $mg/m^3$ (25 to 75 ppm). Methemoglobinemia and more severe symptoms such as somnolence, feelings of intoxication, syncope, and loss of consciousness, occur after prolonged exposure.[53] Toxicity results from asphyxiation. Perception of the vapor of this gas by olfactory cues occurs around 225 $\mu g/m^3$ or 0.12 ppm, a lower concentration than that which produces neurological symptoms.[54]

**J. Nitrogen dioxide** Changes in sensory perception are the most sensitive indicators of the presence of nitrogen dioxide. Exposure of normal volunteers to the vapor of nitrogen dioxide produced changes in olfactory recognition thresholds, mainly increasing the thresholds, as well as a metallic taste in the mouth and pharyngeal roughness or dryness.[55,56] These effects occurred at concentrations ranging from 0.1 to 30 ppm. Irritative symptoms lessened with repeated exposures and even disappeared at concentrations as high as 20 ppm. Increased humidity produced a sharp increase in odor perception and in irritation of the mucous membranes of the respiratory tract.

Exposure to low concentrations of nitrogen dioxide, 0.075 ppm, may also impair dark adaptation.[56,57] The threshold for producing this sensory defect may be as much as 40% lower than that which produces changes in odor perception.[56]

## VAPORS OF VOLATILE OR VOLATILIZABLE LIQUIDS AFFECTING THE NERVOUS SYSTEM

**A. Sulfuric acid** Exposure to vapors of sulfuric acid occur in some industrial situations and concentrations of 0.8 to 17 $mg/m^3$ produce symptoms of irritation or tickling of the upper respiratory tract as well as sneezing and coughing. At lower concentrations irritation of the mucous membranes of the eye also occurs. In experimental situations when subjects were exposed to sulfuric acid aerosols at concentrations of 0.6 to 0.85 $mg/m^3$, throat tickling and scratching occurred.[58] At 1.1 to 2.4 $mg/m^3$ all subjects noted irritation of some sort. At 2.4 to 6.0 $mg/m^3$ all subjects noted acute irritation of several mucous membranes and a reflex cough. In other studies exposure to concentrations of 0.73 $mg/m^3$ elicited a 19% prolongation of chronaxy.[59]

**B. Benzene** Inhalation of benzene at a concentration of 100 ppm for several hours has caused headache and a sense of fatigue. Inhalation of concentrations of 1,500 ppm for an hour has produced marked depression, while 3,000 ppm for a few minutes produced irritation to the eyes and nose. The vapor of benzene is recognized by the olfactory system at concentrations of 1.5 to 5 ppm, far below the level of any toxic effects.[60]

Alkyl benzenes, homologues of benzene, are less damaging than benzene to cells, but may produce more depression of the central nervous system. Toluene irritates the mucous membranes of the eyes and upper respiratory tract at concentrations ranging from 100 to 500 ppm. Ethyl benzene is only slightly irritating to the eyes at 1,000 ppm, but it is intolerably irritating to eyes and the upper respiratory tract at concentrations of 2,000 ppm. The odor threshold for toluene is 1 to 5 ppm. These substances particularly affect the nervous system by producing central nervous system depression, headache, anorexia, fatigue or muscular weakness, neuromuscular incoordination, nausea, vertigo, paresthesias, mental confusion, drowsiness, and loss of consciousness.[61]

**C. Carbon tetrachloride** Exposure to this atmospheric pollutant usually occurs in industrial situations and in the chemical industries. Since it is heavier than air, it tends to accumulate in high concentrations at floor or ground level from which it slowly diffuses. When exposure to this vapor is combined with the consumption of alcohol, the toxicity increases two- to threefold. Daily inhalation of 45 to 100 ppm for several hours has been associated with symptoms of irritability, anorexia, headache, giddiness, and emotional instability. With acute or chronic exposure there is brain edema, especially affecting the myelin of the white matter, and degeneration of brain cells, especially those of the cerebellum. Optic neuritis and polyneuritis have also been observed.[62] The odor threshold in air is 10 to 50 ppm, a level that overlaps that at which some toxic effects have been observed.

**D. Carbon disulfide** This heavy, volatile liquid has a characteristic, unpleasant odor and is mainly an industrial poison, although it is also widely used as a rodenticide. Because of its lipid solubility it quickly attacks the nervous system. Initial neurological symptoms are headache, pains in the extremities, excitement, narcosis, and finally motor and respiratory paralysis. In high concentrations rapid loss of consciousness may occur. Upon recovery symptoms of emotional instability, anorexia, and epileptiform seizures may persist for months. Chronic exposure to smaller amounts of this vapor can result in headache, malaise, fatigue, muscle weakness, anorexia, weight loss, visual disturbances, and paresthesias. Polyneuritis is a frequent occurrence and psychic changes have been commonly observed, as well as daytime somnolence, nocturnal insomnia, and partial blindness.[63,64]

Carbon disulfide is absorbed and stored in body fat and in other lipids including myelin of the nervous system. Areas of degeneration have been found throughout the central nervous system. The gray matter is affected more than the white and the cerebellum more than the rest of the brain. Chronic exposure has been associated with atherosclerotic changes in the smaller arteries of the brain. Acute death is due to central nervous system depression accompanied by respiratory failure. Chronic poisoning may lead to death from inanition.

## AEROSOLS AFFECTING NERVOUS SYSTEM FUNCTION

Aerosols of various substances in the environment produce adverse effects in neural function. These substances include metals, pesticides, biological aerosols, and radioactive substances.

### A. Metals

**1. Beryllium** The acute symptoms of contact of beryllium with the mucous membrane of the eye and upper respiratory tract include itching, burning, and rhinorrhea,[65] while chronic pulmonary exposure produces fatigue, anorexia, and weakness.[66] Peak concentrations of 25 $\mu g/m^3$ or greater are necessary for these effects to occur.

**2. Boron** The boron compounds that are associated with air pollution are mainly boron hydrides (boranes), which are considered highly toxic.[67] Intended as high-energy fuels, they are released into the atmosphere during combustion and cause severe central nervous system irritation that can produce permanent injury or death.

The boranes consist mainly of pentaborane, decaborane, and diborane. Others are known or still being developed. Pentaborane is the most toxic, whereas decaborane and diborane are slightly less toxic. In 1951 a syndrome similar in some ways to metal fume fever was described after exposure to diborane, while exposure to pentaborane produced marked central nervous system irritation.[46] Diborane produced symptoms of headache, drowsiness, and dizziness that appeared soon after exposure and were of short duration.[68] Pentaborane produced the same symptoms but the vertigo, headache, and drowsiness were more pronounced. Convulsions, semicoma, and disorientation were among the other central nervous system symptoms reported.

**3. Cadmium** Cadmium has been released into urban air and drinking water from a variety of sources including cadmium ore smelting stacks and cadmium tailings. In Japan, smelter workers exposed to cadmium fumes in a concentration of 75 to 240 $\mu g/m^3$ for varying periods of time[69] complained of a variety of symptoms. Among the early effects of cadmium toxicity was decreased olfactory acuity, which may be used as an index for elevated levels of this metal in other tissues. Whether these changes correlate closely with the proteinuria that is also associated with this exposure is not known. The syndrome of Itai-Itai (Ouch-Ouch) disease in women and the many neurological symptoms, including disturbed mineral metabolism and renal damage, have been diagnosed as being caused by cadmium poisoning.[70] Feeding cadmium[71] to rats has led to vestibular disturbances, paralysis, convulsions, rolling fits, and death. Another study with animals showed that a single inhalation of cadmium oxide aerosol in amounts of 13 to 18 $\mu g/l$, as well as repeated exposures to 1.8 to 2 $\mu g/l$, produced disturbances of conditioned reflex activity.[72] After both acute and chronic exposures similar results followed prolonged latency periods and were accompanied by lower reflex responses and, in some cases, by hypnotic activity.[72]

**4. Manganese** In 1837 inhalation exposure to manganese prompted the first clinical description of manganism in 5 workers at a pyrolinsite mill. This study led to the conclusion that manganese could provoke a syndrome similar to paralysis agitans.[73] These symptoms were rediscovered in 1901 by Joksch[74] and Emden.[75] As described by Stokinger:[76]

> . . . the onset of this syndrome is insidious, with apathy, anorexia, and asthenia. The Mn psychosis . . . has certain definitive features: unaccountable laughter, euphoria, impulsiveness, and insomnia, followed by overpowering somnolence. Headache is often present, as are recurring leg cramps and sexual excitement, followed by impotence. Following or concomitantly with these manifestations are speech disturbances with slow and difficult articulation, incoherence, even complete muteness. Mask-like facies sets in, general clumsiness of movement, noticeable in altered gait and balance, which may develop into severe propulsive and retropulsive movements, and ultimate development of "hen's gait." Micrographia is a consistent finding. As the poisoning progresses, rigidity is marked, frequent falls occur, and tremor sets in, which becomes exaggerated by stresses such as fatigue or emotion. Absolute detachment, broken by sporadic and spasmodic laughter, ensues and, as in extrapyramidal affections, salivation and excessive sweating. Despite the severe incapacitations imposed by the disease, the patient survives, although permanently disabled unless treated; chronic Mn poisoning is not a fatal disease.
>
> The characteristic pathological lesion in man is destruction of the ganglion cells of the basal ganglia, followed by scarring and shrinking. Perivascular degenerative areas in the striatum and pallidum occur and to lesser extent in the frontal and parietal cortex. Others[77] believe Mn effects are not confined to the corpus striatum but involve the vegetative and vasomotor centers of the midbrain and also the medulla and cord. Unfortunately, owing to the customs in the countries in which manganism is currently most prevalent, autopsies cannot be performed and thus no recent histological examinations have been reported.
>
> The prognosis of manganism depends on the duration of the disease; many of the symptoms will regress or disappear if the worker is removed from exposure shortly after the appearance of symptoms, although there may be some residual disturbances in speech and gait. Well-established manganism, however, is a crippling disease with permanent disability particularly in respect to the use of the legs . . . Some of these patients may improve slowly with occasional remissions of tremor, weakness, and muscular cramps. Seriously poisoned individuals have been considered lifelong cripples.

**5. Mercury** The major factors that determine the effects of inhaled mercury vapor relate to the physicochemical properties of the compounds inhaled and to the individual's susceptibility.[78] Acute toxicity is usually manifested by renal and gastrointestinal changes, but chronic toxicity is usually indicated by neurological changes. Acute toxicity following inhalation of mercury in concentrations of 1,200 to 8,500 g/m$^3$ is characterized by a metallic taste, nausea, abdominal pain, and headache. Facial tics, sleeplessness, headache, digital tremors, delirium, and hallucinations may follow.[79] In mild cases of acute poisoning patients who recover, usually within 10 to 14 days, may develop chronic symptoms such as muscular tremors or an abnormal increase in nervous system irritability. In severe cases death can occur.

Chronic toxicity produced by mercury vapor and inorganic mercury compounds have similar symptomology except that inorganic mercury aerosols may be less toxic than mercury vapor because of differences in their uptake and deposition.[80] Chronic toxicity is a more common form of mercury poisoning and primarily affects the nervous system, with the usual symptoms of erethism and muscular tremors.[78,81] The mildest symptoms are psychological in nature and include anxiety, anorexia, and insomnia. Mild changes in personality occur, as well as erethism characterized by excessive, undue embarrassment, timidity, depression, discouragement, irritability, resentfulness, or excitability.[78,82] Other symptoms relate to loss of ability to concentrate and to remember, inappropriate fear, and indecision. Tremors are the most frequently reported physical manifestation, varying from slight movements of the hands, eyelids or tongue to disabling, intense trembling that may affect the whole body.[78,81] Early toxic effects consist of trembling of the lips and tongue followed by tremors of the fingers and hands.[83] These signs are followed by ataxia, usually expressed initially as stammering, by dysarthria and dysphagia, and later by poor coordination of the arms and legs. Severe toxic effects are manifested by entirely uncoordinated movements, impairment of hearing, taste, and smell and an inability to write or speak with clarity. Other neurological disturbances include paresthesia, hypogeusia and hyposmia with or without dysgeusia or dysosmia, neuralgia, and dermographism.[82]

Inhalation of alkyl mercury derivatives predominantly affects motor and sensory nerves and leads to permanent impairments of the nervous system. The effects may include gross ataxia, peripheral hypesthesia, impairments of vision and hearing, personality changes, and loss of intellectual capacity.[78] On the other hand, the neurological signs associated with inhalation of inorganic mercury compounds are more likely to be erethism.[84] Inhalation of other mercury compounds, such as methoxyethyl mercury, a fungicide, has also produced the neurological symptoms of anorexia and fatigue.

Nearly all of the mercury vapor inhaled by man in the concentration range of 50 to 350 $\mu$g/m$^3$ is absorbed. Although all mercury compounds bind tightly to cells and are slowly metabolized and excreted, the alkyl derivatives appear to concentrate more rapidly in the brain. The mechanisms for these predilections are not clearly established.

**6. Phosphorus** Phosphine is a toxic gas which ignites spontaneously in air and has an odor threshold of 2 ppm. This gas produces chronic effects similar to those observed with phosphorus. The acute effects on the nervous system include weakness, vertigo, convulsions, and death[85] after exposure to concentrations of 100 ppm for periods longer than one hour. Phosphorus trichloride vapors are irritating to the mucous membranes of the eyes and upper respiratory tract.

Inhaled organic phosphorus compounds act on the nervous system by inhibition of cholinesterase activity. This inhibition is irreversible with some compounds and can lead to muscular weakness and toxic-chronic convulsions due to the accumulation of acetylcholine at myoneural junctions. These compounds are found most prevalently as insecticides and include, for example, Parathion, Phosdrin, Thinet, and Trithion. Since farm workers frequently come in contact with these insecticides, they could become prey to the toxic effects of these phosphorus compounds.

**7. Lead** Inhalation of toxic amounts of airborne lead can give rise to several neurological disturbances that include acute and chronic encephalopathy and peripheral neuropathy. Prolonged or recurrent, uncontrolled, excessive exposure can result in a slow, progressive cerebral incompetence. Although the symptoms of acute lead poisoning may abate on cessation of exposure, a single exposure in a young child may be followed by brain damage, varying in severity from subtle learning defects to profound mental deficiencies. Early neurological symptoms of lead poisoning may be subtle, subjective, and nonspecific and, therefore, may be poorly recognized in children and perhaps even less so in adults.

Encephalopathy is the most severe acute clinical neurological problem associated with inhalation of toxic amounts of lead. This problem may arise acutely with intractable seizures followed by coma. Fulminant encephalopathy usually develops within a week. Apathy progresses to drowsiness and stupor.[88,89] About 20% of the patients give a history of recent onset of clumsiness and exhibit frank ataxia. Prior to convulsions adults may report mental depression, persistent headache, vertigo, and tremor.[90] The prodrome of acute encephalopathy consists of subtle changes in mental attitudes, sluggishness, poor memory, inability to concentrate, restlessness, and hyperirritability.

Chronic exposure to airborne lead may lead to progressive mental deterioration in children. Historically, development may be normal for the first 18 months of life followed by a steady loss of motor skills or speed. Children may also exhibit hyperkinetic activity, aggressive behavior disorders, and convulsive disorders that may be poorly controlled with conventional therapy. The blood concentration of lead is usually greater than 60 $\mu$g/100 ml of whole blood.

The most important clinical feature of the peripheral neuropathy of lead poisoning is the predominance of motor involvement with minimal or absent sensory abnormalities. There is a tendency for extensor muscles of the hands and feet to be involved. Three clinical forms have been described: In one there is initially pain and tenderness in the muscles of the limbs. As the pain and tenderness subside, weakness may emerge with a gradual, slow recovery. In the second form, the neuropathy is a painless, peripheral extensor weakness.[91] This is the most common form of the toxicity. In the third form, neuropathic and myopathic features are difficult to separate from one another. Muscle fasciculations and proximal girdle atrophy are present with findings clinically similar to amyotrophic lateral sclerosis.[92]

At least one-fourth of the young children who survive an attack of acute encephalopathy due to lead poisoning sustain severe permanent neurological sequelae.[93] These sequelae are similar to those following a diffuse cerebral injury incurred during childhood and include cortical atrophy, hydrocephalus ex vacuo, convulsive states, mental retardation, and in severe forms, blindness. In less severe forms motor incoordination and impairment of sensory perception occur. Short attention spans are common, and the affected children are easily distracted.

Pathological changes in the peripheral nervous system have been demonstrated in experimental studies of lead poisoning.[94,95] These changes include segmental demyelination associated with injury to the Schwann cell, including injury to its mitochondria. Regeneration can occur but "onion-bulb" formation and scarring of the nerve fibers may be present. It is of interest to note that lead crosses the placenta,[96] so that excessive exposure of the mother to lead during pregnancy has resulted in neurological disorders in children.[97]

Of considerable importance, but clouded by present controversy, is the issue of whether or not low-level lead absorption adversely affects the neuropsychological function of school-age children. Bryce-Smith and Waldron in a report about Spanish-American children living near a lead smelter in El Paso, Texas, who had no overt symptoms but with blood lead levels between 40 and 68$\mu$g/100 ml suggested "that low levels of lead can adversely affect the central nervous system during its maturation and thereby produce behavioral abnormalities" (Bryce-Smith, D. and Waldron, H.A., *Lancet* I, 1166, 1974). The results of these studies differ from the results of other workers (Lansdown, R.G., Shepherd, J., Clayton, B.E., Delver, H.T., Graham, P.J., and Turner, W.C., *Lancet* I, 538, 1974; McNeil, J.L. and Ptosnik, J.A., International Symposium on Recent Advances in the Assessment of the Health Effects of Environmental Pollution, June 24, 1974, Paris). McCabe also raises questions about some of the conclusions drawn by the previous investigators (McCabe, E.B. Bloodlead levels, behavior and intelligence. *Lancet* II, 896, 1974). In spite of these disagreements this important and difficult area of investigation requires further study in order to understand more fully the potential role of low-level lead absorption on learning and intelligence in children.

**8. Arsenic** Arsine is produced whenever certain compounds of arsenic come into contact with nascent hydrogen such as is produced by an active metal (iron, zinc, tin) reacting with dilute acid or water. Many coloring materials, mordants, rodenticides and insecticides are arsenical in nature; hence, arsine and diethyl arsine are apt to be released by the action of various fungi acting on such materials.[98,99]

As little as 1 part of arsenic in 25,000 parts of air can produce serious neurological symptoms. Initial symptoms of acute toxicity relate to a metallic taste in the mouth and a feeling of mouth dryness along with irritation of the mouth and throat. Pain in the limbs, headache, convulsions, muscular weakness, and loss of consciousness are central nervous system effects of inhalation of this chemical. Actual degeneration of areas of the central nervous system and of the optic nerves have occurred. The pathological lesions that occur are increased permeability and dilation of small arterioles and capillaries and subsequent edema.

In chronic toxicity, numbness, burning, tingling, or itching have been reported. These symptoms are commonly followed by fasciculation or gross tremors with muscle atrophy and paralysis. These signs are commonly associated with cerebral edema with patchy necrosis or hemorrhagic encephalitis; this latter change is most often produced by sulfursphenamine or other arsphenamines. Peripheral neuropathy also occurs with demyelination and degeneration of the axis cylinder.[100]

## B. NEUROLOGICAL EFFECTS OF BIOLOGICAL AEROSOLS

The neurological effects of biological aerosols reflect the neurological symptoms produced by the specific organism. Airborne organisms include bacteria, fungi, and viruses as causative agents for various diseases. These agents generally enter the respiratory tract, are absorbed there and produce their specific effects. Hypersensitivity reactions with neurological manifestations also can occur after infection with airborne fungi and algae.[101]

## C. NEUROLOGICAL EFFECTS OF AIRBORNE PESTICIDES

Pesticides that are intended to be toxic to specific forms of life other than man exhibit some degree of toxicity for man. Indeed, cases of poisoning, either accidental or occupational, have been reported for practically every insecticide.[102]

**1. Insecticides** Insecticides, primarily the chlorinated hydrocarbons and the organophosphates, are the two most prevalent groups of synthetic organic insecticides that give rise to toxicity in man. As a group the chlorinated hydrocarbons are relatively stable in the environment, are not easily biodegradable or readily metabolized, and persist in the body for long periods of time, being stored in body fat tissues. From a toxicological standpoint, these hydrocarbons can be divided into three groups: (1) DDT, lindane, methoxychlor, etc.; (2) aldrin, dieldrin, chlordane, etc.; and (3) simple chlorinated aliphatic hydrocarbons. The mode of entry for each group in man includes the respiratory tract, the digestive tract, and the skin or mucous membranes. Each insecticide produces a spectrum of severity of effects on the nervous system. Mild toxicity is generally characterized by headache, dizziness, numbness and weakness of the extremities, apprehension, and hyperirritability. More severe symptoms, relating to absorption of larger amounts of the toxic agents, include fine muscular tremors that spread from the head to the extremities, jerking movements that involve whole muscle groups, loss of equilibrium and, finally, tonic and clonic convulsions.[103] The severity of symptoms appears directly related to the concentration of the insecticide in the nervous system, primarily in the brain.[104] Death is invariably due to cardiac failure or respiratory collapse.[105] These symptoms have been reported for DDT,[102] aldrin,[102] chlordane,[99] dieldrin,[102] benzene hexachloride,[106] lindane,[102] and p-dichlorobenzene, an irritant that can also cause cataracts.[102]

The first neurological symptoms of all organophosphate poisonings are anorexia, nausea, and headache. These are followed by salivation and papillary constriction. In severe cases, convulsions, coma, and death may occur. Usually body temperature does not rise and there are no signs of meningeal irritation. Onset of symptoms is rapid and death has been known to occur at any time from 5 minutes to 12 hours after exposure. These agents owe their toxicity to their ability to inhibit acetylcholinesterase, that results in an accumulation of acetylcholine which, in turn, causes excessive stimulation of the parasympathetic nervous system.[107] The insecticidal organophosphates generally are not poor acetylcholinesterase inhibitors per se. However, they are transformed metabolically to other products which do act as acetylcholinesterase inhibitors. This conversion occurs relatively rapidly in man, although it is much more rapid in invertebrates. These substances include many commonly used chemicals such as parathion and diazinon.

**2. Herbicides** In general, herbicides have been observed to produce local mucous membrane irritation, including the eyes, and some muscular disturbances. They may also produce fever. These substances include such commonly used agents as 2,4-D, silvex, and dinitrophenols.

**3. Fungicides** Fungicides produce toxic effects similar to those observed with herbicides. These substances include maneb and ferbam.

## D. NEUROLOGICAL EFFECTS OF ODOROUS COMPOUNDS IN THE ATMOSPHERE

Unpleasant odors are one of the first and frequent manifestations of air pollution and frequently arouse extreme emotional reactions in people. Offensive odors are capable of producing nausea, vomiting, headache, anorexia, and decreased fluid intake. They also may disturb sleep, produce hyperirritability, emotional disturbances, and mental depression.[82]

## E. NEUROLOGICAL EFFECTS OF RADIOACTIVE SUBSTANCES IN THE ATMOSPHERE

Respiratory tract contamination is the most direct and important effect of radiation pollution. However, cataracts can occur after exposure of the lens of the eye to heavy doses of x-rays, gamma rays, beta-particles, and especially neutrons. Although lens changes have been reported from doses as low as 200 rads, the average dose of gamma radiation that is believed to produce cataracts in adults is about 800 rads, but somewhat less in children.[109] Other less specific effects may include anorexia and loss of taste and smell.[110]

## SUMMARY

Vapor phase air pollutants are common and significant factors that affect neural and sensory functions in man. These pollutants arise from both natural and man-made sources and act either singly or in combination. Synergism is not uncommon when some pollutants, such as sulfur dioxide, occur in combination with other vapor phase organic air pollutants.

Vapor phase pollutants may be classified with respect to the form of their presentation in air; these consist of true gases, vapors of volatile or volatilizable liquids, and aerosols. Contact with man may be enhanced by environmental factors such as excessive humidity, precipitation, and temperature inversion, and by occupational factors that require the subject to be in close contact with these vapors for prolonged periods of time. The effects on man may be due to direct contact with the air pollutant on the skin or mucous membrane or to inhalation.

Effects on the neural and sensory functions in man vary widely. Odorous pollutants can cause minor annoyance which, if persistent, can produce irritation, emotional upset, anorexia, and mental depression. True gas phase pollutants also produce specific responses. Eye irritations can occur as a secondary consequence of direct exposure to substances such as sulfur dioxide, whereas death may ensue secondarily to the depression of central nervous system respiratory centers as a result of carbon monoxide poisoning. The former effects are acute and of relatively short duration due to direct contact of the gas with the exposed nerve endings of the mucosal surfaces. The effects of carbon monoxide persist until the carboxyhemoglobin concentrations in the tissues return to normal. Those agents that acutely depress central nervous system function can produce acute respiratory failure. This asphyxia is commonly associated with central nervous system edema and petechial hemorrhage. If asphyxia is more protracted, as with exposure to nitrous oxide, there is often perivascular round cell infiltration and chromatolysis, particularly of cortical brain and basal ganglia cells. With repeated and prolonged exposure to carbon monoxide and other substances of this type, neural and sensory changes occur that can alter various aspects of sensory perception including visual acuity, temporal perception, and higher mental function. These effects may be long lasting if exposure is protracted. Vapors in air that originate from liquids and aerosols made up of metals, dust, or pesticides can similarly produce acute and chronic effects. Aerosols that are highly lipid soluble can enter the body and be absorbed into the lipids of the nervous system where they may persist and produce effects long after the initial stimulus contact has been removed. Such long-term chronic effects can occur from organophosphate pesticides and from vapors carrying metals such as lead, mercury, and cadmium.

Sensory changes may be early and useful diagnostic signs of the toxic tissue levels of these substances in man. Dysgeusia, usually phantogeusia (a phantom, usually metallic, taste present in the oral cavity without the presence of food) may occur after toxic exposure to lead or mercury. Decreased visual acuity may occur after toxic exposure to carbon monoxide, while hyposmia may be an early sign of cadmium toxicity. Toxic exposure to lead and mercury aerosols may produce both hyposmia and hypogeusia during the early stages of intoxication.

Changes in neural and sensory function after exposure to vapor phase air pollutants may be difficult to determine. Changes in personality or in sensory function may also be difficult to quantitate and may be considered to be subjective changes. Yet, these changes may be properly assessed provided tests are instituted and information made available on potentially hazardous symptoms before other, more accepted and more clinically obvious signs are present. Since sensory and neural functions are easily tested and since they may be early signs of disease, it is up to the astute clinician to evaluate these changes and to relate them to occupational exposure if such hazards exist. In addition, among those subjects who exhibit pathological or other changes which might render them more susceptible to toxic effects of certain vapor phase air pollutants, repeated testing of sensory function may be of value if early signs of toxicity are to be observed. Qualitative measurements of various sensory and neural functions can aid in diagnosis and in evaluation of effects of treatment in various states of intoxication from vapor phase air pollutants.

## REFERENCES

1. Community Air Quality Guides — Aldehydes. *Am Ind Hyg Assoc J* 29:505, 1968.

2. Fassett DW: Aldehydes and Acetals. In *Industrial Hygiene and Toxicology* vol. II, FA Patty (ed), New York, Interscience, 1963.

3. Sim M, Pattle RE: Effect of possible Smog Irritants of Human Subjects. *JAMA* 165:1908, 1957.

4. Sax NI: *Dangerous Properties of Industrial Materials*, ed 3. New York, Reinhold, 1968.

5. Formaldehyde, Its Toxicity and Potential Dangers. Supplement. Public Health Report US 181, 1945.

6. Morrill EE Jr: Formaldehyde Exposure From Paper Process Solved by Air Sampling and Current Studies. *Air Cond Heat Ventilating* 58:94, 1961.

7. Melekhina VP: Maximum Permissible Concentration of Formaldehyde in Atmospheric Air. USSR Literature on Air Pollution and Related Occupational Diseases. 3:135, 1960.

8. Leonardos G, Kendall DA, Barnard NJ: Odor Threshold Determinations of 53 Odorant Chemicals. *J Air Pollut Control Assoc* 19:91, 1969.

9. Prentiss AM: *Chemicals in War*. New York, McGraw Hill, 1937.

10. Silverman L, Whittenberger JL, Miller J: Physiological Response of Man to Ammonia in Low Concentration. *J Ind Hyg Toxicology*. 31:74, 1949.

11. Hemeon WCL: The Estimation of Health Hazards From Air Pollution. (AMA) *Arch Ind Health* 11:307, 1955.

12. Amdur MO, Corn M: The Irritant Potency of Zinc Ammonium Sulfate of Different Particle Sizes. *Am Ind Hyg Assoc J*. 24:326, 1963.

13. Polezhaev NG: Determination of Active Chlorine in the Air. USSR Literature on Air Pollution and Related Occupational Diseases 3:29, 1960.

14. Kramer CG: Chlorine. *J Occup Med* 9:193, 1967.

15. Heyroth FF: Chlorine, C12. In *Industrial Hygiene and Toxicology*, vol II, ed 2, New York, Interscience, 1963.

16. Chasis H, Zapp JA, Bannon JH, et al: Chlorine Accident in Brooklyn. *J Occup Med* 4:152, 1947.

17. Joyner RE, Drel EG: Accidental Liquid Spill in a Rural Community. *J Occup Med* 4:152, 1962.

18. Beck H: Experimental Determination of Olfactory Threshold for Some Important Irritant Gases (chlorine, sulphur dioxide, ozone, nitrosylsulfuric acid) and Their Manifestations of the Effect of Low Concentrations on Man. *Thesis Wurzburg*, 1959.

19. Stern AC: Air Pollution, vol. II, pp 99, 325. New York, Academic Press, 1968.

20. Denmead CF: Air Pollution by Hydrogen Sulfide From a Shallow Tidal Inlet. Presented at the Clean Air Conference, University New South Wales, Auckland, New Zealand, 1962.

21. Permissible Emission Concentrations of Hydrogen Sulfide. Sub-committee on effects of hydrogen sulfide of the committee on effects

of dusts and gas of the Verein Deutsches Ingenieure Committee on Air Purification. VDI, 2107, 1960.

22. Patty FA: Industrial Hygiene and Toxicology, vol. I, p 896 New York, Interscience, 1958.

23. Petri H: The Effects of Hydrocarbon Sulfide and Carbon Disulfide. *Staub* 21:64, 1961.

24. Halley PD: Hazards of Hydrogen Sulfide. *Med Bull* 27:219, 1967.

25. Technical Report of California Standards for Ambient Air Quality and Motor Vehicle Exhaust. Suppl.#2. State of California, Department of Public Health, Berkeley, California, 1962.

26. Lehmann KB: Experimentelle Studien Über Den Einfluss Technisch und Hygienisch Wichtiger Gase und Dampfe auf den Organismus Teil VI. Schineflige Saure. Arch Hyg 18:180, 1893.

27. Greenwald I: Effects of Inhalation of Low Concentrations of Sulfur Dioxide Upon Man and other Mammals. *Arch Ind Hyg Occup Med* 10:455, 1954.

28. Buechley RW, Riggan WB, Hasselblad V, et al: $SO_2$ Levels and Perturbations in Mortality. A study in the New York-New Jersey Metropolis. *Arch Environ Health* 27:134, 1973.

29. Cornwall CJ, Raffle RAB: Bronchitis-sickness Absence in London Transport. *Br J Ind Med* 18:24, 1961.

30. Drinker CK: *Carbon Monoxide Asphyxia*. New York, Oxford University Press, 1938.

31. Beard RR: Carbon Monoxide as a Community Air Pollutant. *Calif Med* 113:53, 1970.

32. Gilbert GJ, Glaser GH: Neurological Manifestation of Chronic Carbon Monoxide Poisoning. *N Engl J Med* 261:1217, 1959.

33. Snyder RD: Carbon Monoxide Intoxication with Peripheral Neuropathy. *Neurology* 20:177, 1970.

34. Bowr H, Ledingham I: *Carbon Monoxide Poisoning*. New York, Elsevier Publ. Co., 1967.

35. Groll-Knapp EH, Wagner H, Hanck H, et al: Effects of Low Carbon Monoxide Concentrations on Vigilance and Computer-analyzed Brain potentials. *Staub Reinholt-Luft* 32:64, 1972.

36. Fodor GG, Winneke G: Effect of Low CO Concentrations on Resistance to Monotony and on Psychomotor Capacity. *Staub Reinholt-Luft* (Engl. Ed) 32:46, 1972.

37. Horvath SM, Dahms TE, O'Hanlon JF: Carbon Monoxide and Human Vigilance. A deleterious effect of present urban concentrations. *Arch Environ Health* 23:343, 1971.

38. Beard RR, Wertheim GA: Behavioral Impairment Associated with Small Doses of Carbon Monoxide. *Am J Public Health* 57:2012, 1967.

39. McFarland RA: Low Level Exposure to carbon monoxide and driving performance. *Arch Environ Health* 27:355, 1973.

40. Weir FW, Rockwell TH: An Investigation of the Effects of Carbon Monoxide in Humans in the Driving Task. First Report — The Ohio State Univ. Res. Found., Columbus, Ohio, 1973, 170 pp.

41. McFarland RA: The Effects of Exposure to Small Quantities of Carbon Monoxide on Vision. *Ann NY Acad Sci* 174:301, 1970.

42. Beard RR, Grandstaff N: Carbon Monoxide Exposure and Cerebral Function. *Ann NY Acad Sci* 174:385, 1970.

43. Ramsey JM: Effects of Single Exposures of Carbon Monoxide on Sensory and Psychomotor Response. *Am Ind Hyg Assoc J* 34:212-216, 1973.

44. Wright G, Randell P, Shephard RJ: Carbon Monoxide and Driving Skills. *Arch Environ Health* 27:349, 1973.

45. Schulte JH: Effects of mild carbon monoxide intoxication. *Arch Environ Health* 7:524, 1963.

46. Stewart RD, Peterson JE, Baretta ED, et al: Experimental Human Exposure to Carbon Monoxide. *Arch Environ Health* 21:154, 1970.

47. Mikulka P, O'Donnell R, Heinig P, et al: The Effect of Carbon Monoxide on Human Performance. *Ann NY Acad Sci* 174:409, 1970.

48. Bender W, Gothert M, Malorny G: Effects of Low Carbon Monoxide Concentrations on Psychological Functions. *Staub Reinholt-Luft* 32:54, 1972.

49. Bender W, Gothert M, Malorny G, et al: Wirkungsbildniedriger Kohlenoxidkonzentrationen Bein Menschen. *Arch Toxikol* 27:142, 1971.

50. Sayers RR: Physiological Effects Due to Rescue Apparatus. *Cool Trade Bull* 45:28, 1921.

51. Renzetti NR, Gobran V: Studies of Eye Irritation Due to Los Angeles Smog, 1954-1956. San Marino, California Air Pollution Foundation. July, 1957.

52. Griswold SS, Chambers LA, Motley HL: Report of a Case of Exposure to High Ozone Concentrations for Two Hours. (AMA) *Arch Indust Health* 15:108, 1957.

53. Von Oettingen WF: The Toxicity and Pollution Dangers of Nitrogen Fumes. US Public Health Service, Washington, DC, PHS Bull. N 272, 1941.

54. Shalamberidze OP: Reflex Effects of Sulfur and Nitrogen Dioxides. *Hyg Sanit* 32:10, 1967.

55. Henschler D, Stier A, Beck H: Gerochsschwellen Einiger Wichtiger Reizgase (Schinefeldioxyd, Ozon, Stichstaffdioxyd) und Erscheinungen bei den Einwirkung Geringer Konzentrationen auf den Menschen. *Arch Gewerbepath Gewerbehyg* 17:547, 1960.

56. Shalamberidze OP: Reflex Effects of Mixtures of Sulfur and Nitrogen Dioxides. *Hyg Sanit* 32:10, 1967.

57. Bondareva EN: Hygienic Evaluation of Low Concentrations of Nitrogen Oxides Present in Atmospheric Air. *USSR Lit Air Pollut Rel Occup Dis* 8:98, 1963.

58. Bushtueva KA: The Determination of the Limit of Allowable Concentration of Sulfuric Acid in Atmospheric Air. In *Limits of Allowable Concentrations of Atmospheric Pollutants*, Bk. 3. US Department of Commerce, Washington, DC, 1957, pp 20-36.

59. Bushtueva KA: Threshold Reflex Effect of $SO_2$ and Sulfuric Acid Aerosol Simultaneously Present in Air. In *Limits of Allowable Concentrations of Atmospheric Pollutants*, Bk. 4. US Public Health Service, Washington, DC, 1961, pp 72-79.

60. Hygiene Guide Series. Benzene. *Am Ind Hyg Assoc J* 31:383, 1970.

61. Gerade HW, Aklstrom DB: Toxicological Studies on Hydrocarbons. *Toxicol Appl Pharmacol* 9:185, 1966.

62. Harden BL: Carbon Tetrachloride Poisoning. A review. *Ind Med Surg* 23:3, 1954.

63. Brieger H: Chronic Carbon Disulfide Poisoning. *J Occup Med* 3:302, 1961.

64. Browning E: Toxic Solvents. A review. *Br J Ind Med* 16:23, 1959.

65. Cholak J, Kehoe RA, Miller LH, et al: Toxicity of Beryllium. US Air Force System Command Report. ASD-TR-62-7-665, April 1962.

66. Hall TC, Wood CH, Stoeckle JD, et al: Case Data From the Beryllium Registry. (AMA) *Arch Ind Health* 19:100, 1959.

67. Sax HI: Dangerous Properties of Industrial Materials, ed 3. New York, Reinhold, 1968.

68. Rozendaal HM: Clinical Observations on the Toxicology of Boron Hydrides. (AMA) *Arch Indust Hyg Occup Med* 4:257, 1951.

69. Tsuchya K: Proteinuria of Workers Exposed to Cadmium Fume. *Arch Environ Health* 14:875, 1967.

70. Fukuyama Y: Itai-itai Disease and Proteinuria Caused by Cadmium Poisoning. *Med Biol* (Tokyo) 84:41, 1972.

71. Schroeder HA, Balassa JJ, Vinton WH Jr: Chromium, Cadmium and Lead in Rats: Effects on Life Span, Tumors and Tissue Levels. *J Nutr* 86:51, 1965.

72. Vasilion P: Pressure-volume Relations of Rat Lung Exposed to Repetitive Irritation. US DHEW National Center for Air Pollution Control, Health Effects Res. Prog., Cincinnati, Ohio, 1966.

73. Couper J: On the Effects of Black Oxide of Manganese When Inhaled into the Lungs. *Brit Ann Med Pharm Vit State Gen Sci* 1:41, 1837.

74. Emden H: Ueber eine Nervenkrankheit nach Manganvergiftung. *Munch Med Wochenscher* 48:1852, 1901.

75. Jaksch RV: Cited in *Progress Med Wochenscher* 18:213, 1902.

76. Stokinger HE: The Metals (excluding lead), pp 987-1217 FA Patty, (ed) *Industrial Hygiene and Toxicology* (2nd rev. ed) DW Fossett, DD Irish (eds) vol. II *Toxicology* pp 1087-1089, New York, Interscience, 1962.

77. Voss H: Über das Vorkommen von Beruflichen Manganvergiftungen Stahlindustrie. *Arch Gerwerbepath Gewerbehyg* 9:453, 1939.

78. Battigelli MC: Mercury Toxicity from Industrial Exposure, a Critical Review of the Literature. Part 1. *J Occup Med* 2:337, 1960.

79. Demidov AV, Mokhov LA: Rapid Methods for the Determinations of Harmful Gases and Vapors in the Air. USSR Literature on Air Pollution and Related Occupational Disease 10:114, 1962.

80. Recommendations on maxium allowable concentrations of mercury and its compounds; report of an international committee. Symposium at Institute of Hygiene, Karolinska Inst., Stockholm, Sweden, 1969.

81. Duffield DP: The Mercury Hazard in Industry and Its Control. *Chem Ind* No. 7:197, 1968.

82. Stokinger HE: Mercury, $HG^{237}$. In *Industrial Hygiene and Toxicology*, vol. II, ed 2, FA Patty (ed)., New York, Interscience, 1963, p 1090.

83. Hook O, Lundgren KD, Swensson A: On Alkyl Mercury Poisoning. *Acta Med Scand* 140:131, 1954.

84. Brown JR, Kulkarni MV: A Review of the Toxicity and Metabolism of Mercury and Its Compounds. *Med Ser J Can* 23:786, 1967.

85. Jacobs MB: *The Analytical Toxicity of Industrial Inorganic Poisons*. New York, Interscience, 1967.

86. Kosolopoff GM: Organophosphorus Compounds. New York, J Wiley and Sons, 1950.

87. Mackworth JF, Webb EC: The Inhibition of Serum Chlorinesterase by Alkyl Fluorophosphonates. *Biochem J* 42:91, 1948.

88. Amer Acad Ped, Subcommittee on Accidental Poisoning. Prevention, Diagnosis and Treatment of Lead Poisoning in Childhood. *Pediatrics* 44:291, 1969.

89. Chisolm JJ Jr: Lead Poisoning. In *Pediatrics* (14th Edition) HL Barnett (ed). New York, Appleton-Century-Crafts, 1968, pp 313-319.

90. Arib JC, Fairhall LT, Minot AS, et al: Lead Poisoning. *Medicine Monographs*, vol 7. Baltimore, Williams & Wilkins Co., 1926.

91. Catton MJ, Harrison JJG, Fullerton PM, et al: Subclinical Neuropathy in Lead Workers. *Br Med J* 2:80, 1970.

92. Simpson JA, Seaton DA, Adams JF: Response to Treatment

With Chelating Agents of Anaemia, Chronic Encephalopathy and Myelopathy Due to Lead Poisoning. *J Neurol Neurosurg Psychiatry* 27:536, 1964.

93. Perlstein MA, Attala R: Neurologic Sequelea of Plumbism in Children. *Clin Pediatr* 5:292, 1966.

94. Lampert PW, Schochet SS Jr.: Demyelination and Remyelination in Lead Neuropathy. Electron microscopic studies. *J Neuropathol Exp Neurol* 27:527, 1968.

95. Fullerton PW: Chronic Peripheral Neuropathy Produced by Lead Poisoning in Guinea Pigs. *J Neuropathol Exp Neurol* 25:214, 1966.

96. Barltrop D: Transfer of Lead to the Human Fetus. In *Mineral Metabolism in Pediatrics*, D Barltrop, WL Burland (eds). Philadelphia, FA Davis Co, 1969. pp 135-151.

97. Angle CR, McIntire MS: Lead Poisoning During Pregnancy. *Am J Dis Child* 108:436, 1964.

98. Pinto SS, McGill CM: Arsenic Trioxide Exposure in Industry. *Indust Med* 22:281, 1953.

99. Dinman HB, Arsenic. Chronic Human Intoxication. *J Occup Med* 2:137, 1960.

100. Chuttani PN, Chawla LS, Sharma TD: Arsenical Neuropathy. *Neurology* 17:269, 1967.

101. Burrows W: Textbook of Microbiology, 19th Edition. Philadelphia, Saunders, 1968.

102. Davingnon LF, St. Pierre J, Charest G: A Study of the Chronic Effects of Insecticides in Man. *Can Med Assoc J* 92:597, 1965.

103. The Merck Index, ed 8. Rahway, New Jersey, Merck & Co, 1968.

104. Dale WE, Gaines TB, Hayes WJ Jr.: Poisoning by DDT: Relationship Between Clinical Signs and Concentration in Rat Brain. *Science* 142:1474, 1963.

105. Spencer JN: Pesticide poisoning: The Insecticides. *S Dakota J Med* 20:25, 1967.

106. Sax NI: Dangerous Properties of Industrial Materials, ed 3. New York, Reinhold Book Corp, 1968.

107. Zavon MR: Diagnosis and Treatment of Pesticide Poisoning. *Arch Environ Health* 9:615, 1964.

108. Interstate Air Pollution in the Selbyville, Delaware-Bishop, Maryland Area. US Public Health Service, Division of Air Pollution, Washington, DC, 1965.

109. Merriam GR, Focht EF: A Clinical Study of Radiation Cataracts and Relationship to Dose. *Am J Roentgenol Radium Ther Nucl Med* 77:759, 1957.

110. Henkin, RI: Prevention and Treatment of Hypogeusia Due to Head and Neck Irradiation. *JAMA* 220:870, 1972.

# 17 Effects of Moderate CO Dose on the Central Nervous System — Electrophysiological and Behaviour Data and Clinical Relevance

M. Haider, M.D., Ph.D.
E. Groll-Knapp, Ph.D.
H. Höller, Ph.D.
M. Neuberger, M.D.
H. Stidl, Sc.D.

## INTRODUCTION

Carbon monoxide continues to be not only a potential industrial but also general environmental and social hazard, mainly because of the increasing use of automobile transport in cities. Various institutes have conducted experiments on the effects of exposure to CO on human and animal behaviour. In this connection we may mention the works of Forbes et al.,[8] McFarland et al.,[21] Halperin et al.,[12] Schulte,[26] Beard and Wertheim,[3] Ray and Rockwell,[25] Stewart et al.,[30] Mikulka et al.,[22] Beard and Grandstaff,[1,2] Hanks,[13] Lewis,[19] and Horvath et al.,[14] who carried out human experiments, and the reviews published by the National Academy of Science and the National Academy of Engineering,[35] the New York Academy of Science,[34] and the U.S. Department of Health, Education and Welfare.[33]

These investigators arrived at partly contradictory results, probably due largely to differences in the planning and performing of tests, but mainly to their reliance on different parameters of behaviour.

The reported research is financially supported by the Austrian "Fonds zur Förderung der wissenschaftlichen Forschung" and the Austrian "Allgemeine Unfallversicherung."

Many aspects of human performances are apparently affected even by low CO concentrations, while others show an impairment only considerably later. The capacity for observation in situations involving comparatively weak stimulation could be a very sensitive indicator, while complex psychomotor dexterity is only slightly affected. In the presence of significant changes in behavioral parameters, all investigators infer an impairment of the central nervous system by an increased CO concentration.

This problem may be elucidated by recording spontaneous cerebral activity by electroencephalography, by determining the cerebral electric reactions to signals (evoked potentials) through electronic data processing, and by determining the specific slow changes in brain potentials in situations of anticipation and expectation (expectancy wave or contingent negative variation, CNV). Little relevant data on these aspects are yet available.

In our Institute of Environmental Health we have performed a series of investigations with special emphasis on the electrophysiological aspects (EEG, evoked potentials, slow brain potentials) during the last years.

## METHODOLOGY

There were three series of experiments, each conducted with 20 subjects, nonsmokers, of both sexes. Each subject participated in each test under all experimental conditions, but the sequence was rotated systematically. The CO concentrations in the first study were 50, 100, and 150 ppm, corresponding to 3%, 5.4% and 7.6% COHb after two hours exposure. In the second experiment 200 ppm CO were used for four hours with a calculated 15% COHb and an actually measured mean of 13% COHb at the end of the experiment. The third experiment included 7 hours sleep in a control condition and under 100 ppm CO with an estimated 12% COHb after the sleep period.

All experiments were performed under a double blind condition. In the second experiment the administration of Pyrithioxin (pyritinol)* and a placebo were added to test the possible countereffect of this substance against CO. During the first experiment subjects were submitted to a 1.5-hours acoustic vigilance test with pairs of tones in 1.1-second intervals. In random intervals the second tone was slightly weaker, constituting a critical signal, to be answered with a motor reaction. In the second experiment a similar but more difficult vigilance test was used together with other tests of mental performance (e.g., attention tests, memory tests, and a mental arithmetic task). Finally, in this experiment, the subjects were exposed to a 105 dB octave noise band with a middle frequency of 2,000 Hz for 15 minutes. After the noise exposure, the temporary threshold shift (TTS) was measured after 4, 8, 16, 32 and 64 minutes. As physiological parameters,

---

*pyritinol = 3.3[1]-(Dithiodimethyl)-bis-(5-hydroxy-6-methyl-4-pyridyl-methanol).

the pulse rate was measured and the EEG was recorded from vertex to mastoid. In order to analyze slow brain potentials we used Ag/AgCl electrodes. All electrophysiological data were registered on an analog magnetic tape for subsequent computer analyses.

## PULSE-RATES

As shown in Figure 1, in the second experiment the pulse-rate was slightly depressed during the first hour during CO exposure that corresponded to COHb levels below 5%. Then it increased during the vigilance task and finally exceeded the control level when the CO corresponded to 9% to 10% COHb. In the last part of the experiment, the pulse rate during CO exposure was statistically significantly higher than in the control condition.

From animal experiments (Mills and McEdwards[23]) we know that high COHb values (18% to 40%) increase the impulse traffic in the carotid sinus nerves two- to fivefold. Chevalier et al.[4] found mean heart rates to be lowered for COHb levels below 4%; on the other hand, Shephard[27] demonstrated that pulse rates were not influenced by CO in humans for COHb levels from 2.4% to 5.4%. Our results seem to fill the gap between such studies and show that COHb levels around 10% may accelerate the pulse, but low levels may depress it.

## ELECTROENCEPHALOGRAM IN WAKING AND SLEEPING

The EEGs were frequency analyzed and the integrated values were used for different computations. In Figure 2 the frequency analysis of control recordings (0) and those for CO exposure (CO) are shown in the upper part. In the lower part the results under the same conditions together with premedication of Pyrithioxin are demonstrated. CO produces a slight drop in the alpha-wave frequency accompanied by augmentation in the theta- and beta-range. Similar results were noted by Grudzinska[10] and Sluijter.[29] Pyrithioxin had an activating and stabilizing effect in the alpha-range. Similar results were described by Künkel and Westphal.[18] This effect seems to be able to counteract the slight depressing effect of CO in the alpha-range.

In Figure 3 a very interesting result of CO on sleep patterns is shown. According to these data, which are derived from only six persons up to now, CO has a depressing effect on REM-stages and stage 2 sleep. On the other hand it leads to a greater amount of deep sleep (stage 3 and 4) and also a greater amount of wakefulness during the night. If this result can be verified it would add CO as one important environmental factor for sleep disturbances. Some animal experiments (Fodor et al.[6] and Colmant[5]) have demonstrated REM-depressions in rats with CO concentrations above 200 ppm in the air. In our human experiments only 100 ppm were used.

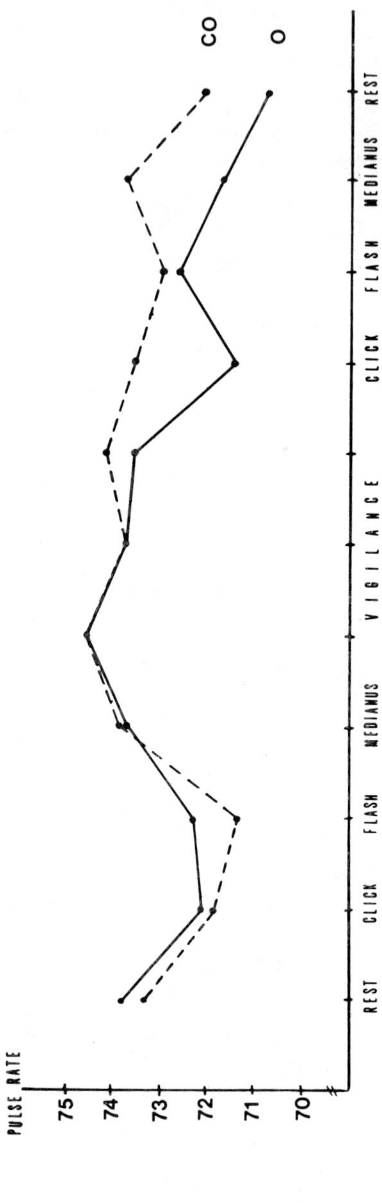


**Figure 1.** Pulse rates over a time of two hours under 200 ppm carbon monoxide (CO) and a control condition (0).

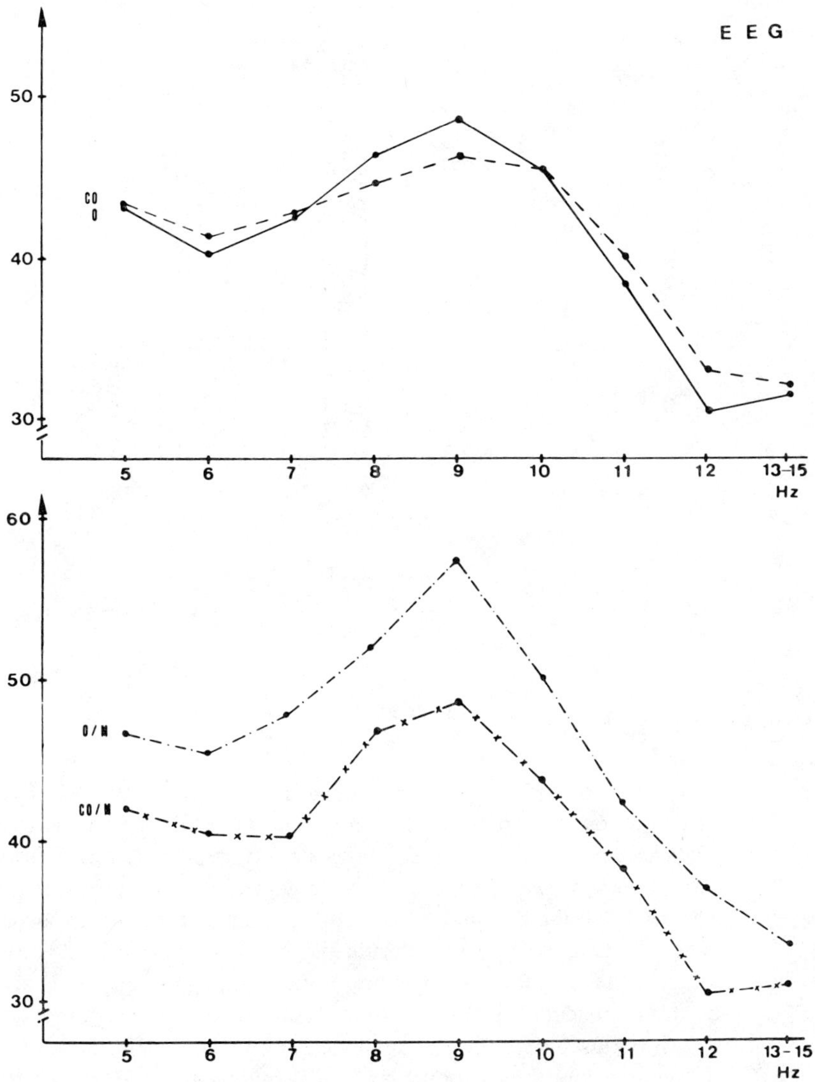


**Figure 2.** Upper part: Frequency analysis of the EEG after four hours 200 ppm CO and a control condition (0) Lower part: Same conditions with premedication of Pyrithioxine (M).

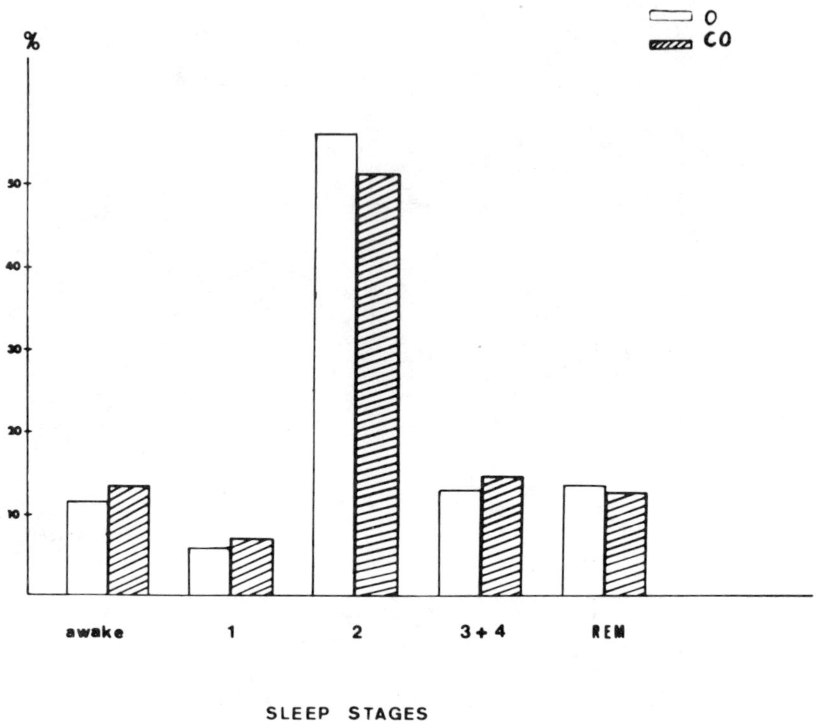


**Figure 3.** Sleep patterns of 7 hours sleep under 100 ppm carbon monoxide (CO) and a control condition (0)

## EVOKED POTENTIALS

The analysis of signal processing in the brain permits the investigation of processes that are more specific than those expressed by spontaneous electric activity. Experiments have already been performed to determine the extent to which the variations in potential produced by signals (evoked potentials) in the human and animal brain are affected by exposure to CO. The activity specific to the stimulus is very low and overlaid by permanent, much higher spontaneous activity in the brain. Normally, this activity can hardly be detected in the encephalogram. Its analysis requires computers. Synchronous activity can be separated from spontaneous activity by the repeated presentation of stimuli and the summation of the brain activities occurring after the signals. The activity specific to the stimulus, which is related in time to the signal, sums up and can thus be recorded and analyzed, while the spontaneous activity is annulled by summation, because it has no time relationship to the signal.

In this connection Xintaras et al.[32] and Ikeda[16] found changes in evoked potentials under influence of CO in animal experiments. The primary components were enlarged whereas the late components displayed a definite reduction of amplitudes. Tests on humans yielded similar results. Stewart et al.[30] noticed changes in the evoked brain responses after a four-hour exposure to 200 ppm CO while Hosko[15] also described changes in evoked potentials at 20% COHb.

In Figure 4 some evoked potentials to clicks and to electric stimulation of the median nerve are demonstrated. The evoked potentials to the median nerve stimulation especially show a definite reduction in the amplitude in the late components. Surprisingly, this reduction occurred in the very first evoked potential, recorded after 30 to 40 minutes of 200 ppm CO with an estimated COHb value from 4% to 5%. Pyrithioxin enhances the evoked potentials and in this respect is also able to counteract the CO effect to a certain degree, as shown in Figure 5.

## SLOW BRAIN POTENTIALS (EXPECTANCY WAVE OR CNV)

In recent years experiments have shown that besides the above-mentioned stimulus evoked potentials there exist other potential changes

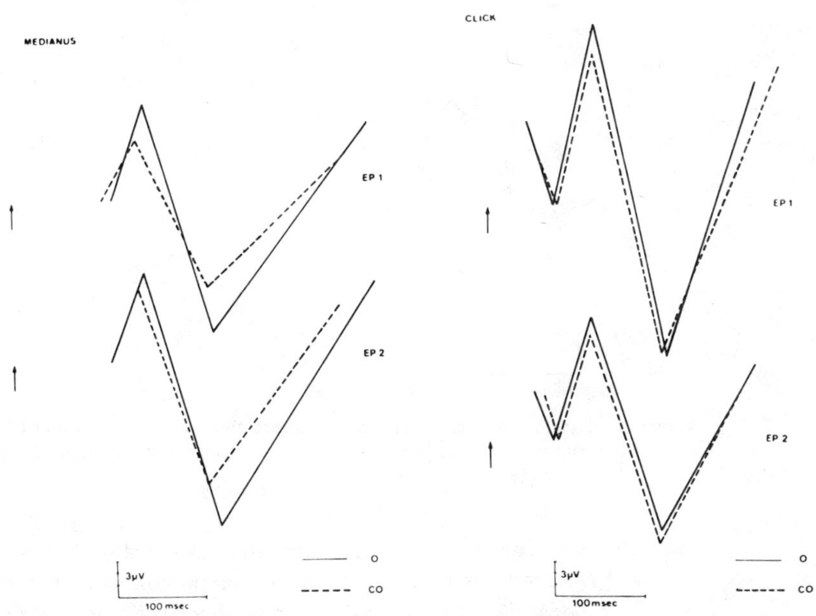


**Figure 4.** Mean values for peak to peak amplitudes and latencies of computer analyzed evoked potentials to electrical medianus shocks and clicks, after 40 min (EP 1) and 120 min. 200 ppm CO (EP 2), compared with controls (0).

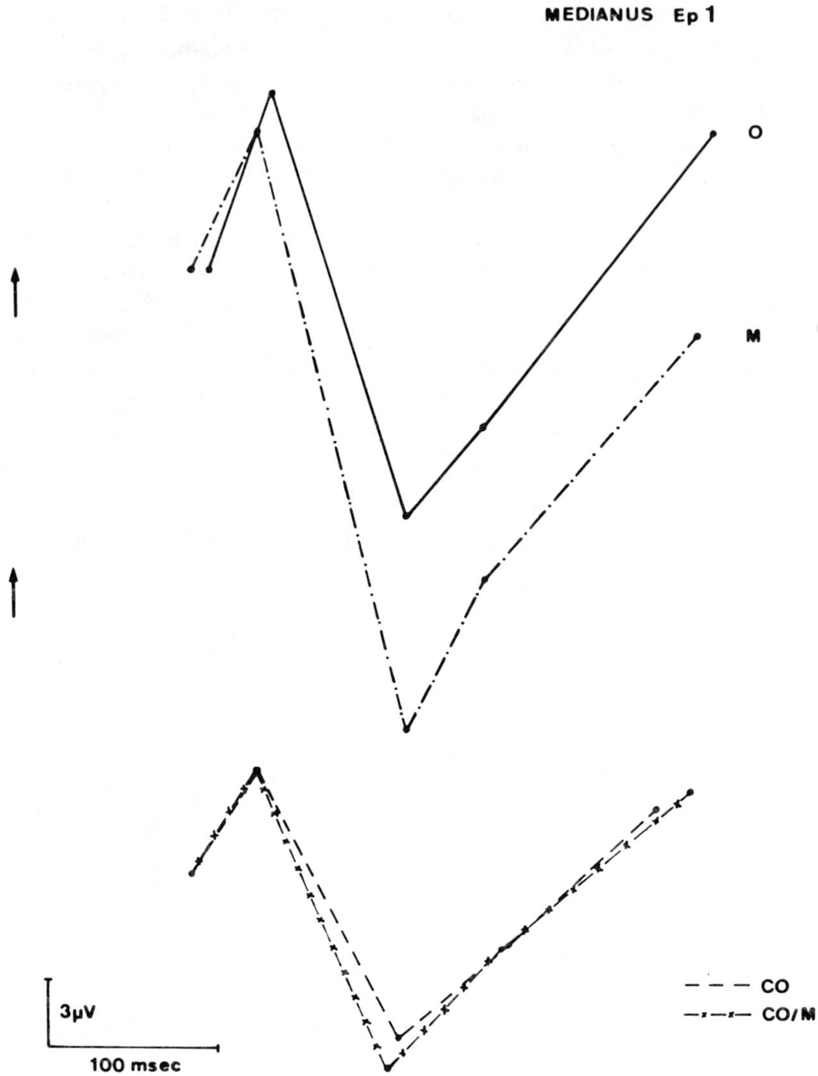


**Figure 5.** Evoked potentials after electrical medianus shocks with 40 min CO (lower part) and controls (upper part), both with premedication of Pyrithioxin compared to placebo.

that are associated with expectancy and anticipation situations (Mostofsky,[24] Haider[11]). In contrast to the evoked potentials they consist mainly of a large negative component that can be recorded over wide areas of the brain. Like the previously described evoked potentials, these slow potentials are much lower than those of spontaneous activity; the same computerized evaluation techniques must therefore be used for their analysis.

No research has yet clarified to what extent the slow brain potentials are affected by exposure to CO.

In two of our experiments we used an experimental design that enabled us to study these phenomena. The EEG was computer analyzed between the warning signals and the imperative signals on the vigilance tasks described in the section on methodology. Selective analysis of relevant pairs of signals and nonrelevant signals as well as differentiation between slow potential changes before correct responses and before misses could be performed by special trigger devices. The main results are demonstrated in Figure 6, which shows original expectancy waves of two subjects. The diminution of the CNV under influence of CO can be seen clearly.[9]

Figure 7 gives an overview of the computations of both experimental series. The results of our first experiment are shown in the upper part. It can easily be seen that CO application caused a reduction of the negativity as well as the following positivity of the slow potential changes in this experiment. The degree of diminution clearly seems to be related to the amount of CO administered. In the second part of Figure 7, the results of the selective analysis of relevant signals (RS) and nonrelevant signals (NRS) from the second experiment are shown. A finer analysis reveals differences between correctly detected relevant signals (RSrr) and nondetected relevant signals (RSnr). In all analyzed signal groups the results are in the same direction as in the first experiment, but they are much less clear cut. One of the causes may be that we used a more difficult vigilance test and other tests in the same situation. This could have changed the activation state of the subjects and thus influenced the results.

## BEHAVIOUR DATA

The results of psychological tests are poor in our investigations. We could not demonstrate an effect on attention tests, memory tasks, or mental arithmetic. One positive result concerned vigilance tests in our first experiment. There we found a clear correlation between missed signals and CO concentration. But we were unable to repeat this result in the second experiment with higher activation state and attentiveness of the subjects. Some authors (Fodor and Winneke[7]) reported a similar experience. Nevertheless there are some reports in the literature with clear-cut results of CO influence on mental performance. Obviously there is a complex interaction between task and situational variables and the detrimental influences of CO on behaviour. Beard and Grandstaff[1,2] have discussed this very ably.

## CLINICAL RELEVANCE

When the clinical relevance of our data is considered, the first question should be: does CO exert any measurable psychophysiological or even

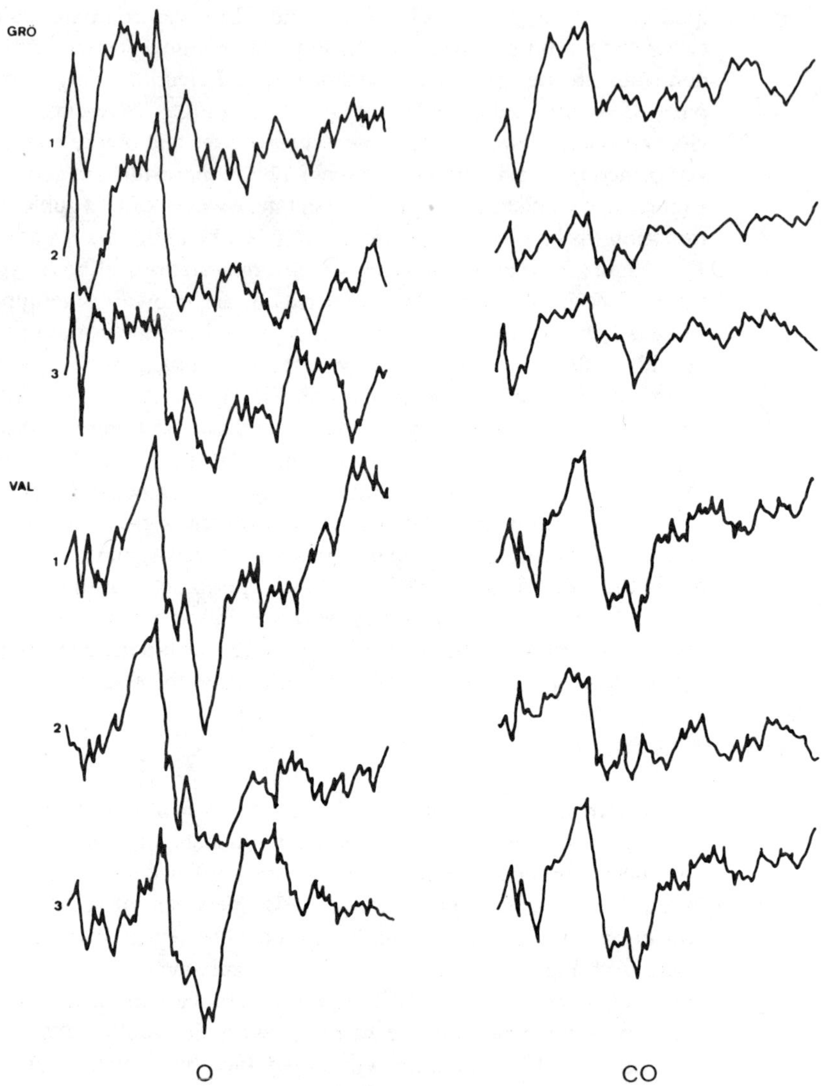


**Figure 6.** Computer analyzed expectation waves (CNVs) for two subjects under 150 ppm carbon monoxide (CO) and control conditions (0).

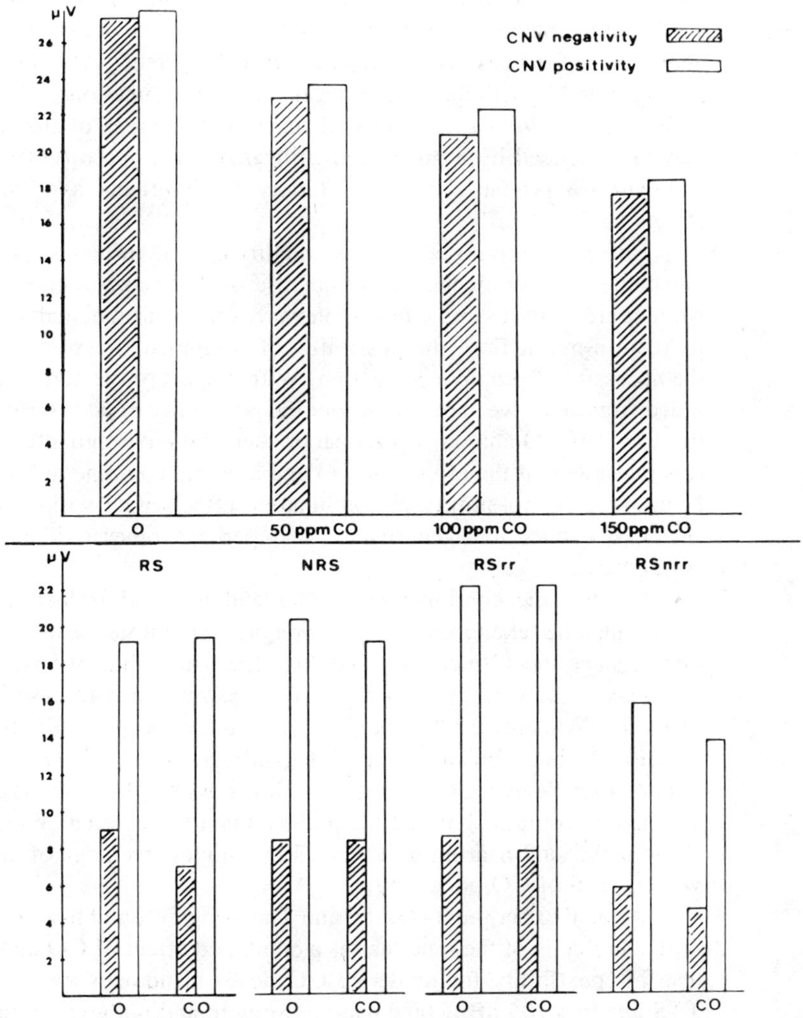


**Figure 7.** Upper part: Mean values for negativity and positivity of expectancy waves (CNVs) from controls (0) and 50, 100, and 150 ppm CO. Lower part: CNV negativity and positivity of the second experiment with 200 ppm CO selectively analyzed for relevant signals (RS), non relevant signals (NRS), correctly responded relevant signals (RSrr) and not responded relevant signals (RSnr).

pathological effects in moderate doses? Research workers until some years ago and clinicians even nowadays would probably answer no. But today one must say that there is obvious evidence that the cardiovascular system and the central nervous system are influenced even by moderate doses of CO. For the cardiovascular system the possibility that CO might be involved in the pathogenesis of atherosclerosis must be considered. But how about the influences on the central nervous system? Although we have demonstrated functional disturbances, do these functional changes deal with disease? We must admit that functional changes of the same order may be produced by a pill or tranquilizer or by a cup of coffee. For the clinician we probably have to classify these effects as "non-disease-effects."

But we are exposed to these environmental influences without any personal choice or mechanism of defense — that makes a great difference. Moreover, in discussing clinical relevances of environmental data we have to keep in mind, first, the possibility of combined stressors and, second, the necessity of prophylactic action. With respect to the concept of "combined stressors" we should consider the possibility of additive or synergistic effects of CO and other physical or chemical environmental influences. It is possible that the effects of CO and hypoxia combine at high altitude. Moreover, the possibility of combination effects with various air pollutants like ozone, nitrogen oxides, and perhaps other oxidants must be considered.

Finally, the combination of CO and physical influences, such as noise, must be taken into account. The ototoxic influence of CO has been known since the 17th century and there are many clinical observations on combined effects of CO and noise. For example, a man described in the literature (Wagemann[31]) worked 12 years under noise conditions of 80 to 90 dB(A), but clinical symptoms (audiogram, vestibular signs, etc.) seemed to indicate a CO intoxication. It is possible that both noxious stresses had a combined effect. Scandinavian reports after World War II (Lumio[20]) found hearing deficiencies in a high percentage of patients that were typical of CO poisoning.

Animal experiments (Kittel and Theissing[17]) found histologic damage in the hair cells of the inner ear as a combined effect of CO and noise. To test this possibility for moderate CO doses in humans we measured the TTS due to a 105 dB octave band noise with and without a four-hour CO exposure of 200 ppm. Fortunately no synergistic effects could be demonstrated (Fig. 8). But we need further investigations to find out at which level the effects possibly start.

As I mentioned earlier, even the demonstration of functional changes only should be taken as a special stress for the organism and should necessitate prophylactic action. In the field of environmental hygiene the most

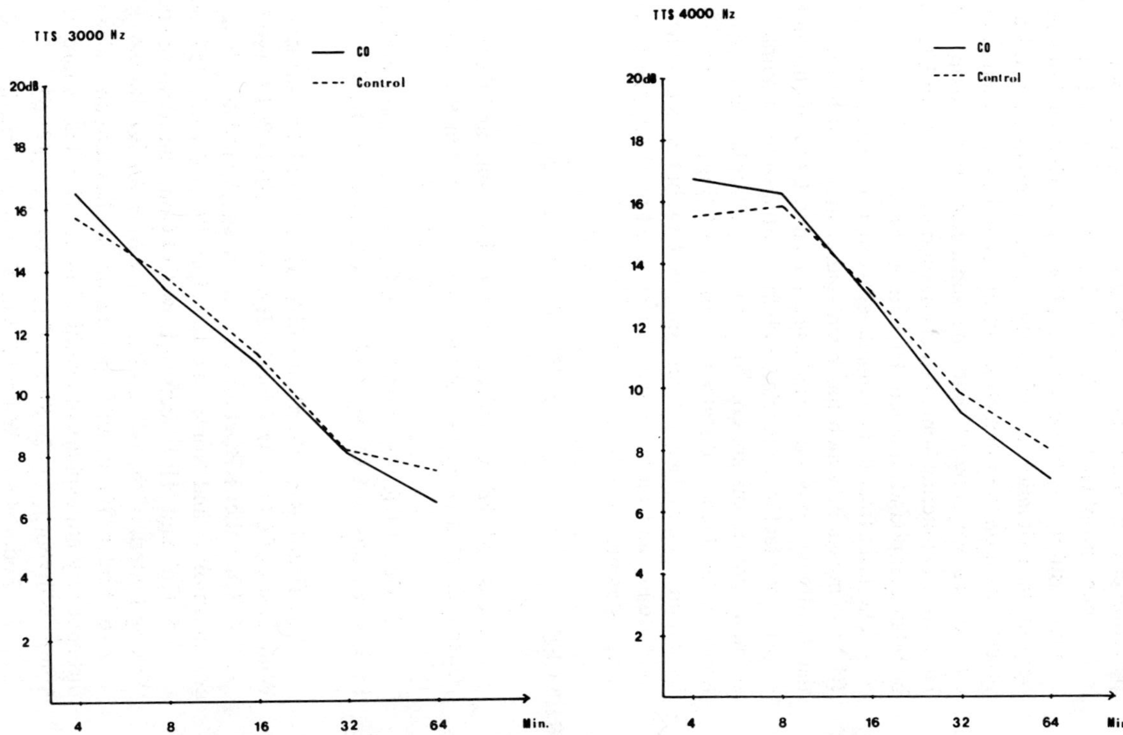


**Figure 8.** TTS after noise exposure (105 dB, 15 min) with four hours 200 ppm carbon-monoxide (CO) and control condition (0). Left part: test tone 3000 c/s Right part: test tone 4000 c/s.

important kind of preventive action is to prevent the individual from exposure to environmental noxious influences. In this connection physicians should take special care of such environmental influences in their own surroundings (e.g., in hospitals). For data like ours may help to produce air pollution criteria and finally accepted international air pollution standards.

Chemoprophylaxis will only be the last resort in environmental hygiene. But it may be better to find out now what kind of substances may help us in a situation of necessity. Here it might be good to have more investigation on the interactions of drugs and environmental influences. In our studies we showed that one substance (Pyrithioxin, i.e., pyritinol HCl) may counteract some effects of moderate doses of CO. There might be other compounds that still have to be tested.

The last effect to be considered is that of CO on sleep patterns. The REM depression, which has been found, may well be one of the effects that finally has to be classified as pathogenic. CO is in this and some other respects one factor, connected with the "urbanization trauma," with which so many physicians are confronted today. Sleeplessness, irritability, restlessness, headaches, disturbances of well-being, etc., are symptoms that have many causes. One of these may be CO, but each case has to be tested individually and further research is needed to clarify the extent to which such a statement can be verified.

## REFERENCES

1. Beard RR, Grandstaff N: CO Exposure and Cerebral Function. Paper presented at the NY Acad. of Sci. Conf. on Biol. Effects of CO, 1970.

2. Beard RR, Grandstaff N: CO Exposure and Cerebral Function. In Biol. Effects of CO, RF Coburn (ed), *Ann NY Acad Sci* 174 (1) 385, 1970.

3. Beard RR, Wertheim GA: Behavioral Impairment Associated With Doses of CO. *AM J Public Health* 57:2011–2012, 1967.

4. Chevalier RB, et al: Circulatory and Ventilatory Effects of Exercise in Smokers and Nonsmokers. *J Appl Physiol* 18(2):357–360, 1963.

5. Colmant HJ: Effects of Low CO Concentrations on the Sleep and Waking Pattern of the Albino Rat. *VDI Berichte Nr* 180:84–87, 1972.

6. Fodor GG, et al: Experimental studies into animals on sleep disturbances by atmospherical chemical noxes. In: The nature of sleep. Int. Symp. Würzburg, 23–26 Sept 1971, *Abstr.* No. 87, 65.

7. Fodor GG, Winneke G: Wirkungen niedriger CO-Konzentrationen auf Monotonie-Resistenz und psychomotorisches Leistungsvermögen. Staub-Reinhalt. *Luft* 32:169–175, (1972).

8. Forbes WF, et al: The Influence of Moderate CO Poisoning Upon the Ability to Drive Automobiles. *J Ind Hyg Tox* 19:598–603, 1937.

9. Groll-Knapp E, et al: Effects of Low CO Concentrations on Vigilance and Computer Analyzed Brain Potentials. *VDI Berichte* No. 180, 116–120, 1972.

10. Grudzinska B: Electroencephalographic Patterns in Cases of Chronic Exposure to CO in Air. *Folia Med Cracov* 5:493–515, 1963.

11. Haider M (ed.): *Neuropsychologie*. Huber, Bern, 1971.

12. Halperin MH, et al: The Time Course of the Effects of CO on Visual Thresholds. *J Physiol* 146:583–593, 1959.

13. Hanks TG: Human Performance of a Psychomotor Test as a Function of Exposure to CO. In: Biol. Effects of CO, RF Coburn (ed), *Ann of the NY Acad of Sci* 174–421, 1970.

14. Horvath S, et al: CO and Vigilance: Potential Danger From Existing Urban Concentrations. In: Project Clean Air, Vol. 2, Santa Barbara Univ of California Res Proj S 4, 11: 1, 1970.

15. Hosko MJ: The Effect of CO on the Visual Evoked Potential in Man. *Arch Environ Health* 21:174–180, 1970.

16. Ikeda T: Experimental CO Poisoning: Its Electrophysiological Effects in Cats. *Folia Psychiatr Neurol Jap* 23:135–142, 1969.

17. Kittel G, Theissing G: Histologische Untersuchungen der Cochlea an Häutchenpräparaten and Treppenserienschnitten Nach Hochgradiger Protrahierter Hypoxidose. *Arch klin exp Ohren Nasen Kehlkopfheilk* 191:534, 1968.

18. Künkel H, Westphal M: Quantitative EEG-analysis of Pyrithioxine action. *Pharmakospsychiat* 1:41–49, 1970.

19. Lewis J: Traffic Pollution and Mental Efficiency. *Nature* 225: 95–97, 1970.

20. Lumio JS: Otoneurological Studies of Chronic CO Poisoning in Finland. *Acta oto-laryngol,* Suppl. 71. (Stockholm), 65–112, 1948.

21. McFarland RA, et al: The Effect of CO and Altitude on Visual Thresholds. *Aviation Med* 15:381–394, 1944.

22. Mikulka P, et al: The Effect of CO on Human Performance. Paper pres. at the 5th Ann. Conf. on Atmospheric Contamination in Confined Spaces. Air Force Aerospace Med Res Lab, Wright Patterson Air Force Base, Dayton, Ohio, 1969.

23. Mills E, McEdwards IW: Stimulation of Aortic and Carotid Chemoreceptors During CO Inhalation. *J Appl Physiol* 25:494–502, 1968.

24. Mostofsky D (ed): Attention: Contemporary Theory and Analysis. Appleton Century Crofts, 1970.

25. Ray AM, Rockwell TH: An Exploratory Study of Automobile Driving Performance Under the Influence of Low Levels of COHb, Ohio State Univ Dept of Ind Engineer, Columbus, Rep No. IE-i, Aug 1971.

26. Schulte JH: Effects of Mild CO Intoxication. *Arch Environ Health* 7:524–530, 1963.

27. Shephard RJ: The Influence of Small Doses of CO Upon Heart Rate. *Resp.* 29:516–521, 1972.

28. Shulga TM: New Data for Hygienic Evaluation of CO in Air. *USSR Lit* 9:73–81, 1964.

29. Sluijter ME: The Treatment of CO Poisoning by Administration of Oxygen at High Atmospheric Pressure. *Prog Brain Res* 24: 123–182, 1967.

30. Stewart RD, et al: Experimental Human Exposure to CO. *Arch Environ Health* 21:154–164, 1970.

31. Wagemann W: Das otologische Bild der Kohlenoxidvergiftung. *Z Laryngol Rhinol Otol* 38:691, 1960.

32. Xintaras C, et al: Brain Potentials Studies by Computer Analysis. *Arch Environ Health* 13:223–232, 1966.

33. Air Quality Criteria for CO. US Dept of Health, Education and Welfare, Washington, 1970.

34. Biological Effects of CO. RF Coburn (ed). *Annals of the NY Acad of Sci* 174(1), 1970.

35. Effects of Chronic Exposure to Low Levels of CO on Human Health, Behaviour and Performance. Natl. Acad. of Sci, Natl. Acad. of Engineering, Washington Lib, Stand. Book No. 309-01735-1, 1969.

# 18 The Effect of Ozone on Brain Function

Barry L. Johnson, Ph.D.
J.G. Orthoefer, D.V.M.
T.R. Lewis, Ph.D.
C. Xintaras, Sc.D.*

## INTRODUCTION

Ozone ($O_3$) is a major air pollutant formed by photochemical reaction between oxygen and oxides of nitrogen in the presence of ultraviolet radiation. The major source of these effluents in ambient community air is motor vehicle emissions. Ozone is also present in such industrial processes as high-voltage ozonizers for treatment of sewage, for water purification, and for control of molds and bacteria in cold storage plants. Appreciable levels of ozone are also found near inert gas-shielded arc welding devices, x-ray machines, high-voltage spectrographs, insulators, and the brushes of electric motors.[1]

The primary biologic effects of exposure to high but non-lethal concentrations of ozone are irritation of the mucous membranes, acute lung injury characterized by pulmonary congestion and edema, and augmentation of morbidity effects of respiratory infection. Both human and animal

*All of the work described in this paper was performed while the authors were with the U.S. Environmental Protection Agency.

studies have focused on the pulmonary effects of ozone with the result that comparatively little information is available regarding ozone effects on other organs.

A review of the literature indicates few studies concerned with the effects of $O_3$ on the central nervous system. Skillen et al.[2] demonstrated a significant decrease in brain serotonin after exposure of rats to 6 ppm for 4 hours. Xintaras et al.[3] reported the effect of a one-hour 0.5 to 1.0 ppm $O_3$ exposure on electrical responses recorded from the rat visual cortex and the superior colliculus in response to photic stimulation. Ozone increased the latency of negative components of the visual evoked response and decreased the overall amplitude of the response. Trams et al.[4] reported the effects of $O_3$ on the biochemistry of the cerebral cortex in 44 dogs exposed for 18 months to $O_3$ levels ranging from 1 to 3 ppm. Catechol-O-methyltransferase activity showed a steady decline as the ozone exposure increased in severity. Concurrently, the catecholamine content of cerebral cortex also decreased.

The acute effects of $O_3$ on vision have been investigated by Lagerwerff,[5] who reported that visual function (visual acuity, peripheral vision, night vision, convergence, and visual fields) was diminished by an $O_3$ exposure of as little as 0.20 ppm $O_3$ for three hours. Hore and Gibson[6] examined the effect of a 0.2 to 0.3 ppm $O_3$ exposure of 70 minutes duration on mental functioning by administering intelligence tests to 99 university students. Ozone was found to have no noticeable effect on mental functioning, as measured by intelligence tests, during the exposure period. Pedersen et al.[7] exposed 13 male subjects to a combination of air pollutants ($O_3$, CO, $NO_2$), together with elevated ambient temperature. Ozone exposures of 0.5 ppm for four to five hours yielded only marginal effects on subjects' peripheral vision and performance on a task requiring divided attention.

With the exception of the report by Trams et al.,[4] all the aforementioned reports have examined only the effects of acute ozone exposure on the central nervous system. The purpose of the present study was to investigate, by means of electrophysiological techniques, the effects on the brain of chronic exposure to ozone and to relate any observed effects to the neurochemical findings by Trams et al.,[4] who utilized the same experimental animals described in the present study.

## METHODS AND MATERIALS

Female purebred beagles were randomly assigned to 100 $ft^3$ (2.83 $m^3$) exposure chambers identified as "five-feet-square" chambers by Hinners et al.[8] Six groups of eight animals each comprised the study groups. Animals were housed eight per exposure chamber for the duration of the study, which was 18 months. All dogs were obtained from a single commerical source and were approximately 6 months old at the start of the study. All chamber atmospheres were maintained at a mean temperature of 73° to 76°F (22.8–24.4°C) and a mean relative humidity of 43% to 50%. All air delivered to the six chambers was from a common air source that was filtered through CBR filters as described by Hinners et al.[8] Ozone was generated by passing medical-grade oxygen through a neon tube ozone generator and was added to the filtered air delivered to each chamber. Three groups were exposed to 1 ppm of $O_3$ in air: group A was exposed continuously for 8 hours per day, group B for 16 hours daily, and group C for 23.5 hours daily (0.5 hour was used for chamber maintenance). Group D was exposed to 2 ppm and Group E to 3 ppm, each for 8 hours per day. Group F, the control group, was maintained in a chamber that contained filtered air only. All exposures were maintained for 7 days per week for the duration of the study. Ozone chamber concentrations were continuously monitored by total oxidant meters (Mast Development Co., Model 724-2).

Twice during the course of the study all dogs were removed from the exposure chambers and electroencephalograms and visual evoked potentials recorded. The first test was given after 9 months of ozone exposure and the second test was at the termination of the study (i.e., after 18 months of $O_3$ exposure). Each dog (with the exception of those in Group E at 18 months) was tested twice during each testing period. Group E was tested only once at the second test period because of the poor physical condition of these animals. Dogs were taken from their chambers on a random schedule immediately prior to EEG recording. All dogs were fasted 12 hours prior to testing. Clipping of head hair and weighing of each animal were performed within minutes prior to data collection. The hair was clipped from the occiput anteriorly to the supraorbital process and laterally to the base of the pinnae. Each animal was given Sucostrin (succinylcholine chloride) intravenously at the rate of 0.5 mg per kg of body weight. The animals were immediately intubated with a cuffed endotracheal tube and artificially respirated by a demand type mechanical respirator with a mixture of 95% $O_2$ and 5% $CO_2$. The respiratory rate was constant, but was adjusted individually for each animal in order to maintain a well oxygenated state, as indicated by a pink tongue and mucous membranes.

Four monopolar subdermal needle electrodes were used for EEG recordings from each dog. The electrodes were approximately 0.5 inch (1.27 cm) in length. Two electrodes were placed at parietal locations $P_3$ and $P_4$ while the other two electrodes were placed in the occipital area at $O_1$ and $O_2$. Electrodes placed in each ear flap and tied together electrically served as the reference electrode. The parietal and occipital areas were chosen for EEG recording because of the reported effects of $O_3$ on the visual system. Symmetrical electrode placements were used in order to investigate the possibility of assymmetric cortical damage due to ozone. The EEGs were recorded in a quiet, well-lighted room with a Sanborn (Model 350-1500) electroencephalograph and recorded on a FM analog tape recorder (Honeywell, Model 7600). The bandwidth of the EEG amplifier-recording system was essentially flat from 0 to approximately 300 Hz.

The EEG recordings lasted 10 minutes for each dog. The first 7 minutes of data acquisition consisted of recording a background EEG during minimal sensory stimulation. The remaining 3 minutes of data acquisition consisted of recording visual evoked potentials recorded from $P_3$, $P_4$, $O_1$, and $O_2$ that were generated in response to photic stimulation from a Grass (Model PS-2) photic stimulator. The stimulator was positioned at dog eye level and positioned 15 inches from the animal's nose. The stimulator was set at minimum intensity (Intensity = 1). Sound absorbant material was placed within the photic unit in order to minimize any auditory click coincident with the gas tube discharge. The visual stimuli were presented at a constant rate of 1 flash every 2 seconds.

The EEGs recorded were ignored for each dog during the first 3 minutes of data recording. During the remaining four minutes four epochs of EEG were subjected to autospectral density analysis[9] by use of an analysis program (DECUS 21-62) written for a PDP-12 computer. The autospectral density analysis resolved the EEG into spectra amplitudes with a resolution of 0.25 Hz. The four spectra estimates computed for each EEG lead were then averaged in order to yield one spectral density estimate for each EEG lead each time a dog was tested.

Averaged evoked potentials were obtained through use of the Advanced Averager program supplied with a PDP-8 computer. For each dog, the visual evoked potentials for the first 75 photic stimuli were averaged for each EEG lead by the aforementioned computer program and saved on punched paper tape for subsequent analysis.

Each dog was tested twice at 9 months and at 18 months of exposure by the aforementioned procedure. The results for each test period were averaged for each dog. At the end of 18 months of exposure, all animals were sacrificed and samples of cerebral cortex from parietal and occipital areas were removed for biochemical assay by Trams et al.[4]

The EEGs for each dog were characterized by six measures obtained from the autospectral density anaysis. These six measures were as follows:

1. Peak frequency, the frequency at which the autospectral density amplitude was maximum
2. Root mean square value of the electrical power represented by the autospectral estimate
3. Percent of total electrical power contained within the delta EEG band (0.25 to 3.75 Hz)
4. Percent of the total electrical power contained within the theta EEG band (4 to 7.75 Hz)
5. Percent of total electrical power contained within the alpha EEG band (8 to 13.75 Hz)

6. Percent of total electrical power contained within the beta EEG band (14 to 31.75 Hz).

The averaged evoked potentials (AEP) were characterized by the following ten measures:

1. Latency of the first negative peak ($N_1$)
2. Amplitude of $N_1$
3. Latency of the first positive peak ($P_1$)
4. Amplitude of $P_1$
5. Ratio of $P_1/N_1$
6. The root mean square (RMS) value of the AEP
7. The area under the positive portions of the response ($A_+$)
8. The area under the negative portions of the response ($A_-$)
9. Ratio of $A_+/A_-$
10. The Pearson product moment correlation coefficient between contralateral EEG leads.

The latter measure was computed in order to investigate possible effects of $O_3$ on symmetric cortical electrical activity.

## RESULTS

The results of this study are divided into two parts: findings at 9 months and findings at 18 months. Tables 1 and 2 summarize the EEG findings at 9 and 18 months, respectively; while Tables 3 and 4 contain the summary of salient visual evoked potential results at 9 and 18 months, respectively.

The EEG results contained in Tables 1 and 2 are all measures derived from autospectral density analyses of each EEG lead. Since no evidence was found to indicate any EEG asymmetry due to ozone exposure, only data from EEG leads $O_1$ and $P_3$ are listed. A review of the data in Table 1 indicates that the primary effect of 9 months of ozone exposure was to decrease the amount of beta (14 to 31.75 Hz) activity in both occipital and parietal leads and to increase the amount of alpha (8 to 13.75 Hz) and theta (3 to 7.75 Hz) activity. This observation was statistically significant ($P<0.05$) for the occipital lead ($O_1$) for both alpha (treatment group E, Dunnett's test) and beta (treatment groups B and E) activity, but did not reach statistical significance for the parietal lead ($P_3$). No indication of any effect of ozone on EEG amplitude or low frequency content (i.e., delta activity) was found at 9 months of exposure. The pattern portrayed by the data of Table 1 is that ozone has effected an EEG shift from the higher frequencies toward the midrange frequencies, primarily alpha, with little effect on overall EEG amplitude.

The EEG data after 18 months are given in Table 2 and illustrate again the shift from beta activity toward alpha activity in both parietal and

**Table 1**
**Means and Standard Deviations for EEG Measures after 9 Months of $O_3$ Exposure**

| | Ozone Treatment | | | | | | | |
|---|---|---|---|---|---|---|---|---|
| EEG Spectral Measure | A (1/8)* | B (1/16) | C (1/23.5) | D (2/8) | E (3/8) | F (0/24) | F Value§ | P |
| Peak frequency | 4.5±0.9† | 5.1±1.2 | 6.0±2.6 | 6.3±3.3 | 7.9±2.7 | 6.1±2.8 | 1.63 | N.S. |
| | 6.3±3.4‡ | 6.9±2.2 | 7.3±3.2 | 7.6±4.1 | 6.8±3.0 | 6.7±4.2 | 0.13 | N.S. |
| RMS power | 1654±629 | 2123±482 | 1652±622 | 1598±914 | 1469±338 | 1278±302 | 1.81 | N.S. |
| | 1503±187 | 2284±313 | 2033±675 | 1786±278 | 1485±336 | 2079±297 | 5.89 | 0.01 |
| Percent delta | 16.6±5.0 | 14.1±4.3 | 11.7±2.2 | 14.9±3.3 | 11.2±2.9 | 12.7±4.1 | 2.17 | N.S. |
| | 8.9±2.1 | 10.5±3.3 | 9.8±2.9 | 9.2±4.3 | 9.9±2.9 | 8.1±1.3 | 0.56 | N.S. |
| Percent theta | 26.7±5.2 | 27.7±3.7 | 26.6±4.4 | 22.7±6.0 | 26.5±3.8 | 22.8±4.4 | 1.59 | N.S. |
| | 23.1±5.6 | 25.2±4.1 | 22.2±4.1 | 19.9±5.4 | 22.0±2.6 | 21.2±4.8 | 1.17 | N.S. |
| Percent alpha | 27.1±3.6 | 29.9±4.9 | 31.6±3.3 | 30.2±4.0 | 36.9±7.5 | 28.2±6.2 | 3.06 | 0.025 |
| | 30.0±5.2 | 30.2±6.6 | 29.3±8.3 | 30.7±5.5 | 30.9±5.7 | 27.6±4.3 | 0.29 | N.S. |
| Percent beta | 29.5±5.0 | 28.1±5.6 | 30.2±6.2 | 32.0±7.0 | 25.5±3.5 | 35.9±5.8 | 2.85 | 0.05 |
| | 38.1±6.0 | 34.2±6.0 | 38.7±6.3 | 39.9±5.9 | 37.3±3.6 | 43.1±4.0 | 2.12 | N.S. |
| Animals tested | 8 | 8 | 8 | 7 | 6 | 8 | | |
| | 8 | 8 | 6 | 8 | 6 | 7 | | |

*Denotes 1 ppm $O_3$ for 8 hours per day.
†Mean ± S.D. for occipital lead $O_1$.
‡Mean ± S.D. for parietal lead $P_3$.
§Univariate analysis of variance.

**Table 2**
**Means and Standard Deviations for EEG Measures after 18 Months of $O_3$ Exposure**

| EEG Spectral Measure | Ozone Treatment A (1/8)* | B (1/16) | C (1/23.5) | D (2.8) | E (3/8) | F (0/24) | F Value[4] | P |
|---|---|---|---|---|---|---|---|---|
| Peak frequency | 5.3±4.1† | 4.2±0.6 | 7.4±4.7 | 7.6±4.4 | 3.7±0.6 | 5.4±4.8 | 1.08 | N.S. |
| | 5.6±3.7‡ | 5.0±3.2 | 5.1±3.4 | 9.2±4.9 | 3.4±0.5 | 10.9±6.1 | 2.41 | N.S. |
| RMS power | 1363±316 | 1524±518 | 1810±427 | 1644±405 | 1194±135 | 1306±572 | 1.58 | N.S. |
| | 2020±405 | 2136±516 | 2219±947 | 2163±570 | 2231± 69 | 2573±866 | 0.47 | N.S. |
| Percent delta | 16.0±5.1 | 16.5±4.5 | 11.6±2.4 | 13.0±3.3 | 16.9±3.3 | 17.3±6.3 | 1.88 | N.S. |
| | 11.2±1.5 | 15.2±3.2 | 12.6±5.1 | 8.8±3.9 | 16.2±2.2 | 9.4±3.2 | 3.83 | 0.025 |
| Percent theta | 22.2±3.4 | 24.2±3.0 | 23.1±5.7 | 22.1±4.9 | 24.2±1.4 | 24.5±7.2 | 0.36 | N.S. |
| | 17.6±5.5 | 23.8±3.7 | 21.0±5.6 | 18.8±3.8 | 23.1±0.3 | 19.8±5.5 | 1.73 | N.S. |
| Percent alpha | 27.5±3.4 | 27.8±3.6 | 29.0±4.5 | 29.3±6.7 | 28.1±2.3 | 25.6±2.9 | 0.64 | N.S. |
| | 29.0±3.4 | 27.0±4.4 | 28.2±4.7 | 28.1±6.1 | 25.9±1.6 | 23.2±2.7 | 1.23 | N.S. |
| Percent beta | 34.4±3.4 | 31.5±6.3 | 36.3±7.9 | 35.7±7.0 | 30.5±7.1 | 32.6±8.2 | 0.71 | N.S. |
| | 42.3±3.8 | 34.0±4.5 | 37.8±5.5 | 44.4±4.4 | 34.9±4.1 | 47.2±9.0 | 5.77 | 0.01 |
| Animals tested | 8 | 8 | 7 | 8 | 3 | 7 | | |
| | 8 | 8 | 7 | 8 | 2 | 5 | | |

*Denotes 1 ppm $O_3$ for 8 hours.
†Mean ± S.D. for occipital lead $O_1$.
‡Mean ± S.D. for parietal lead $P_3$.

occipital leads. The decrease in parietal EEG beta activity was statistically significant ($P<0.05$) for treatment groups B, C, and E. The trend of increased alpha activity was not statistically signficant for either the parietal or occipital leads.

The effects of ozone on visual evoked potentials are summarized in Tables 3 and 4. Figure 1 illustrates a typical averaged visual evoked potential and illustrates the primary component waves at $P_1$ and $N_1$. A review of the data in Table 3 indicates that the mean effects of $O_3$ after 9 months are upon the primary positive wave, $P_1$. The effect of $O_3$ was to lengthen the latency of $P_1$ in both occipital and parietal leads. The latency of $P_1$ in EEG lead $O_1$ increased from a control (Group F) mean of 98.4 ms to 111.3 ms for Group E, the most severely exposed group. This increase in the latency of $P_1$ was statistically significant ($P<0.05$) for treatment groups A, B, and E. The same result was noted in parietal leads where the mean latency for $P_1$ for the control dogs increased from 94.3 ms to 111.3 ms for Group E, which was statistically significant for treatment groups B, D, and E. The latencies of both $P_1$ and $N_1$ for the control dogs are in agreement with values given by Lee et al.,[9] who report a mean for the primary negative component ($N_1$) of 43 ms and 100 ms for the primary positive component ($P_1$). These values can be compared to the present values of 58.4 ms and 98.4 ms obtained from the control dogs after 9 months. No significant increase in the latency of $N_1$ was found after 9 months of ozone exposure. The data of Table 3 also suggest that the amplitude of $P_1$ increases as a consequence of $O_3$ exposure, but this observation was not statistically significant. In fact, no other evoked potential measure achieved statistical significance, although some trends are evident. These trends suggest that the overall amplitude of the average evoked potential increases due to $O_3$ exposure and this increase is due primarily to an increase in the positive portion of the response.

Table 4 contains the evoked potential measures obtained at 18 months. As with the data at 9 months, the latency of $P_1$ is greater for $O_3$-exposed dogs than for the control dogs; however, the effect was not statistically significant.

## DISCUSSION

The purpose of this investigation was to ascertain through use of electroencephalographic methods the effects on the brain of chronic exposure to ozone. The results obtained suggest a neurotoxic effect of $O_3$ on brain function evident at 9 months, but with a possible tolerance present after 18 months of $O_3$ exposure.

The results indicate that after 9 months of $O_3$ exposure a decrease in the frequency bandwidth of the EEG has occurred and that this decrease

**Table 3**
**Visual Evoked Potential Measures after 9 Months of $O_3$ Exposure**

| Evoked Potential Measure | Ozone Treatment A (1/8) | B (1/16) | C (1/23.5) | D (2/8) | E (3/8) | F (0/24) | F Value | P |
|---|---|---|---|---|---|---|---|---|
| $P_1$ latency (ms) | 107.0±2.3* | 109.4±9.0 | 103.5±8.8 | 106.9±8.0 | 111.3±4.4 | 98.4±4.6 | 3.41 | 0.025 |
| | 101.9±6.2† | 107.6±9.6 | 102.1±7.5 | 104.1±7.7 | 111.3±5.5 | 94.3±4.1 | 4.89 | 0.01 |
| $N_1$ latency (ms) | 61.9±5.9 | 60.9±3.0 | 58.0±2.1 | 57.9±2.2 | 61.5±2.7 | 58.4±2.6 | 2.31 | N.S. |
| | 60.8±5.4 | 60.9±5.7 | 56.3±3.1 | 57.0±3.7 | 56.3±6.6 | 57.6±3.7 | 1.55 | N.S. |
| $P_1$ amplitude (mv) | 12.3±9.5 | 11.0±6.4 | 11.1±5.0 | 10.6±4.3 | 12.9±6.6 | 8.6±4.4 | 0.43 | N.S. |
| | 12.9±8.4 | 8.1±5.1 | 9.6±4.1 | 9.0±5.5 | 11.7±5.7 | 6.5±3.7 | 1.31 | N.S. |
| $N_1$ amplitude (mv) | 10.9±5.5 | 10.7±4.0 | 13.0±7.4 | 11.9±6.8 | 13.5±6.2 | 11.2±5.8 | 0.27 | N.S. |
| | 8.6±4.0 | 6.2±3.3 | 8.2±5.6 | 5.4±2.3 | 6.4±4.2 | 6.8±5.5 | 0.65 | N.S. |
| Ratio $N_1/P_1$ | 1.23±0.74 | 1.16±0.53 | 1.23±0.36 | 1.14±0.51 | 1.10±0.24 | 1.42±0.75 | 0.29 | N.S. |
| | 0.86±0.48 | 0.97±0.57 | 0.85±0.32 | 0.71±0.19 | 0.56±0.33 | 1.12±0.64 | 1.31 | N.S. |
| RMS value | 5.0±3.6 | 4.6±2.2 | 5.2±2.6 | 4.7±2.4 | 5.1±2.2 | 3.9±1.6 | 0.26 | N.S. |
| | 5.4±4.6 | 3.4±2.0 | 3.8±2.2 | 3.1±1.7 | 4.6±2.9 | 2.9±1.6 | 0.96 | N.S. |
| Animals tested | 8 | 8 | 8 | 8 | 6 | 8 | | |
| | 8 | 8 | 8 | 8 | 6 | 8 | | |

*Mean ± S.D. for occipital lead $O_1$.
†Mean ± S.D. for parietal lead $P_3$.

**Table 4**
**Visual Evoked Potential Measures after 18 Months of $O_3$ Exposure**

| Evoked Potential Measure | Ozone Treatment A (1/8) | B (1/16) | C (1/23.5) | D (2/8) | E (3/8) | F (0/24) | F Value | P |
|---|---|---|---|---|---|---|---|---|
| $P_1$ latency | 101.1±5.4* | 101.3±2.9 | 106.3±7.0 | 104.6±5.2 | 108.3±9.7 | 100.3±2.4 | 2.08 | N.S. |
| | 96.7±6.3† | 102.8±9.0 | 103.6±8.0 | 104.3±6.5 | 107.7±8.6 | 99.0±3.5 | 1.76 | N.S. |
| $N_1$ latency | 60.3±3.0 | 61.5±2.1 | 60.9±4.5 | 62.4±1.8 | 65.7±3.8 | 61.3±4.3 | 1.36 | N.S. |
| | 59.9±3.4 | 58.6±2.5 | 59.1±3.8 | 60.6±2.6 | 61.0±7.5 | 59.9±3.6 | 0.37 | N.S. |
| $P_1$ amplitude | 25.2±12.3 | 15.6±6.9 | 18.7±6.3 | 23.8±9.9 | 14.7±1.8 | 24.3±8.8 | 1.70 | N.S. |
| | 13.8±4.9 | 10.4±3.7 | 11.1±5.1 | 12.1±3.8 | 9.1±4.0 | 13.4±2.9 | 1.08 | N.S. |
| $N_1$ amplitude | 22.2±10.7 | 15.6±11.6 | 14.3±8.7 | 21.8±5.6 | 13.4±1.3 | 18.9±10.5 | 1.04 | N.S. |
| | 12.8±5.9 | 9.9±6.3 | 8.2±4.3 | 9.6±6.2 | 4.6±2.9 | 9.3±4.4 | 1.17 | N.S. |
| Ratio $N_1/P_1$ | 1.24±0.84 | 1.09±0.54 | 0.92±0.61 | 1.28±0.91 | 0.92±0.03 | 0.79±0.41 | 0.59 | N.S. |
| | 1.12±0.56 | 0.95±0.58 | 0.78±0.44 | 0.89±0.64 | 0.52±0.37 | 0.71±0.32 | 0.86 | N.S. |
| RMS value | 11.3±3.2 | 6.6±4.2 | 7.0±2.2 | 9.5±3.0 | 4.7±1.0 | 9.5±4.2 | 2.96 | 0.05 |
| | 6.1±2.8 | 5.1±2.9 | 4.4±1.7 | 5.1±2.1 | 3.0±1.5 | 5.0±1.2 | 1.01 | N.S. |
| Animals tested | 8 | 8 | 7 | 8 | 3 | 7 | | |
| | 8 | 8 | 7 | 8 | 3 | 7 | | |

*Mean ± S.D. for occipital lead $O_1$.
†Mean ± S.D. for parietal lead $P_3$.

occurs as the result of loss of beta activity into the alpha range. In addition, the latency of the primary positive component, $P_1$, of the averaged visual evoked potential is increased. The findings indicate that the ozone exposures of 1 ppm given for 16 hours per day was the minimum ozone concentration required to effect abnormal EEG activity. It is interesting to note that the 2 ppm ozone exposure given for 8 hours per day, which yields the same time weighted average as the 1 ppm, 16 hour exposure, was much less effective in causing abnormal EEG activity. However, the 3 ppm 8 hour per day exposure was clearly more neurotoxic than was the 1 ppm exposure given for 23.5 hours per day, both treatments yielding a time weighted exposure of approximately 1 ppm.

The decrease in EEG frequency bandwidth (i.e., the slowing of the EEG) is in the direction of a mild depression of cortical function. As such, the effect resembles observations[10] of slowed EEGs due to hypoxia. For example, Brazier[11] reported a progressive slowing of alpha activity in human EEGs as a consequence of progressive hypoxia. Skillen et al.[2] reported that brain serotonin levels decreased as a result of ozone exposure and Trams et al.[4] found that catecholamine content of cerebral cortical tissue decreased as a consequence of ozone exposure. Both these reports indi-

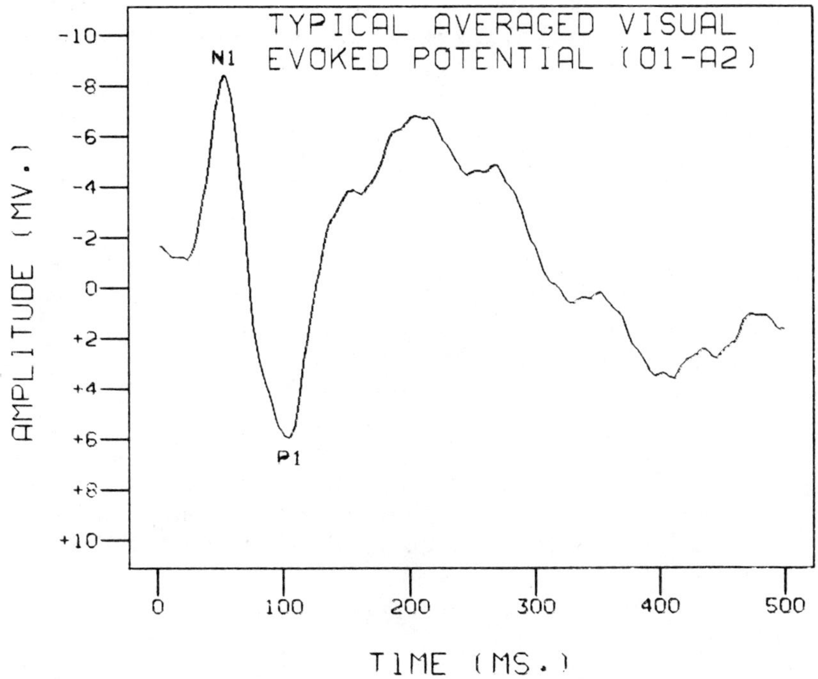


**Figure 1.** Typical averaged visual evoked potential from occipital lead $O_1$.

cate a decrease in neurotransmitters that could be reflected as a slowed EEG and increased latencies in evoked potentials. Robins and Hartman[12] point out that the catecholamines are generally considered to be excitatory neurotransmitters. A decrease in serotonin content is described by Smythies[13] as leading to functional loss of perception and cognitive behavior.

The lack of effect of ozone on brain electrophysiology at 18 months of $O_3$ exposure is not in agreement with Trams' finding of decreased catecholamine content. It is suggested that the decrease in catecholamine content was not sufficiently great to be reflected as significant changes in the EEG.

The question of whether ozone exerts a chronic effect on brain function appears to be answered in the affirmative as the result of Trams' report that used biochemical indicators of toxicity and by our finding that brain electrophysiology is affected after 9 months of ozone exposure. The question of whether these effects are the direct result of a neurotoxic reaction of ozone with brain tissue or are the by-products of damage to other body organs remains unanswered.

## REFERENCES


1. Jaffe LS: The Biologic Effects of Ozone on Man and Animals. *Am Ind Hyg Assoc J* 28:267–277, 1967.
2. Skillen RG, et al: Brain 5 — Hydroxytryptamine in Ozone-Exposed Rats. *Proc Soc Exp Biol Med* 108:121–122, 1961.
3. Xintaras C, Johnson BL, Ulrich CE, et al: Application of the Evoked Response Technique in Air Pollution Toxicology. *Toxicol Appl Pharmacol* 8:77–87, 1966.
4. Trams EG, Lauter CJ, Brown EA, et al: Cerebral Cortical Metabolism After Chronic Exposure to Ozone. *Arch Environ Health* 24:153–159, 1972.
5. Lagerwerff JM: Prolonged Ozone Inhalation and Its Effects on Visual Parameters. *Aerosp Med* 34:479–483, 1963.
6. Hore T, Gibson DE: Ozone Exposure and Intelligence Tests. *Arch Environ Health* 17:77-79, 1968.
7. Pedersen EE, Breisacher P, Patterson J, et al: Experimental Studies on Human Health Effects of Air Pollutants: Psychophysiological and Psychomotor Assessment. Final Report ARB 2-372, Rancho Los Amigos Hosp, Inc, Downey, Calif, 1974.
8. Hinners RG, Burkart JK, Punte CL: Animal Inhalation Exposure Chambers. *Arch Environ Health* 16:194–206, 1968.
9. Lee SI, Klass DW, Wakim KG: Changes in Visual Evoked Potentials Induced by Electrolyte Alterations in Dogs. *Electroencephalogr Clin Neurophysiol* 27:677, 1969.

10. Van Liere EJ, Stickney JC: Hypoxia. Univ of Chicago Press, Chicago, 1963.

11. Brazier MAB: Effect of Hypoxia on the EEG. *J. Neurol Neurosurg Psychiatry* 11:118–125, 1948.

12. Robins E, Hartman BK: Biochemical Theories of Mental Disorders. In: *Basic Neurochemistry,* RW Albers, GJ Siegel, R Katzman, et al (eds), Little, Brown & Co, Boston, 1972.

13. Smythies JR: *Brain Mechanisms and Behaviour*. Academic Press, New York, 1970.

# 19 The Relative Contributions of Inspired and Dietary Lead to Body Lead Burden*

Joel D. Kopple, M.D.
Michael B. Rabinowitz, Ph.D.
George W. Wetherill, Ph.D.

A number of studies have now established that inspired air may contribute to the body lead burden. However, there is little information concerning the quantitative relationships between atmospheric lead and respired intake of lead under typical urban conditions. We have, therefore, conducted a series of metabolic studies to evaluate normal human lead metabolism and the relative contributions of inspired and dietary lead to the body lead burden. In these investigations the absorption, excretion, and sizes and transfer coefficients of physiological pools of lead were assessed with the use of stable isotope tracers administered under conditions of controlled dietary and atmospheric exposure.

Three healthy adult male volunteers lived in the metabolic unit at Wadsworth Hospital Center, which is situated in the western part of Los Angeles. They resided there for periods ranging up to six months and ingested rigidly controlled constant diets consisting of foods selected to be

*Thanks are due to the nursing, dietetic, and laboratory staff of the metabolic unit of the Veterans Administration, Wadsworth Hospital Center.

Work was supported by the National Science Foundation Research Applied to National Needs (RANN). VA Project Number 5016-02.

low in lead content. The diets were supplemented at each meal with a stable isotope tracer, usually lead-204 as nitrate. Thus, the subjects' total lead intake approximated their pre-study levels, but they ingested a type of lead that was distinct from other inputs of lead (i.e., atmospheric).

The concentration and isotopic composition of lead were determined in all feces and urine, which were pooled continuously into 10-day collection periods; in duplicate diets, prepared at 10-day intervals; and in blood, hair, nails, desquamated skin and sweat, saliva, bone, bile, and gastric and pancreatic secretions, which were collected periodically. Lead was measured by mass spectrometric isotope dilution analysis. Intake of nitrogen, potassium, sodium, calcium, phosphorus, magnesium, calories, and vitamins was designed to provide an abundance of each of these nutrients and was kept constant throughout the study. Simultaneous metabolic balance studies were carried out for nitrogen, potassium, calcium, phosphorus, and magnesium. The study has been completed in two subjects and is currently in progress in a third individual (Subject C).

Two subjects smoked eight cigarettes per day of a brand containing 1 $\mu$g of lead per cigarette; these cigarettes provided approximately 1 $\mu$g of inspired lead intake per day.[1] Lead input from cigarette smoking in Subject C was slightly greater since he smoked about one pack per day of cigarettes higher in lead content. Two subjects were Caucasian; Subject B was Negro.

The techniques used in these studies have the advantage of allowing one to perform kinetic studies without employing an abnormally high lead intake or administering radioactive lead isotopes. Another benefit of this approach is that the tracer lead that is employed in these studies may be analytically separated from lead contamination arising during the metabolic or laboratory procedures because of its markedly different isotopic composition. Thus, the tracer lead data are independent of any blank or yield corrections. In contrast, the lead balance data were calculated from measurements of lead that was largely of normal isotopic composition. Although small yield and blank corrections were made (approximately 5%), there is inherently less certainty with the balance data than with data derived solely from the tracer measurements.

Three independent sets of data from these studies indicate that approximately 10 to 18 $\mu$g per day of lead were incorporated from respiration.

1. The first set of data is derived from the lead balance studies in the three subjects (Table 1). Lead balance is calculated by subtracting total lead output from total dietary lead intake. The tabulated numbers represent total lead, which includes common lead plus the isotope tracer. The major output is in the stool, which includes a small quantity of endogenous lead lost from internal sources. The second major output of lead is in urine. A small amount of lead is also lost from hair, nails, and desquamated skin

and sweat. It is apparent that all three subjects were in continuously negative lead balance (i.e., they excreted more lead than they ingested in the diet). However, during the balance periods, which lasted for 65 to 116 days, their nitrogen, potassium, calcium, phosphorous, and magnesium balances were neutral or positive; and their blood lead levels did not fall. We therefore believe that the subjects were in neutral lead balance, and the magnitude of the excessive output represents the quantity of lead intake from respiration.

2. The second set of data indicating that the subjects were incorporating approximately 10 to 18 $\mu$g of lead from air each day is derived from the observation that for each subject the isotopic composition of his blood lead levels would never become identical with the isotopic composition of his dietary input. Figure 1 shows for Subject A both the total blood lead levels (the upper curve) and the concentration of lead in blood that has an isotopic composition identical to the diet (the lower curve). During the initial 116 days of the study, when the subject ingested a low lead diet supplemented with lead-204 to give a total dietary intake of 362 $\mu$g per day, total blood lead levels were relatively constant. However, the quantity of lead in blood that had an isotopic composition identical to the diet rose rapidly (lower curve). The difference between the upper and lower curves represents unlabelled common lead that is derived either from deeper pools within the subject's body or from an external source, such as inspired atmospheric lead. It is apparent that even though the duration of this study exceeded several mean-lives of the blood compartment, the blood lead for Subject A never became more than 55% labelled with dietary lead. From these data a strict upper limit of about 50% can be set on the fraction of blood lead that is derived from unlabelled sources. Similar data are available for Subjects B and C.

The question then is: How much of this unlabelled lead is derived from the atmosphere and how much is transferred from a source within the body? We attempted to solve this problem by developing a kinetic model to

**Table 1**
**Lead Balance Data***

| Subject | Age (years) | Days | Input | Output ($\mu$g/day) | | | Balance |
|---|---|---|---|---|---|---|---|
| | | | Diet | Diet-Feces | Urine | Others† | |
| A | 53 | 116 | 362 | 26 | 38 | 4 | −16±4‡ |
| B | 50 | 108 | 366 | 12 | 27 | 4 | −19±3 |
| C | 25 | 65 | 183 | 23 | 30 | 5 | −12±6 |

*Mean values.
†Lead in hair, nails, desquamated skin, and sweat.
‡Standard Deviation.

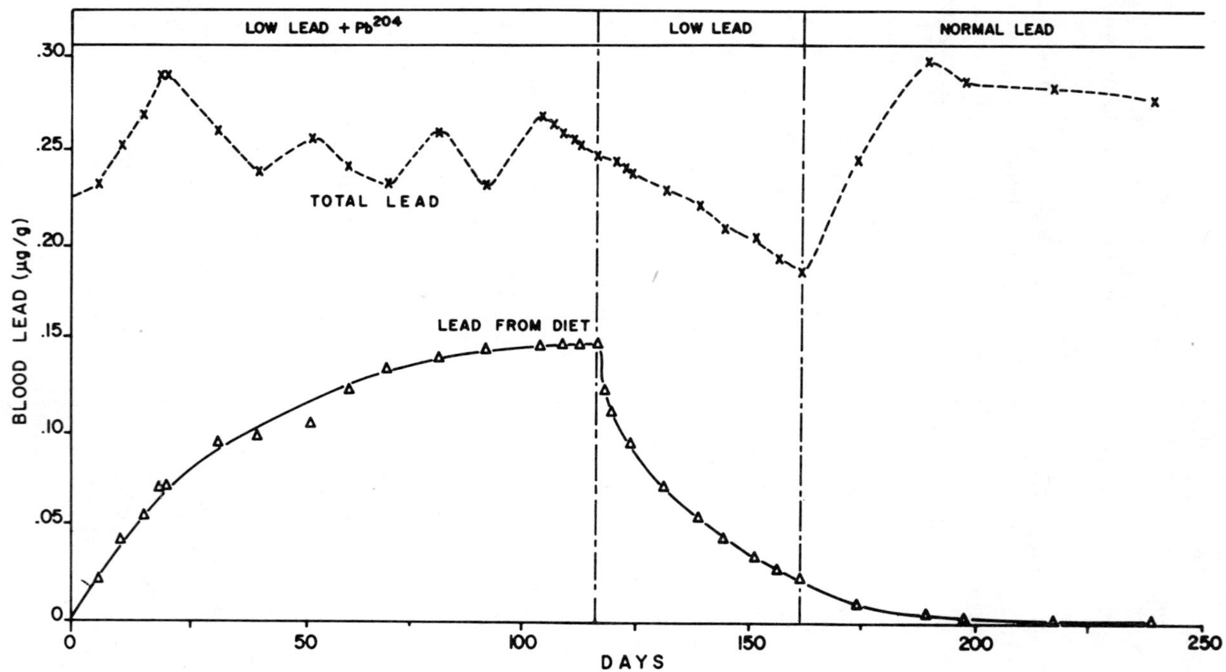


**Figure 1.** Blood lead levels in subject A who was fed a low-lead diet supplemented with lead-204 nitrate, the same diet without the lead-204 tracer, and a normal lead diet.

assess transfer rates of lead from various body pools. The experimental data suggest a three-compartment model; this model, derived from the averaged data from Subjects A and B is shown in Figure 2. Data from Subject C is excluded from this figure as his study is still in progress. Blood lead is contained in Pool 1 and comprises approximately one-half to two-thirds of this pool. This pool contains about 1.8 mg of lead and has a mean-life of about 27 days. It is considered to exchange with two pools. One of these pools resides at least largely in soft tissue (Pool 2), is somewhat smaller than the first pool, and has a mean-life that is similar to this latter pool. Evidence for its existence is derived from the relatively slow rates of labelling of lead in hair, gastrointestinal secretions, and other outputs from soft tissues. The third pool is much larger and encompasses most of the body lead. It resides primarily in bone although it includes most of the lead that is situated in soft tissue. It contains approximately 200 mg of lead and has a much longer mean-life, approximately 10,000 days. The exchange rates of these two pools with the blood pool can be determined from the tracer data and the kinetic model. With these data one can project the growth curve for the tracer lead in blood to a time when a steady state is achieved in both labile body pools (Pools 1 and 2). When equilibrium between Pools 1 and 2 is achieved, about 60% of the lead in blood would

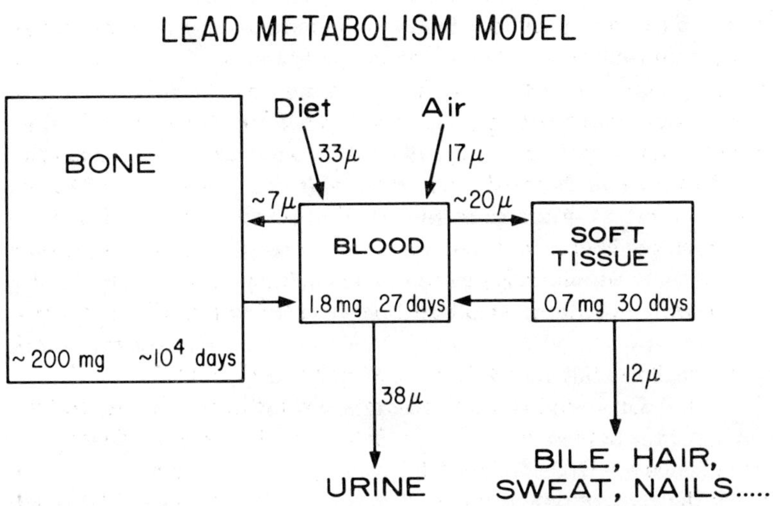


**Figure 2.** A three-compartmental model for human lead metabolism from analyses of tracer and balance data. The lead content and mean life of each pool, the exchange constants (λ), and the rates of flux (micrograms per day) are shown. The first pool contains blood lead; the second pool is comprised of lead at least primarily from other soft tissues; the third pool contains skeletal lead and most of the lead in soft tissues. Loss of lead from the second pool is via integumentary structures (hair, nails and sweat) and the alimentary tract (bile and gastric and pancreatic juice). The numbers are average values derived from the kinetic data of Subjects A and B.

eventually have an isotopic composition similar to the dietary lead and 40% would continue to resemble the isotopic composition of lead from an unlabelled source, either internal or external to the body.*

Analysis of the transfer coefficients of lead from blood for the three subjects indicates that approximately 4 to 9 $\mu$g per day of lead enters Pool 3 from the blood pool. Since medical evaluation of these subjects revealed no evidence of bone loss and since calcium, phosphorus, and magnesium balances were not negative, it may be inferred that the skeletal mass was in a steady state. These findings suggest that skeletal lead, and hence the third pool, was in a steady state. The first (blood) pool was also probably unchanging as the total lead concentration in blood was relatively constant. Hence, it can be inferred that these deeper pools yield approximately the same amount of unlabelled lead back to the blood; 9, 5, and 4 $\mu$g per day for Subjects A, B, and C, respectively. The remaining quantity of the unlabelled lead transferred into the first pool can be concluded to be inspired from the atmosphere 17 $\pm$ 5 $\mu$g (mean $\pm$ S.D.) per day for Subject A, 16 $\pm$ 6 $\mu$g per day for Subject B, and 8 $\pm$ 4 $\mu$g per day for Subject C.

The data from Figure 1 also indicate the sensitivity of blood lead levels to changes in intake. When dietary lead was decreased by approximately 50% by discontinuing the lead-204 tracer, both total blood lead and the quantity of lead from the diet fell rapidly. When the subject was allowed to eat a self-selected diet which had a normal isotopic composition and a normal lead content, total blood lead rose while the blood lead from the labelled source continued to decrease.

3. The third independent method indicating that there is a respired input of about 10 to 18 $\mu$g per day is derived from an experiment in which the atmospheric content of lead was reduced, and blood lead levels and lead balances were monitored. This study was carried out in Subject B (Figure 3). During the first 108 days of the study, this subject resided in the metabolic unit and ingested a constant low-lead diet that was supplemented with lead-204 to achieve a total daily intake of 366 $\mu$g per day. During this time blood lead concentrations were relatively constant while the fraction of lead in blood that was of normal isotopic composition continued to fall gradually.

On the 109th day, the subject moved into a single-bed room in the metabolic unit in which the lead content in the atmosphere was reduced. This was accomplished by using two air purifiers, each consisting of a blower that pulled room air through a fiberglass prefilter, an HEPA fiber filter, and a bed of activated charcoal. These purifiers ran continuously and recirculated the air in the room every three minutes. The room entrance was provided with an air lock consisting of two doors arranged in series. Both particulate lead salts in the atmosphere as well as organic lead vapor

---

*This calculation assumes no input of labelled lead from the third pool as data concerning its size and residence time are too imprecise for this analysis.

were monitored and averaged 0.07 and 0.05 $\mu$g per cubic meter, respectively. The subject spent at least 23 hours each day in this room for a total of 40 days.

During the first 15 days of the experiment in this room, the subject received an additional dietary supplement of lead-207. This increase in his

**Table 2**
**Respired Lead Intake Determined by Three Methods ($\mu$g/day)**

| Subject | Lead Balance Studies | Blood Labelling with Tracer Lead | Response of Blood Lead and Lead Balance to Low-Lead Air |
|---|---|---|---|
| A | 16 ± 4* | 17 ± 5 | — |
| B | 19 ± 3 | 16 ± 6 | 18 ± 3 |
| C | 12 ± 6 | 8 ± 4 | — |

*Mean ± Standard Deviation.

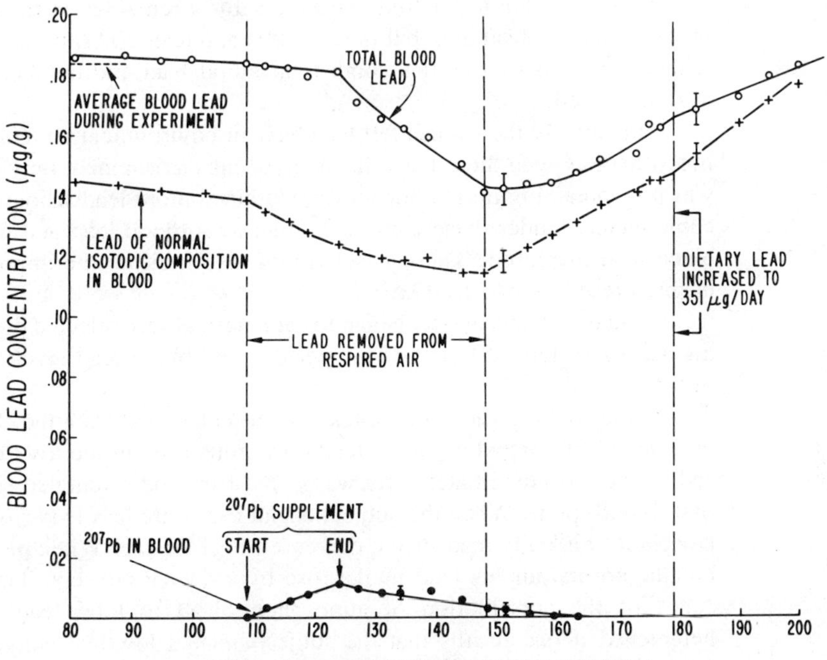


**Figure 3.** Blood lead levels in subject B who was fed, initially, a low-lead diet supplemented with lead-204 nitrate. On day 109 a dietary supplement of lead-207 was added to this intake, and simultaneously atmospheric lead was removed by filtering. The total lead dropped slightly indicating that the additional lead-207 did not exactly balance the loss of atmospheric lead. On day 124, both the lead-204 and lead-207 supplements were stopped and the low-lead diet was continued. Both total lead and labelled lead concentrations in blood fell. On day 148 the subject began to breathe atmospheric lead, and both total and unlabelled blood lead began to rise. On day 179 a normal lead diet was begun, and total and unlabelled blood lead increased further.

dietary lead intake was designed to counterbalance the anticipated decrease in inspired lead so that his total daily lead intake would remain relatively constant. This was done in order to prevent a decrease in blood lead. The possibility was considered that a decrease in blood lead might activate homeostatic mechanisms which could mobilize lead from other pools, such as the skeleton. Since skeletal lead was unlabelled, it would be virtually impossible to distinguish between the contribution of mobilized skeletal lead and of inspired atmospheric lead to the blood. Thus, by adding lead-207 to the diet, total blood lead should not decrease and no potential homeostatic mechanism would be activated.

It is apparent from Figure 3 that, during this period, total blood lead decreased only slightly. However, the content of unlabelled lead in blood fell substantially. Lead balance, which had been consistently negative previously, became neutral or slightly positive during this period. These observations indicate that lead inspired from the atmosphere does make a substantial contribution to total blood lead levels and body lead burden.

From the 16th to the 40th day, the subject remained in the room and ingested his low-lead diet, but the lead-204 and lead-207 supplements were withdrawn. During this time both total blood lead and unlabelled blood lead decreased.

On day 149 the subject left the clean air environment but continued to ingest his low-lead diet. Thus, he changed his environment only in that he was now breathing normal air which contained more lead, about 2 $\mu$g per cubic meter. Under these conditions both total blood lead and unlabelled blood lead increased. These observations also indicate the importance of inspired lead to blood lead levels.

On day 178 the subject began to eat a normal self-selected diet providing approximately 351 $\mu$g of lead per day and blood lead levels rose still further.

Subject B's respired lead intake can be calculated from the changes in the content of unlabelled blood lead during his time in the low-lead room and the period immediately afterwards and from the calculated size of his first (blood) pool. When the subject began to inspire less lead from the atmosphere, his daily lead intake decreased by 14.6 $\mu$g. When he began to breathe normal air his lead intake rose by 18.3 $\mu$g per day. The significance of the contribution of atmospheric lead to total lead intake is heightened if one recalls that the subject inhaled low-lead air and never lead-free air. Moreover, throughout this "clean-air" part of the experiment, he continued to leave the room for up to one hour each day. It is estimated that during his stay in the low air-lead room, he still inspired 2.5 $\mu$g of lead per day. Hence, his daily inspired lead intake when breathing normal urban air is calculated to be 18 $\pm$ 3 $\mu$g per day.

The results of these studies indicating apparently negative lead balance, incomplete labelling of blood lead with a dietary tracer, and changes in blood lead and lead balance with varying concentrations of atmospheric lead suggest that normal men incorporate a substantial quantity of lead from inhalation of the ambient urban atmsophere. A summary of these results for Subjects A and B and the preliminary data for Subject C are shown in Table 2.

It is of interest that the quantity of inspired lead intake calculated from these studies is similar to the respired lead intake estimated from the observed concentrations of air lead in the subjects' living quarters (about 2 $\mu$g per cubic meter), generally accepted values for ventilatory and particle retention rates, and the quantity of lead incorporated from their cigarette smoke.[1]

What then are the clinical or epidemiological implications of these findings? It is apparent that blood lead levels and body lead burden are very sensitive to lead concentrations in the atmosphere as well as the lead content of the diet, and under normal conditions inspired lead makes a major contribution to lead intake. Moreover, the contribution of airborne lead to the daily lead intake clearly exceeds its effect upon respired input since significant quantities of lead from atmospheric sources are also ingested. Airborne lead adheres to leaves, seeds, and fruit of edible plants, is taken up by their root systems, and is transported into their internal structures. Lead also enters food during processing, is inspired or ingested by cattle and other animal sources of food, and is swallowed by humans.

It is important to note, however, that despite the relationship between atmospheric lead and body lead burden, it has not yet been established that in the normal urban environment, lead levels in air represent a health hazard. This question will only be answered by continued attempts to define precisely the relationships between blood lead levels and body lead burden and health.

## REFERENCE

1. Lehnert G, et al: *Int Arch Gewerpath.* Gewerbehyg: 23, 358–363, 1967.

# 20 Clinical Implications of Air Pollution Research for Diseases of the Central Nervous System

Frank E. Speizer, M.D.*

The papers presented in this session have discussed a variety of issues related to pollutants in the atmosphere and their effects on the nervous system as well as other organ systems in the body. In this summary I will try to use the information that was presented and bring the data into a context for clinicians who not only have to provide care for individual patients but also may be called upon to provide expertise on health related matters for the communities in which they work and live. This issue becomes important because all physicians are more likely to come into contact with issues of long-term exposure to environmental contaminants as these exposures become of greater concern to citizens at large.

Two personal experiences will be given to begin the discussion, after which more detailed information about some of the more general pollutants that have been under discussion will be presented. One cannot possibly cover all the effect of all pollutants, so the discussion will concentrate on the central nervous system effects of carbon monoxide, with some brief comments about other pollutants, as time permits.

*Supported in part by a Career Development Award No. ES33332 from the National Institute Environmental Health Sciences and by research grant No. ES503.

Several years ago, I had a discussion with the head of the traffic division of a major United States city police force concerning the health problems related to being a policeman. One of the particular problems raised was what to do with the partly disabled policeman. Certainly, the policeman injured on duty is taken care of financially, at least. But, the discussion focused on what was to be done with the policeman who, for example, suffered a mild heart attack, when he was medically cleared to return to work. Not all of these men are ready to return to the street and not all can be given inside desk jobs. The force has to find a reasonable alternative. In the case of heart attack victims, it was decided in good faith that for at least a three-month period after returning to work, these policemen would be given the relatively "soft" job of guarding the large number of towed vehicles that were stored by the city. The only trouble was that the vehicles were stored in an underground parking garage, where the averaged measured ambient carbon monoxide levels were sometimes as high as 170 ppm. In effect, the police force, trying to act on behalf of the employees, had exposed some of their possibly most susceptible personnel to a particular environmental contaminant that increased rather than decreased their occupational hazard. Fortunately, once identified, the situation was relatively easy to correct, and no adverse effect of the excess exposure on what were probably highly susceptible individuals, was known to have occurred.

The second incident relates more directly to a clinical situation. A house officer who within the previous hour had just admitted the fourth firefighter who had collapsed at the scene of a fire in a plastics warehouse asked me for a consultation. The house officer had just been notified that two more firefighters from the same fire were being brought in, and he wanted to know what material in plastics produced a yellow smoke that would cause the sudden collapse of these men. The story was exactly the same for each man. These very experienced firefighters had all gone up the ladder to a second floor window, had stepped through the window casing, and within a minute had collapsed and had to be brought down. Fortunately, an easy method for estimating carboxyhemoglobin was available. Each man had levels ranging between 35% to 50% and obviously the treatment of choice was oxygen. When the situation was discussed with the recovered men, it was discovered that a firefighter judges whether or not it is safe to enter a building by the blackness of the smoke. The blacker the smoke, the higher the carbon monoxide level and, thus, the greater the danger. Something in the plastic that was burning in this fire caused the smoke to be quite yellow, and the firefighters judged the building to be safe to enter. One other clue that was available was that all the men were quite pink and complained of headaches. In addition, since cyanide was the only other chemical that might produce sudden collapse, the fact that the men were all alive helped in the diagnostic considerations.

This second example has been presented to point out that although carbon monoxide at ambient levels may have some toxic effects, it is generally considered an unimportant pollutant. Situations do occur, however, when CO exposure can be life-threatening. In fact, most deaths associated with fires are not related to smoke inhalation per se but to hypoxia produced by CO exposure.

With reference to the papers presented, without going over the details of the presentations, some general clinical comments are in order.

The wide variety of materials and symptoms and signs discussed by Dr. Henkin might suggest to some persons that one would need to take a year off to study toxicology of air contaminants — and even then one would not be able to master the field. However, most of the materials discussed by Dr. Henkin do not occur as general environmental pollutants; and one cannot stress enough the usefulness in clinical medicine not only of finding out the current or last occupation of the individual, but also of probing in depth the work history. The latter should include questions regarding materials handled and, from the respiratory point of view, whether or not masks or respirators were suggested or required for the job; how long one worked at a particular job; and other past jobs. One should also inquire about hobbies and at-home jobs, activities, or repairs in which a large number of exposure problems may occur. For the large-scale disasters which do occur — such as the explosion of a hydrogen disulfide tank — it is sufficient to know that these are dangerous chemicals, for it is probable that public health officials, state police, and local pollution control authorities will take over the management of the problem very quickly.

Although Dr. Haider showed some interesting physiologic data on the effects of moderate levels of carbon monoxide and Dr. Johnson's data on chronic exposure of animals with ozone produced evidence of similar effects in dogs, one must look to behavioral effects data for understanding what these changes might mean. What little data are available suggest that carbon monoxide affects individuals in the performance of tasks that might best be characterized as dull, boring, or routine and CO seems to have less effect on complex intellectual tasks. It is also apparent that the behavior effects are small and the data are quite variable. Thus, the data are difficult to interpret.

The behavior effects of exposure to carbon monoxide can be divided into several categories that include vigilance testing, reaction time, time discrimination and estimation, coordination and tracking, sensory processes, and complex intellectual behaviors, including automobile driving. The fact that automobile driving is included in the category of complex intellectual behavior suggests just how simple some of the tests in the other categories are.

Psychologists define a vigilance test as one that determines how well an individual can detect small changes in his environment that occur at unpredictable intervals.[1] A less than clinical situation, but one that is readily understandable, is the task of a submarine radio man assigned to watch a sonar screen and identify the changes in patterns of signals that might literally reflect impending danger. This is a monotonous task and a common finding is that men would, after a while, miss signals that they would not have missed when first starting their vigil.

One can use visual or auditory signals occurring randomly during an exposure period to determine if exposure to carbon monoxide at varying concentrations causes any change in vigilance. Figure 1 shows the results of one such study. The data are taken from the study by Horvath et al. on carbon monoxide and human vigilance.[2] In this study, the subjects, who were all healthy male volunteers aged 21 to 32, were pre-exposed for one hour to the concentration of carbon monoxide they were going to receive during the test period. This was done to increase the likelihood that a

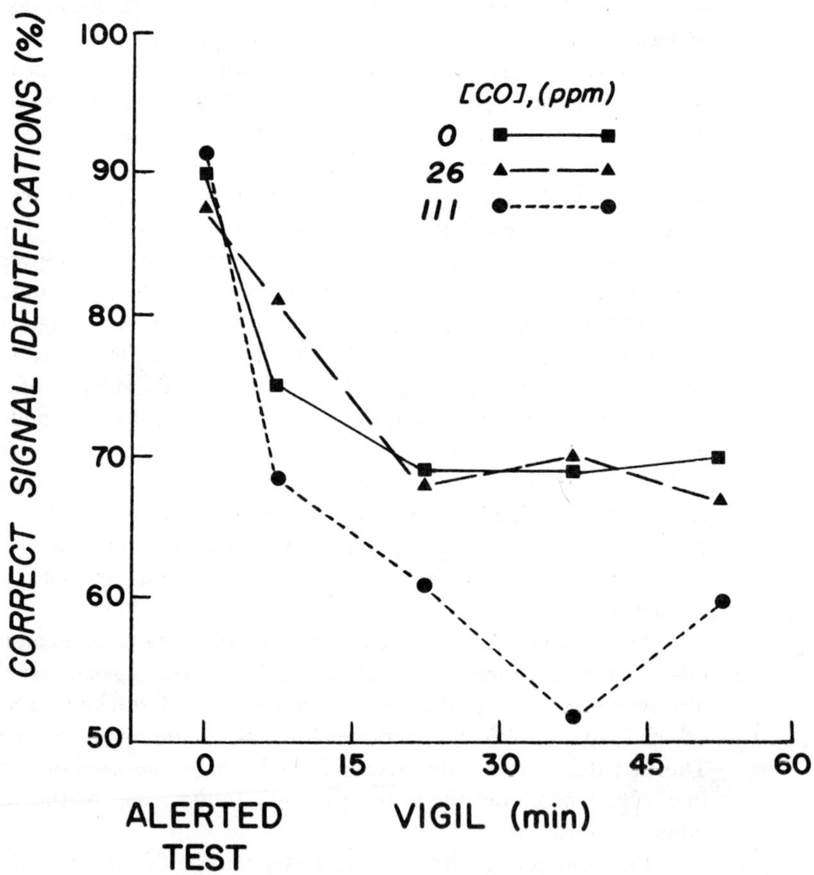


**Figure 1.** Test of vigilance after exposure to varying levels of carbon monoxide. (By permission of author: Horvath et al.[2])

changed carboxyhemoglobin level would be present at the start of the test and that during the exposure period equilibrium would be reached. Those exposed to 0 ppm or room air, had a COHb level of 0.8% both before and after the test period. Those exposed to 26 ppm had a COHb level at the beginning of the test period of 1.6% and 2.3% after the test hour. Those exposed to 111 ppm had a pre-test COHb of 4.2% and a post-test level of 6.6%.

No effect on vigilance as measured was apparent in those exposed to 26 ppm. There appeared to be a significant decrease in performance in those exposed to 111 ppm. As an indication of how complex this field is, the apparent improvement near the end of the hour is believed to be an example of the so-called end-spurt effect that often occurs in vigilance work when the subjects become aware that their tasks are almost finished. These tests were carried out on non-smokers, but similar tests in smokers, who may chronically have COHb in the range of 3% to 5% showed no change in the vigilance performance when exposed to the same levels of carbon monoxide.[3]

Reaction time studies have often used driving simulation tests in which the subject has been asked to respond to a changed situation. Unfortunately, the reports concerning the effects of carbon monoxide on reaction time do not allow one to come to any firm conclusion. As many studies report positive relationships as report negative results.[4,5] It appears that the studies that used the more simple response patterns were the ones in which a relationship to carbon monoxide exposure was found. In any case, COHb levels under investigation that were associated with positive responses were from about 4.5% to 11.2%.[6,7]

In 1967, Beard and Wertheim reported the effects of carbon monoxide on the ability of human subjects to discriminate between the length of two tones.[8] Figure 2 summarizes some of their results and shows the average length of time at each exposure level for the production of a decrease in performance that was significantly below the performance level in room air. This means that at 90 minutes of exposure to 50 ppm of CO, at 50 minutes of exposure at 100 ppm, at 30 minutes of exposure at 75 ppm, and at 23 minutes of exposure to 250 ppm there were significant decreases in performance in the tone length discrimination test. All of these exposure periods translate into COHb levels ranging between 2.5% and 4%.

Repeated studies using considerably higher levels of carbon monoxide exposure have been unable to confirm or refute these findings.[9,10] Because of this and because of the peculiar circumstances of exposure in the Beard-Wertheim studies, it has been suggested that the effects that were measured must have been related to vigilance rather than to tone discrimination alone.

A large number of studies of dexterity and coordination have been carried out. In many of the studies no effect was seen before the level of

COHb approaches 30%.[4] Similarly, little or no effect in sensory processes — for example, the ability to determine brightness of different lights — was detected below COHb levels of 12%.[6]

Several tests of intellectual function, which include timed readings, arithmetic tests, memory tasks, and dot counting, have shown modest impaired relationships with COHb levels. For example, subjects asked to underline plural nouns in a passage took 150 seconds at COHb levels of 0% to 7% and 210 seconds at COHb levels between 10% to 15%.[11]

When one puts all of the behavior effects into a real test situation, such as driving an automobile, it appears that whatever the effects of carbon monoxide are on behavior, they can be overcome by the necessity of the moment. It is interesting to note that in spite of at least a half a dozen groups working in this field over the last 5 to 10 years, the results of Forbes et al. in 1937, in which they exposed 5 subjects to enough carbon monoxide to produce COHb as high as 30%, still stand.[12] They reported very little effect on a series of reaction times, coordination, and perceptual tasks — all present within the context of a test of driving skills. Although the tests today are considerably more elaborate, the results are approximately the same.

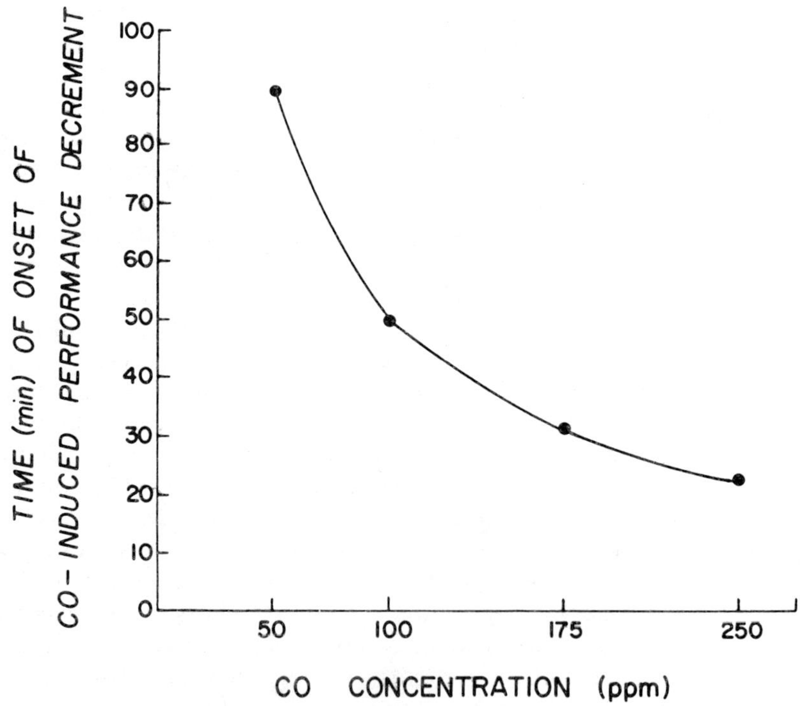


**Figure 2.** Time of decreased performance of time discrimination function after exposure to varying levels of carbon monoxide. (By permission of author: Beard.[8])

On the other hand, as mentioned by Dr. Haider, it is very possible that automobile accident risk may relate to the driver's vigilance. Since vigilance appears to be related to carbon monoxide levels, it would be prudent to be concerned about exposures to levels that produce these changes.

Turning to the question of what populations are at increased risk of central nervous system effects after exposure to carbon monoxide, one must consider the pathway by which carbon monoxide exerts its effects. One might consider the environmental factors in persons who have become disabled by any disease process that impairs the ability to transport oxygen. Ambient levels of carbon monoxide in the range of 25 ppm are not confined to central cities. Either because of where the air intake vents are located or because of inefficient exhaust systems, many automobiles are likely to have carbon monoxide concentrations in the passenger area as high or higher than 25 ppm, even without the added effects of cigarette smoke. Another environmental factor, probably less important in San Francisco than in Boston, relates to the nature of space heating. Space heaters are a potential source of carbon monoxide and depending on the nature of the space being heated can be either an acute hazard or a chronic hazard, particularly for older persons with cerebral vascular disease.

The effect of carbon monoxide is through interference with transport and delivery of oxygen to the tissues. Vital organs having high oxygen demand are especially vulnerable. Obviously, the brain is one such vital organ. Additional diseases that may interfere with oxygenation might also contribute to the effects of hypoxia on the brain. These additional conditions might include anemia, chronic pulmonary diseases, and cerebral vascular disease. Few studies of specific levels of exposure to carbon monoxide have been carried out in these groups of patients to test changes in behavior. However, on theoretical grounds, it is apparent that older persons with these very common conditions need to be protected from excessive carbon monoxide exposure.

One additional burden of carbon monoxide exposure that relates to effect on delivery of oxygen to tissue is worth mentioning — that is, the effect of altitude on oxygen transport. Persons living at 5,000 feet normally have a reduced oxygen hemoglobin saturation. Generally, these persons get along fine. However, if their ability to transport oxygen becomes impaired, they are quick to suffer and on occasion are forced to move to sea level towns. The chronic effect of exposure to carbon monoxide is to simulate an altitude environment. Table 1 indicates the effects of living at environmental levels of carbon monoxide that are not very different from what is present in urban centers for at least part of a 24-hour day. For example, in Boston or San Francisco, although both cities are listed as sea level towns, those parts of the city that have ambient levels of carbon monoxide in the range of 25 ppm are in effect physiologically like 6,000 feet above sea levels. A city like Denver, which is at 5,000 feet, is really like 8,300 feet if the ambient carbon monoxide level reaches 25 ppm.

Since Dr. Kopple discussed the relative contribution of airborne lead to the total body burden, a few comments about ambient exposure to lead are in order. For several years measurements have been made of the biologic incorporation of lead from the environment in a group of workers exposed to automobile exhaust. These have included both traffic police officers and tunnel turnpike workers. Lead levels have been monitored in workers as well as in their spouses to rule out alternative sources of lead exposure. There is no question that lead incorporation, presumably through the respiratory route, occurs in these workers; but, to date no clinical manifestation of lead intoxication has been found in spite of the occasional blood lead levels in excess of 60 $\mu$g%.

The situation in children, however, may be very different than in adults. There may be considerable exposure in small children that may in part relate to automobile exhaust as a source of lead. Household dust, street dirt, and roadside dirt all may contain considerable quantities of lead. Concentration averaging 2,000 $\mu$g/gm of dry dirt are not uncommon.[13] In a recent study of dirt in Detroit, the major source of lead appeared to be from painted frame houses, and there appeared to be a decreasing concentration in dirt samples taken farther away from the house and closer to the road.[14] Street gutter dirt lead content averaged approximately 1,000 $\mu$g/gm compared to 2,000 $\mu$g/gm within two feet of the painted frame houses and 500 $\mu$g/gm from within two feet of brick houses. Studies of dirt wiped from children's hands have shown that a considerable amount of lead is present, and it has been estimated that if only half the concentration found on hands is transferred to the mouth, several times a day, the daily permissible lead intake can be exceeded.[15] Although no cases of lead intoxication in children related directly to road dust have been reported, there is increasing interest and investigation in a number of laboratories on what might be regarded as the effects of subclinical intoxication with lead.[16] Here, in contrast to carbon monoxide, one is attempting to identify chronic rather than

**Table 1**
**Approximate Physiologically Equivalent Altitudes at Equilibrium with Carbon Monoxide Levels**

| Ambient Carbon Monoxide Levels (ppm) | Altitude (feet) | | | |
|---|---|---|---|---|
| | Sea Level | Oxygen Pressure (mm Hg.) | 5000 | Oxygen Pressure (mm Hg.) |
| 0 | Sea Level | 159.1 | 5,000 | 84.8 |
| 25 | 6,000 | 74.1 | 8,300 | 53.9 |
| 50 | 10,000 | 41.6 | 12,000 | 30.5 |

From: *Biological Handbook Respiration and Circulation.* Ed. Altman P.L., and Dittmer, D.S. Federation of American Societies for Experimental Biology. Bethesda, M.D., 1971.

acute effects. Everyone is aware of the tragic, permanent mental retardation that may follow encephalopathy as a result of lead intoxication. What is unknown at present is whether or not there are adverse effects of subclinical lead intoxication on the more subtle behavior and intellectual functions of children.

In summary, in regard to behavior, from the clinical point of view, the only therapeutic tool in most situations is to prevent exposure. To do this, one must recognize that exposure may come in many different forms. It may come not only from general environmental sources but also from specific working and recreational situations. Clinicians must, therefore, always be aware that behavior abnormalities or other changes, particularly in persons handicapped by other chronic diseases, may be the result of environmental stress; and physicians must be prepared to probe for the identification of these exposures. To do this effectively, physicians must incorporate the concept of the multifactorial etiology into their understanding of most chronic disease states.

## REFERENCES


1. Mackworth JF: *Vigilance and Attention,* Baltimore, Penguin Books, Inc, 1970.

2. Horvath SM, Dahms TE, O'Hanlon JF: Carbon Monoxide and Human Vigilance; a Deleterious Effect of Present Urban Concentrations. *Arch Environ Health* 23:343–347, 1971.

3. O'Hanlon JF: Preliminary Studies of the Effects of Carbon Monoxide on Vigilance in Man. In: *Behavioral Toxicology,* B Weiss, VG Laties (eds), New York; Plenium (in Press) as Quoted in Committee Print of Committee of Public Works United States Senate, Air Quality and Automobile Emission Control vol 2, Health Effects of Air Pollutants, Serial No. 93-24, p 76, 1974.

4. Stewart RD, Peterson JE, Baretta ED: Experimental Human Exposure to Carbon Monoxide. *Arch Environ Health* 27:154–164, 1970.

5. Wright G, Randell P, Shephard RJ: Carbon Monoxide and Driving Skills. *Arch Environ Health* 27:349–354, 1973.

6. Ramsey JM: Effects of Single Exposures of Carbon Monoxide on Sensory and Psychomotor Response. *Am Ind Hyg Assoc J.* 28:531–534, 1967.

7. Ramsey JM: Carbon Monoxide, Tissue Hypoxia, and Sensory Psychomotor Response in Hypoxaemic Subjects. *Clin Sci* 42:619–625, 1972.

8. Beard RR, Wertheim GA: Behavioral Impairment Associated With Small Doses of Carbon Monoxide. *Am J Public Health* 57:2012–2022, 1967.

9. O'Donnell RD, Chikos P, Theodore J: Effect of Carbon Monoxide Exposure on Human Sleep and Psychomotor Performance. *J Appl Physiol* 31:513–518, 1971.

10. Stewart RD, Newton PE, Hosko MJ: Effect of Carbon Monoxide on Time Perception. *Arch Environ Health* 27:155–160, 1973.

11. Schulte JH: Effects of Mild Carbon Monoxide Intoxication. *Arch Environ Health* 7:524–530, 1963.

12. Forbes WH, Dill DB, Desilva H: The Influence of Moderate Carbon Monoxide Poisoning Upon the Ability to Drive Automobiles. *J Ind Hyg Toxicol* 19:598–603, 1937.

13. Hardy HL, Chamberlain RI, Maloof CC, et al: Lead as an Environmental Poison. *Clin Pharmacol Therap* 12:982–1002, 1971.

14. Ter Haar G, Arnow R: New Information on Lead in Dirt and Dust as Related to the Childhood Lead Problem. *Environ Health Prospect* No. 7, 83–89, May 1974.

15. Sayre JW: House and Hand Dirt as a Potential Source in Childhood Lead Exposure. *Am J Dis Child* 127:167–170, 1974.

16. Albert RE, Shore RE, Sayers AJ, et al: Follow-up of Children Overexposed to Lead. *Environ Health Prospect* No. 7, 33–39, May 1974.

# DISCUSSION: PART FOUR

**Dr. Henkin:** Two common events among children who are lead poisoned is that they exhibit pica and have decreased acuity of taste and smell. One wonders whether these two factors interact and produce some of the changes one sees. For example, in Iran, where there are mud houses, there are children who eat dirt. Most children are prohibited from doing so by their parents, and after a time they stop. There are, however, some children who continue to eat dirt from the walls and floors of their houses, and these are the ones who commonly appear to be dwarfed and have zinc deficiency, perhaps from eating this dirt, which may chelate zinc in their diet. Could this type of activity be a factor in lead intoxication as well?

**Dr. Kopple:** This is a very interesting possibility. I am not aware of any evidence linking an altered sense of taste and smell with the development of pica and lead intoxication.

**Dr. Lescoe:** Does anyone have information as to whether there is any compensatory factor that allows one to overcome some of the decrements from carbon monoxide?

**Dr. Speizer:** We have looked at hemoglobin concentrations in our groups that are exposed. The problem is that the exposure levels to CO that people are getting on the streets are certainly lower than in chronic cigarette smokers. We have found no evidence of polycythemia in our population of policemen, who seem to be quite healthy and who have high normal hemoglobin concentrations.

**Dr. Goldsmith:** The population data on cigarette smokers do show quite clearly that there is a tendency for smokers to have a higher hematocrit that is probably related to carbon monoxide. I know of several individuals' case studies of this and I believe that Dr. Hammond and others showed this effect when they followed the hematocrit of smokers who stopped and started smoking. In addition, there were occupational studies on the tunnel workers in New York City that showed that those workers with a high occupational carbon monoxide exposure had a high hematocrit.

It is very pleasing to hear these elegant isotopic studies on patients confirm what was more prudently suggested by epidemiologic studies on emission exposures to lead, in which we attempted to relate the blood lead levels to the estimated atmospheric exposure with at least grossly similar quantitative results. We simply used the lowest value of the extrapolated logarithmic regression and made an estimate of what was likely to be contributed from atmospheric sources. Dr. Kopple, from the monitoring data for lead in Los Angeles, which many of us have studied over a period of years, we know that the atmospheric level is somewhat higher during the winter months during which the inversion level is low. I wonder if in your studies you find a seasonal effect of the estimated atmospheric contribution to the body level of lead?

**Dr. Kopple:** We have been unable to establish such a relationship. The center where this study took place is in the western part of Los Angeles several miles from the ocean, on the third floor of a building that is not very far from a freeway. It is possible that these factors tended to obscure the effects of seasonal fluctuations on atmospheric lead levels. We monitored atmospheric lead levels in the living area of the experimental subjects continuously, and we did not observe seasonal changes in atmospheric lead.

**Dr. Henkin:** Dr. Johnson, some investigators felt that there was very little CNS or neural effect of ozone at all, and that the effects of ozone were due to other photochemical oxidants that reacted with ozone. Do you have anything to offer?

**Dr. Johnson:** No, I don't have any data on that point. There are only three or four reports concerning the effect of ozone on the central nervous system; two of these are biochemical studies indicating a decrease in serotonin and in catecholamines in the cerebral cortex.

**Dr. John D. Allen, Seattle:** One of our concerns in the Northwest is the presence of copper smelters that far exceed the allowable sulfur emission level. They also have cadmium particulates mixed with the lead and these escape out of the stack. We also have arsenic levels identified in nearby towns. Would arsenic, cadmium, lead, and sulfur intermixed in a fall-out pattern have a synergistic effect?

**Dr. Speizer:** I can't answer the question, but I would comment that some of these metallic particles may very well become a vehicle for getting some of the gases into the lung. Sulfur dioxide as a pure gas is probably pretty harmless because it just does not get past the upper airway; attached to a particle it may reach an area of the lung that it just otherwise couldn't reach.

**Dr. Henkin:**Certainly with $SO_x$, one gets synergistic effects with many other substances.

**Dr. Kopple:** Since many trace elements inhibit enzymes, it would make sense that a series of trace elements, each present in sub-toxic concentrations, could have an additive or synergistic effect which could be toxic.

**Dr. Henkin:** One of the problems here is that obviously some of those metals are much more active than others. Cadmium, for example, displaces zinc and, where enzymes are zinc-dependent, cadmium can do a lot of damage. Obviously arsenic poses the same kind of problem, although there is not much effect of copper per se. In general there is no direct synergistic effect between copper and arsenic and cadmium. They are quite different in contrast to the situation where one gets antagonistic effects of cadmium and zinc.

**Dr. Johnson:** Some Soviet investigators who have looked at this question have reported that synergistic effects were not found for air pollutants at concentration levels representative of Soviet community air.

**PART FIVE**

# ILLNESSES OF CHILDREN

# 21 Illnesses of Children

Paul F. Wehrle, M.D.
Douglas I. Hammer, M.D.

## INTRODUCTION

With the inevitable urbanization and accompanying industrial development in many countries, environmental contamination has increased substantially. It now involves a greater proportion of the total population, and a greater diversity of compounds is involved. In large urban areas, particularly those with geographic characteristics favoring atmospheric stagnation, the most obvious pollution has been that of ambient air. Initially, concern over increasing air pollution was directed primarily toward the immediate problems of soiling and decreased visibility. More recently, increasing attention has been directed to the acute and chronic adverse effects on health and toward long-term studies to detect presently unrecognized late effects.

The first disaster involving a well-defined population and clearly associated with air pollution occurred during 1930 in a heavily industrialized valley in Belgium. Only after a similar episode occurred in Donora, Pennsylvania in 1948 was any systematic approach attempted to evaluate the real and potential hazards from this form of pollution. It is notable that these disasters, and those reported more recently in London and New York, have involved primarily older individuals, and the emphasis of much of the research effort was directed toward the definition of adverse effects on adults.

The relative freedom of children from the chronic disabling afflictions seen among the elderly and the inherent resilience of young individuals when faced with chemical insults appears to have largely spared them from many of the acute responses recognized among those individuals with chronic lung disease, limited cardiovascular function, and other problems frequently associated with aging. In addition to their perhaps greater resilience, children are less likely to encounter intense and prolonged exposure at close range to industrial sources of air pollution.

The need for data in children is apparent since an adverse effect resulting from exposure may not become clinically or physiologically recognizable for many years. In order to minimize any potential hazard and to reduce the effect of those hazards already identified among older persons, sound information regarding children is essential. It is important in planning the location of schools, playgrounds, and urban housing developments likely to have many children in residence, as well as providing information for air quality standards. It has been difficult to identify the effects attributed to poor air quality and to distinguish them from other hazards such as accidents and surface chemicals, as well as from nutritional deficiency, consequences of infection, social and other factors. Fortunately, some data are now available from studies that have utilized an epidemiological approach permitting the separation of other factors from the effect of air quality.

Due to humanitarian and ethical considerations, volunteer or challenge studies are not possible in children. Thus, establishment of a relationship between the type of concentration of specific air pollutants and adverse health effects among children is dependent upon the ability to define and quantitate various compounds in ambient air. It is reassuring to note that the Council on Environmental Quality in 1972 indicated that progress in developing indices and monitoring techniques for air pollution were further advanced than in most other areas of environmental health.[1] Despite these advances in the technology of monitoring specific sources, as well as those in detecting the chemical content of and characteristics of ambient air, considerable difficulties have been encountered in establishing direct associations between air quality and child health. Nevertheless, significant advances in defining the effect of specific compounds on children have been made in recent years. These effects have included attention to growth and development of children in heavily polluted areas, the occurrence of respiratory illness including asthma, the correlation between various symptoms and specific pollutants, athletic performance, pulmonary function as measured by air flow techniques, and the detection and effects of specific chemicals absorbed during exposure.[2,3,4]

## ACUTE EFFECTS

Although the greatest impact of acute intense exposure to air pollutants has been among adults, particularly those with underlying chronic conditions, excess deaths were noted in infants less than 1 year of age during the severe smog experience in London during 1952.[5,6] During this incident there were approximately 3,000 excess deaths among individuals of 45 years of age and older, 8 among infants of less than 1 year, and no evidence of excess mortality among those individuals between 1 and 24 years of age. In an attempt to discover residual damage from this intense experience, 1,000 children exposed to this smog episode at ages less than 1 year were evaluated during subsequent years, and pulmonary function was compared with appropriate control populations. No evidence of residual damage resulting from this exposure was discovered.[6]

## CHRONIC EFFECTS

Until recent years, much of the evidence on the effect of air pollution in children was based on studies in Great Britain and Japan. As reported by Holland,[6] these studies clearly demonstrate adverse effects on ventilatory function as measured at ages 5, 11, and 14 to 15 years, and indicate a correlation with area of residence, social class, and history of pneumonia or bronchitis during the first 4 years of life. It is interesting to note that the frequency of bronchitis and pneumonia in Great Britain has been correlated with area of residence in a polluted area and with social class.

In Japan, both Toyama[7] and Watanabe[8] have reported changes in ventilatory function, particularly during the winter months, in children exposed to high levels of air pollution. These effects appeared to be reversible when air quality improved. The effect of prolonged exposure is unknown. It seems likely, in view of recent experience in the United States, that these effects represent those of sulfur oxides and particulates. The evidence, though not completely convincing, nevertheless strongly suggests that continued exposure to high levels of air pollution increases the risk of serious respiratory illness in children, a conclusion reached by a committee of the Royal College of Physicians in 1970.[9]

## SPECIFIC COMPOUNDS AND CHEMICALS WITH DEFINITE OR PROBABLE EFFECTS

The Committee on Environmental Hazards of the American Academy of Pediatrics reviewed the available literature concerning the health effects of specific air pollutants on children as reported from more than 100 investigations from many countries.[10] The information tabulated includes that available prior to 1970 and concerns numerous compounds. These are listed in Tables 1, 2, and 3.*

It is apparent that numerous individual pollutants have been associated with some type of health defect. For some, the effect observed is not as clearly asosciated as with others, due to the non-specific changes observed and the frequent association between social class and exposure with respect to industrial sources. As noted in Table 1, fluorine has been associated with clinical evidence of tooth mottling as well as laboratory evidence of a decrease in hemoglobin in children who were exposed intensively in the proximity of particular industrial sources of airborne fluoride. This association seems reasonably certain, due to the specificity of the changes observed, although the comparative effect of ingestion versus inhalation remains unresolved. The toxicity of lead is apparent when paint chips are ingested, or when airborne exposure is intense. As yet there is no clear evidence that clinical disease has resulted from air exposure alone in urban environments, a situation similar to arsenic. Recent experience in Idaho, including two children hospitalized for symptomatic lead poisoning among those residing near a lead smelter, has prompted additional investigation.[11] In this area more than 20% of children residing within 1 mile of the smelter were found to have blood lead levels of at least 80 $\mu$g/ 100 cc. Whether these levels resulted from inhalation alone or from ingestion due to dust fall contamination of the environment remains uncertain. Similarly, proximity to industrial sources has resulted in adverse effects from asbestos, silicates, fatty acids, and chloroprene, although in the latter instance only biochemical abnormalities were reported and no direct clinically recognized health effect was recognized.

Various sources and types of sulfur compounds have been associated with several types of symptomatology, primarily related to irritation and inflammation of the mucous membranes of the upper and lower respiratory tract. Sulfur oxides may originate from industrial sources as a derivative of the manufacturing process involved, or they may be released into ambient air as a result of burning fossil fuel in motor vehicles or in stationary power sources.

The effect of sulfur oxides could be anticipated from the earlier experience in the acute smog episodes described above and from the studies in Great Britain and Japan. Recent studies conducted in several cities in the United States have indicated an excess frequency of acute respiratory illness in children clearly associated with increased atmospheric concentrations of sulfur dioxide and suspended sulfates.[12] Furthermore, evaluation of the ventilatory performance of elementary school children living in neighborhoods of high and low air pollution in three American cities have

*Complete bibliographic references are available upon request to the American Academy of Pediatrics, 1801 Hinman Avenue, Evanston, Illinois 60204.

**Table 1**
**Summary of Reports on Effects of Air Pollutants on Children**

| Industrial or Other Source of Pollutant | Pollutants Incriminated | Clinical Findings | Laboratory Findings |
|---|---|---|---|
| Individual pollutants: | | | |
| Hydroelectric, aluminum, phosphate, fertilizer plants | Fluorine | Teeth mottling; reduced caries | Hemoglobin decrease |
| Lead processing plants* | Lead | No clinical disease resulting from air exposure | X-ray changes as well as increased urine and blood Pb |
| Copper mine | Arsenic | No clinical effect | Arsenic detected in hair and urine; hemoglobin decreased |
| Asbestos plant | Asbestos | Lower respiratory disease and aggravation of pulmonary disease | None reported |

*Metal reclaiming plants and car exhausts other possible sources but not studied.

**Table 2**
**Summary of Reports on Effects of Air Pollutants on Children**

| Industrial or Other Source of Pollutant | Pollutants Incriminated | Clinical Findings | Laboratory Findings |
|---|---|---|---|
| Individual pollutants: | | | |
| Cement plants;abrasive and silicone manufacturing | Silicates | Increased upper respiratory disease | Cytological changes in nasal and ocular mucosae*† |
| Solvent factories | Fatty acids | Irritation of upper respiratory tract with increased respiratory illnesses | None reported |
| Motor vehicles; ferrous, metallurgical, hydroelectric power plants | Carbon monoxide | Anorexia, fatigueability, headache, dizziness, decreased visual threshold | Increased carboxyhemoglobin; increased red blood cell count |
| Neoprene plant | Chloroprene | None | Increased coproporphyrinuria and 17-ketosteroids in urine*† |

*Methodology uncertain.
†Effect on 17-ketosteroids possible due to seasonal changes.

**Table 3**
**Summary of Reports on Effects of Air Pollutants on Children**

| Industrial or Other Source of Pollutant | Pollutants Incriminated | Clinical Findings | Laboratory Findings |
|---|---|---|---|
| Reducing agents: | | | |
| Paper mills, viscose plants | Sulfur dioxide, hydrogen sulfide, carbon bisulfide | Increase in respiratory infections, inflamed eyes, headaches, nausea | Peak expiratory flow decreased |
| Synthetic alcohol plants | Sulfur dioxide, hydrogen sulfide, hydrocarbons | Headache, anorexia, increased upper respiratory infections | None reported |
| Steel mills, coal and oil power plants, petroleum refineries | Sulfur dioxide and particles; hydrocarbons | Growth retardation; increased respiratory infections; increased asthma attacks | Anemia, delayed ossification, decreased peak expiratory flow, accentuated lung markings on chest x-rays |
| Superphosphate, fertilizer producing plants | Sulfur dioxide, sulfuric acid | Increased respiratory infections | None reported |

clearly defined a consistent relationship between impaired ventilatory function and exposure to increased levels of sulfur oxides and particulates in ambient air. These studies also suggested that early exposure to elevated air pollution levels for periods of more than 5 years was accompanied by a prolonged deficiency of ventilatory function.[13] These reports, together with the earlier experience, would now appear to provide convincing evidence that elevated ambient concentrations of sulfur oxides and total suspended particulates have well-defined adverse effects on the normal function of the respiratory tract of children.

With respect to the association between asthma and air pollutants, earlier studies have described an association between specific allergenic pollutants and general increases in multiple known pollutants in ambient air.[14] In the latter studies the specific agents responsible for the effect remain unknown, and other factors such as climatic changes and unrecognized allergens complicate a more specific interpretation. Recently a panel of asthmatics residing near a coal-fueled power plant were followed during intermittent exposure to high levels of pollution. A clear association was found between acute episodes of asthma and elevations in sulfur dioxide, total suspended particulates, suspended sulfates, and nitrates. It was not possible to separate the latter factors in order to determine the relative importance of each.[15] In addition, unpublished data by Hammer et al. from the southeastern United States have clearly related acute episodes of croup among asthmatic children with increased particulate air pollution exposure. These latter studies have also associated morbidity due to lower respiratory tract disease with elevations in the concentration of particulates in ambient air in the presence of low sulfur dioxide concentrations. Thus it would appear that the long suspected association between sulfur compounds and particulate levels with respiratory complaints in children now rests on a more substantial data base.

The association between oxides of nitrogen and both respiratory illness and ventilatory performance has been evaluated in Chattanooga. A large TNT plant in close proximity to one of the study areas provided high ambient $NO_2$ exposure in contrast to another study area nearby with elevated suspended particulate exposure. Ventilatory performance of second grade school children residing in the area with high $NO_2$ concentrations was significantly decreased when compared with children in control areas nearby.[16] No adverse effect was found among children from the area with increased levels of suspended particulate matter.

In addition to the effect on the ventilatory function of children, a relative excess in respiratory illness was observed among members of families residing in the elevated nitrogen dioxide study area.[17] Thus it would appear that elevations in nitrogen dioxide concentrations are associated with adverse effects on respiratory function in children.

Photochemical oxidants have been associated with impaired athletic performance in healthy cross-country runners[18] and with excess headache, eye discomfort, cough, and chest discomfort among healthy young student nurses.[19] Chronic oxidant exposure has shown no relationship with acute respiratory infections, and McMillan et al. found no relationship between oxidant air pollution and peak respiratory flow rates among school children, although the sample size studied was relatively small.[20] This latter experience is in contrast to impairment of ventilatory function in adults, particularly with combinations of ozone and sulfur dioxide.

Evidence of the harmful and even lethal effect of carbon monoxide during exposure in closed spaces is ample, and it is likely that the current energy crisis may be accompanied by an increase in this form of poisoning due to attempts at reduction of heat loss from dwelling units. In contrast, there is little information that levels of carbon monoxide in urban ambient air are associated with symptomatology. While McCarroll et al. reported more headache on "High CO" days than on other days, the levels of CO measured were less than those customarily associated with symptomatology.

## FUTURE STUDIES INDICATED OR IN PROGRESS

Attention should be given to long-term follow-up of children exposed to known or suspected hazards. Although work with children is more difficult because of ethical considerations that prevent challenge or exposure studies, identification of selected locations that lend themselves to appropriate longitudinal epidemiological and clinical evaluation can be expected to yield valuable information. The impairment of function that appears to be defined in the studies described above will be of critical importance if progressive changes become apparent with continued residence in the areas of high concentration of pollutants. Later effects appearing among exposed children will be of equal importance.

In addition to the continued evaluation of those areas of research that appear promising, such as the association of oxides of sulfur and nitrogen to respiratory tract disease, further attention should be given to other compounds. Oxidant exposure appears to have some effect, but further studies are needed for confirmation. One unpublished report concerning athletic performance did not confirm that reported above, although the level of oxidant reported was very low.[21]

An example of the value of continued careful studies is the suggestion that asymptomatic children with blood lead levels in excess of $40\mu g\%$ have diffuse and subtle impairment of fine motor, perceptual, and visual perceptual skills when measured by sensitive tests.[22] Innovative approaches, such as the use of hair for quantitative measurement of heavy metal concentrations resulting from environmental exposure may provide methods for quantitation of exposure to other classes of compounds, in addition to those already noted.[23]

Thought should also be given toward monitoring for human mutagenesis. Several chemicals are well-known teratogens, and it is not inconceivable that some of the compounds involved in current or future atmospheric pollution might be involved in either of these mechanisms.[24]

## CONCLUSION

The data concerning the adverse effects of numerous air pollutants on children are on a more secure foundation than was the case only five years ago. It is anticipated that, with continued study and additional search for more sensitive indicators of disability, the effect of additional compounds may become better defined five years hence. It is reassuring to note that the National Research Council's Commission on Natural Resources in their summary report, "Air Quality and Automobile Emission Control," indicates no substantial basis for changing current federal standards for ambient air quality.[25] It is hoped that additional information concerning child health will provide the basis for standards that will at least maintain or perhaps even improve the quality of air in our environment in future years.

## REFERENCES

1. *Environmental Quality*. The Third Annual Report of the Council on Environmental Quality. US Government Printing Office, No. 411–0011, Washington, DC, 1972.

2. Colley JRT: Respiratory Disease in Childhood, *Br Med Bull* 27:9–14, 1971.

3. Colley JRT, Douglas JWB, Reid DD: Respiratory Disease in Young Adults: Influence of early childhood lower respiratory tract illness, social class, air pollution and smoking. *Br Med J* 3:195–198, 1973.

4. Engle RE, Hammer DI, Horton RJM, et al: *Environmental Lead and Public Health*. Environmental Protection Agency, US Government Printing Office, No. AP–90, Washington, DC 1971.

5. Ministry of Health: *Mortality and Morbidity During the London Fog of December 1952*. Rept. Public Health Med Subj, No. 95, London, Her Majesty's Stationery Office, 1954.

6. Holland WW: Effects of Air Pollution on Children. *Pediatrics* 53 (part II):839–841, 1974.

7. Toyama T: Air Pollution and Its Health Effects in Japan. *Arch Environ Health* 8:153–173, 1967.

8. Watanabe H, Kaneko F, Murayama H, et al: Effects of Air Pollution on Health. Report No. 1 on peak flow rate and vital capacity of primary school children (Rept. Osaka City Inst Hyg), No. 26:32–37, 1964. (Translated from Japanese 12) (Preprint — 58th Annual Meeting of Air Pollution Control Association, Toronto, June 20-24, 1965.)

9. Royal College of Physicians of London. Air Pollution and Health. Summary and report on air pollution and its effect on health by the Committee on Smoking and Atmospheric Pollution. *Pittman Medical,* London, 1970.

10. Pediatric Aspects of Air Pollution. Report of the Committee on Environmental Hazards, *Pediatrics* 46:637–639, 1970.

11. Lead Poisoning, Idaho, *Morbidity and Mortality Weekly Report* 23:323, 1974.

12. French JG, Lowrimore G, Nelson WC, et al: The Effect of Sulfur Dioxide and Suspended Sulfates on Acute Respiratory Disease, *Arch Environ Health* 27:129–133, 1973.

13. Shy CM, Hasselblad V, Burton RM, et al: Air Pollution Effects on Ventilatory Function of US School Children. *Arch Environ Health* 27:124–128, 1973.

14. Cowan DW, Thompson HJ, Paulus HJ, et al: Bronchial Asthma Associated With Air Pollutants From the Grain Industry, *J Air Pollut Control Assoc* 13:546–552, 1963.

15. Cohen AA, Bromberg S, Buechley RW, et al: Asthma and Air Pollution From a Coal-fueled Power Plant, *Am J Public Health* 62:1181–1188, 1972.

16. Shy CM, Creason JP, Pearlman ME, et al: The Chattanooga School Children Study: Effects of community exposure to nitrogen dioxide. I. Methods, description of pollutant exposure, and results of ventilatory function testing, *J Air Pollut Control Assoc* 20:539–545, 1970.

17. Shy CM, Creason JP, Pearlman ME, et al: The Chattanooga School Children Study: Effects of community exposure to nitrogen dioxide. II. Incidence of acute respiratory illness. *J Air Pollut Control Assoc* 20:582–588, 1970.

18. Wayne WS, Wehrle PF, Carroll RE: Oxidant Air Pollution and Athletic Performance. *JAMA* 199:901–904, 1967.

19. Hammer DI, Hasselblad V, Portnoy B, et al: Los Angeles Student Nurse Study. *Arch Environ Health* 28:255–260, 1974.

20. McMillan RS, Wiseman DH, Hanes B, et al: Effects of Oxidant Air Pollution on Peak Expiratory Flow Rates in Los Angeles School Children. *Arch Environ Health* 18:941–949, 1969.

21. McCarroll J: Personal Communication Based on Data Assembled by Koontz CH, Seattle, Washington, 1968.

22. Follow-up Studies on Human Lead Absorption. *Morbidity and Mortality Weekly Report* 23:157–159, 1974.

23. Hammer DI, Finklea JF, Hendricks RH, et al: Hair Trace Metal Levels and Environmental Exposure. *Am J Epidemiol* 93:84–92, 1971.

24. Fraumeni JF: Chemicals in Human Teratogenesis and Transplacental Carcinogenesis. *Pediatrics* 53:807–812, 1974.

25. News Report, National Academy of Sciences, National Research Council, National Academy of Engineering 24:1, 1974.

# 22 Air Pollution and Childhood Ventilatory Function I. Exposure to Particulate Matter in Two Southeastern Cities, 1971–1972

Robert S. Chapman, M.D.
Victor Hasselblad, Ph.D.
Carl G. Hayes, Ph.D.
John V.R. Williams, BS.
Douglas I. Hammer, M.D.

## INTRODUCTION

The determination of the effect of air pollution on the lung function of children is an extremely important goal. The unequivocal finding that ambient pollution promotes impairments in ventilatory function in childhood would constitute a powerful argument for vigorous pollution abatement measures. Furthermore, any such impairments that air pollution may promote must be detected as early as possible in life, while they are small and potentially reversible. The demonstration of a clear association between improvements in air quality and improvements in ventilatory function would fortify the argument for pollution abatement.

The study described here was designed to test the hypothesis that several years of exposure to ambient levels of particulate air pollution promotes impairments in children's ventilatory function. This study was the first of a five-year series of assessments of children's ventilatory function in Charlotte, North Carolina, and Birmingham, Alabama. The series is designed to test the hypothesis that improvements in children's ventilatory function accompany improvements in air quality.

Considerable evidence linking childhood air pollution exposure to decreased ventilatory function has already been gathered.[1,2,3,4,5] Measured differences in function between clean and polluted areas have nearly always been subtle, and it has proven difficult to incriminate one specific pollutant or class of pollutants as the major cause of concern. Conversely, it has proven difficult to exonerate any individual pollutants. Charlotte and Birmingham presented an unusual opportunity to assess the effect on children's ventilatory function of suspended particulate matter, in the relative absence of other pollutants. Within the category of total suspended particulates, two fractions were of particular interest: suspended sulfates, which Shy[6] has associated with decreased ventilatory function, and respirable particulates, of diameter about $3.5\mu$ or less, which can enter and lodge in the pulmonary alveoli.

## METHODS

### Sample Selection

In Charlotte, ventilatory function testing was performed in 12 elementary schools, each within a 2¼-mile (3.6 km) radius of one of two air monitoring stations. In Birmingham, testing was performed in 10 elementary schools, each within 2¼-mile radius of one of three stations. In each school, all classrooms of first through sixth grade students were tested. Neighborhoods of children's residence were selected, on the basis of 1960 Census data and visual inspection, to be socioeconomically and demographically similar in the two cities. On the basis of past air monitoring data, it was expected that neighborhoods in Birmingham would have higher levels of particulate pollution than in Charlotte.

In Birmingham, there was very little interneighborhood busing of children to the schools selected for study. In Charlotte, all schools were subject to busing guidelines that resulted in most school enrollments being about 70% white and 30% black.

### Ventilatory Function Testing

During the 1971–72 school year, testing was performed at each school once in the fall (November 2 to December 17, 1971), once in the winter (January 31 to February 28, 1972), and once in the spring (April 24 to May 18, 1972). In the fall, testing in Charlotte began two weeks later than in Birmingham. In the winter and spring, testing was performed simultaneously in both cities. To insure standardization of data collection methods, testing personnel in both cities were trained and observed by the same EPA staff.

The test performed was the three-quarter second forced expiratory volume ($FEV_{0.75}$), which reflects changes in the patency of large airways more than of small ones. Immediately before testing in each season, all children received a verbal introduction and a test demonstration, which emphasized first maximal inspiration, and then expiration, as forceful and complete as possible, into the spirometer. Each child was tested until he had provided three valid $FEV_{0.75}$ measurements. At the time of testing, children were asked whether they had a cold, cough, sore throat, or asthma.

In the fall of 1971 and the winter of 1972, all tests were performed with the Pulmonary Function Indicator manufactured by the National Cylinder Gas Corporation (NCG).* The NCG spirometer integrates air flow electronically and displays $FEV_{0.75}$ on a digital readout unit. In the spring of 1972, all tests were performed with a 12-liter mechanical bellows-type spirometer manufactured by Cardiopulmonary Instruments, Inc. (CPI),* fitted with an electronic digital readout unit designed by EPA engineers. The investigators thought it wise to replace the NCG instrument with the CPI instrument because linearity of $FEV_{0.75}$ readings on the NCG instrument became difficult to stabilize during the winter of 1972. Testing personnel minimized this problem by resetting the NCG's linearity before each class tested during the fall and winter. In the spring, the linearity of the digital readout unit with the CPI spirometer was checked several times each testing day. Recalibration was performed when necessary.

## Air Monitoring

Two EPA air monitoring stations were installed in Charlotte, and three were installed in Birmingham. All aerometric measurements were obtained daily over the course of the study. Samples of respirable suspended particulates (RSP) were obtained from cyclone samplers, which were designed to trap no particles above $10\mu$ in diameter, 50% of particles of $3.5\mu$ diameter, and all particles of diameters between about $0.1\mu$ and $2.5\mu$. Samples of total suspended particulate (TSP), suspended sulfates, and suspended nitrates were obtained from high-volume samplers. Samples of $SO_2$ were analyzed by a modification of the West-Gaeke method. In Charlotte, one station was atop a one-story fire house, about 16 feet above ground, and the other was at ground level. In Birmingham, all three stations were at ground level. All five stations were installed in suburban residential neighborhoods.

To gain an approximate measurement of the children's lifetime doses of air pollution, estimates of annual average pollution from 1959 onward were constructed for both cities. These estimates were developed from air

*Mention of brand names does not imply EPA endorsement of products.

quality data collected by agencies other than EPA, from point source emissions data, and from meteorologic pollution dispersion models.[7]

### Data Analysis

For each city, mean age- and height-adjusted $FEV_{0.75}$ were computed for each season, and for the school year as a whole. Only results from children tested in all three seasons were used in the analysis of the school year as a whole. Since the $FEV_{0.75}$ is effort-dependent, the maximum of each child's three readings, adjusted to body temperature and pressure, saturated conditions (BTPS), was used in the analysis for each season.[8] To allow intercity comparisons of results, all $FEV_{0.75}$ readings from Charlotte were age- and height-adjusted to Birmingham's values.

Intercity differences in $FEV_{0.75}$ were tested for statistical significance by analysis of covariance techniques. Children who reported asthma were excluded from data analysis. Black and white children were treated separately in data analysis,[9] as were boys and girls.[10] Furthermore, results from children aged nine years or older were analyzed separately from younger children's results. Three factors suggested that age groups should be divided in this way. First, the older children had received a greater lifetime dose of pollution than the younger ones. Second, there is evidence that the formation of new alveoli ceases or decelerates after the age of about eight years.[11] Third, younger and older children might perform the $FEV_{0.75}$ with different proficiencies. Though the investigators had been impressed with the ability of children of all ages to perform the $FEV_{0.75}$ with different proficiencies, there had been a greater element of competition among older children than younger ones.

## RESULTS

### Characteristics of Study Sample

In Charlotte, 2,356 white children and 1,349 black children, a total of 3,705 children, were tested in at least one season. In Birmingham, 1,895 white children and 2,397 black children, a total of 4,292 children, were tested at least once. These figures are distributed by age and sex in Table 1. The numbers of children included in the data analyses for each season are presented in Table 2. As expected, the smallest sample sizes were consistently found in the analyses for the school year as a whole.

In each city, the educational attainment of heads of households was determined by sending self-administered family health questionnaires home with school children. Return rates for these questionnaires were 88% in Charlotte and 95% in Birmingham. As shown in Table 3, both blacks and whites in Charlotte had achieved a somewhat higher educational level than their counterparts in Birmingham. Intercity differences in educational attainment were greater for whites than for blacks. In both cities, white heads of households had achieved a somewhat higher educational level than black ones.

From total school enrollments in the fall of 1971, the proportions of children living outside a 2-mile (3.2 km) radius of the air monitoring station in the neighborhood served by their school were determined. In Charlotte, this proportion was 23.0%. In Birmingham, this proportion was 5.4%. The higher proportion in Charlotte reflects the busing guidelines in that city.

## Outdoor Temperatures

Minimum daily outdoor temperatures for the 1971–72 school year were collected from the municipal airports of Charlotte and Birmingham. During the periods of testing, minimum temperatures in the two cities were very similar, as shown in Table 4. Monthly means of minimum daily temperatures throughout the school year were, on the average, 1.9°F warmer in Birmingham than in Charlotte.

**Table 1**
**Number of Children Taking Pulmonary Function Tests at Least Once in Charlotte and Birmingham in 1971–1972, Distributed by Age, Sex, and Race**

| Race | Sex and Age | Sample Size by City | | | |
|---|---|---|---|---|---|
| | | Charlotte | | Birmingham | |
| | | Number | Percent | Number | Percent |
| Black | Boys 5–8 years | 247 | 18.3 | 545 | 22.7 |
| | Girls 5–8 years | 243 | 18.0 | 542 | 22.6 |
| | Boys 9–13 years | 415 | 30.8 | 676 | 28.2 |
| | Girls 9–13 | 444 | 32.9 | 634 | 26.5 |
| | Total | 1,349 | 100.0 | 2,397 | 100.0 |
| White | Boys 5–8 years | 398 | 16.9 | 456 | 24.1 |
| | Girls 5–8 years | 398 | 16.9 | 421 | 22.2 |
| | Boys 9–13 years | 841 | 35.7 | 562 | 29.7 |
| | Girls 9–13 years | 719 | 30.5 | 456 | 24.0 |
| | Total | 2,356 | 100.0 | 1,895 | 100.0 |

**Table 2**
**Number of Children Included in Pulmonary Function Data Analysis**
**Distributed by Age, Sex, Race and Season: Charlotte and Birmingham, 1971–1972**

| Race | Sex and Age | Sample Size by City | | | | | | | |
|---|---|---|---|---|---|---|---|---|---|
| | | Charlotte | | | | Birmingham | | | |
| | | Fall | Winter | Spring | Whole Year | Fall | Winter | Spring | Whole Year |
| Black | Boys 5–8 years | 205 | 180 | 231 | 166 | 426 | 397 | 395 | 327 |
| | Girls 5–8 years | 208 | 180 | 237 | 161 | 403 | 387 | 396 | 305 |
| | Boys 9–13 years | 349 | 313 | 389 | 287 | 573 | 562 | 549 | 495 |
| | Girls 9–13 years | 363 | 342 | 421 | 306 | 549 | 558 | 537 | 452 |
| | Season Total | 1,125 | 1,015 | 1,278 | 920 | 1,951 | 1,904 | 1,877 | 1,579 |
| White | Boys 5–8 years | 331 | 281 | 383 | 253 | 343 | 320 | 330 | 273 |
| | Girls 5–8 years | 323 | 271 | 390 | 234 | 335 | 307 | 298 | 244 |
| | Boys 9–13 years | 724 | 615 | 809 | 573 | 459 | 430 | 400 | 341 |
| | Girls 9–13 years | 604 | 501 | 693 | 449 | 392 | 375 | 355 | 314 |
| | Season Total | 1,982 | 1,668 | 2,275 | 1,509 | 1,529 | 1,432 | 1,383 | 1,172 |

## Exposures to Air Pollution, 1971–1972

Exposures to air pollution for the period of study in Charlotte and Birmingham are presented in Table 5. In Charlotte, RSP levels fell about 20% from fall to winter (from 40.9 to 32.0 $\mu g/m^3$), and rose about 10% in the spring to 35.7 $\mu g/m^3$. In Birmingham, RSP levels fell about 10% from fall to winter (from 44.4 to 39.9 $\mu g/m^3$), and rose about 25% in the spring to 49.4 $\mu g/m^3$. RSP levels in Birmingham for the whole school year were on the average 35% higher than in Charlotte.

TSP levels in Charlotte fell quite steadily from fall to winter to spring. The mean TSP level during spring testing (67.9 $\mu g/m^3$) was 79% of the level in the fall (85.9 $\mu g/m^3$). In Birmingham, TSP levels rose consistently from one season to the next. The mean level in the spring (119.8 $\mu g/m^3$) was 15% higher than in the fall (104.5 $\mu g/m^3$). For the whole school year, Birmingham's mean TSP level was 34% higher than Charlotte's. In both cities, the ratio of RSP to TSP concentration was generally slightly less than 50%.

In Charlotte, suspended sulfate levels fell 23% from fall to winter (from 11.1 to 8.6 $\mu g/m^3$), and rose 19% in the spring, to 10.2 $\mu g/m^3$. In Birmingham, suspended sulfate levels rose consistently between seasons;

**Table 3**
**Educational Attainment of Heads of Households in Charlotte and Birmingham, 1971–1972, Distributed by Race**

| Race | Highest Educational Level | Percentage Achieving Educational Level | |
|---|---|---|---|
| | | Charlotte | Birmingham |
| Black | Less than high school | 53.7 | 55.8 |
| | High school | 30.0 | 31.3 |
| | Greater than high school | 16.3 | 12.9 |
| White | Less than high school | 39.8 | 47.9 |
| | High school | 34.2 | 35.4 |
| | Greater than high school | 26.0 | 16.7 |

**Table 4**
**Arithmetic Means and Ranges of Daily Minimum Temperatures During Pulmonary Function Testing, Charlotte and Birmingham Municipal Airports 1971–1972, in Degrees Farenheit**

| | Minimum Temperature (°F) by Testing Season | | | | | |
|---|---|---|---|---|---|---|
| | Fall | | Winter | | Spring | |
| | Mean | Range | Mean | Range | Mean | Range |
| Charlotte | 39.6 | 25–65 | 30.1 | 15–48 | 50.8 | 37–60 |
| Birmingham | 38.2 | 24–63 | 30.2 | 21–43 | 52.6 | 41–65 |

**Table 5**
**Means of Daily Mean Air Pollution Exposures in Charlotte and Birmingham, September 1971 to June 1972, in $\mu g/m^3$**

| Pollutant | Time Period | Pollution Exposures by City | | |
|---|---|---|---|---|
| | | Charlotte | Birmingham | Birmingham/ Charlotte |
| | Whole school year | 33.4 | 45.0 | 1.35 |
| | Fall test period | 40.9 | 44.4 | 1.09 |
| RSP | Winter test period | 32.0 | 39.9 | 1.25 |
| (Geometric Mean) | Spring test period | 35.7 | 49.4 | 1.38 |
| | Geometric mean of test periods | 36.0 | 44.4 | 1.23 |
| | Whole school year | 76.8 | 103.2 | 1.34 |
| | Fall test period | 85.9 | 104.5 | 1.22 |
| TSP | Winter test period | 80.3 | 107.0 | 1.33 |
| (Geometric Mean) | Spring test period | 67.9 | 119.8 | 1.76 |
| | Geometric mean of test periods | 77.7 | 110.2 | 1.42 |
| | Whole school year | 0.43 | 0.44 | 1.02 |
| | Fall test period | 0.48 | 0.42 | 0.88 |
| RSP (Geom. Mean) / | Winter test period | 0.40 | 0.37 | 0.93 |
| TSP (Geom. Mean) | Spring test period | 0.53 | 0.41 | 0.77 |
| | Geometric mean of test periods | 0.46 | 0.40 | 0.87 |

**Table 5** *(continued)*
**Means of Daily Mean Air Pollution Exposures in Charlotte and Birmingham, September 1971 to June 1972, in $\mu g/m_3$**

| Pollutant | Time Period | Pollution Exposures by City | | |
|---|---|---|---|---|
| | | Charlotte | Birmingham | Birmingham/ Charlotte |
| | Whole school year | 10.4 | 13.0 | 1.25 |
| | Fall test period | 11.1 | 11.8 | 1.06 |
| Suspended Sulfates | Winter test period | 8.6 | 13.2 | 1.53 |
| (Arith. Mean) | Spring test period | 10.2 | 13.7 | 1.34 |
| | Arithmetic mean of test periods | 10.0 | 12.9 | 1.29 |
| | Whole school year | 1.8 | 2.2 | 1.22 |
| | Fall test period | 2.9 | 2.4 | 0.83 |
| Suspended Nitrates | Winter test period | 1.8 | 2.2 | 1.22 |
| (Arith. Mean) | Spring test period | 1.2 | 1.9 | 1.58 |
| | Arithmetic mean of test periods | 2.0 | 2.2 | 1.10 |
| | Whole school year | 21.0 | 8.5 | 0.40 |
| | Fall test period | 24.4 | 7.9 | 0.32 |
| | Winter test period | 30.9 | 16.5 | 0.53 |
| $SO_2$ | | | | |
| (Arith. Mean) | Spring test period | 14.5 | 8.3 | 0.57 |
| | Arithmetic mean of test periods | 23.3 | 10.9 | 0.47 |

the level during spring testing (13.7 $\mu g/m^3$) was 16% higher than during fall testing (1.8 $\mu g/m^3$). For the whole school year, suspended sulfate levels were 25% higher in Birmingham than in Charlotte.

Levels of suspended nitrates were less than 3$\mu g/m^3$ in both cities throughout the school year. In both cities, levels of nitrates declined steadily as the school year progressed, though this decline was more marked in Charlotte than in Birmingham. Concentrations of $SO_2$, while low in both cities, were consistently higher in Charlotte than in Birmingham.

In summary, exposures to all particulate fractions were higher in Birmingham than in Charlotte during the 1971–1972 school year. Intercity exposure differences were generally smallest in fall, intermediate in winter, and largest in spring.

## Estimates of Past Pollution Exposures

Estimates of past pollution exposures for Charlotte and Birmingham are presented in Table 6. The table is divided into five time periods, 1959–1962, 1963–1966, 1967–1970, 1971, and 1972. The first division represents the period in which most children in the younger age group had yet been born. The table indicates that estimated exposure levels of all particulate fractions were higher in Birmingham than in Charlotte from 1959 through 1972. It is also evident from the table that both TSP and RSP levels in Charlotte and Birmingham underwent a gradual and steady decline from the late 1950s to the period of study. Suspended nitrate levels in both cities rose until the late 1960s, when they began a gradual decline. Levels of $SO_2$ were quite low, and fairly constant, in both cities throughout the lives of the children under study.

It is noteworthy that suspended sulfate levels were estimated to have risen in both cities between 1959–1962 and 1972. In Charlotte, the sulfate level was estimated to have risen about 90%, from 5.2 to 9.7 $\mu g/m^3$ over the 13-year period. The sulfate level in Birmingham was estimated to have risen about 30% from 10.4 to 13.3 $\mu g/m^3$. The reason for the increases in sulfate levels is not at all clear, especially since levels of particulates and $SO_2$, which may well be sulfate precursors, were estimated to have fallen or remained constant in the two cities.

Table 6 also indicates that, in the first four years of life, older children in Birmingham and Charlotte received very similar cumulative doses of RSP. In later years, Birmingham's levels of RSP were between 20% and 50% higher than Charlotte's. Thus, long-term effects on $FEV_{0.75}$ attributable to different RSP doses would probably have manifested themselves most strongly in the younger children, since they had received different RSP doses in Charlotte and Birmingham throughout their lives.

In Birmingham, both younger and older children had received 50% to 75% higher lifetime doses of TSP than had children in Charlotte. Thus intercity differences in FEV attributable to different TSP doses would probably not have been more apparent in one age group than in the other. Long-term effects of suspended sulfates, however, might well have been most evident in the older children, since intercity differences in suspended sulfate levels were greatest before the younger children were born. Effects attributable to suspended nitrates or to $SO_2$ would have been difficult to observe in either age group, since absolute levels of these pollutants were estimated to have been quite low, and intercity differences were estimated to have been quite small, throughout the lives of the children under study.

## Pulmonary Function Testing — Charlotte

In Charlotte, mean age- and height-adjusted $FEV_{0.75}$ values tended to remain fairly constant or to rise slightly from one season to the next (Table 7 and Figures 1 and 2). For all age- and sex-specific groups, whites had a

**Table 6**
**Estimates of Past Pollution Exposures, in $\mu g/m^3$, Charlotte and Birmingham, 1959–1972**

| Pollutant | Time Period | Pollutant Level ($\mu g/m^3$) by City | | |
|---|---|---|---|---|
| | | Charlotte | Birmingham | Birmingham / Charlotte |
| Respirable Suspended particulates (Geometric Mean) | 1959–62 | 61 | 63 | 1.03 |
| | 1963–66 | 53 | 61 | 1.15 |
| | 1967–70 | 45 | 56 | 1.24 |
| | 1971 | 41 | 57 | 1.39 |
| | 1972 | 39 | 55 | 1.41 |
| Total Suspended particulates (Geometric Mean) | 1959–62 | 110 | 167 | 1.52 |
| | 1963–66 | 96 | 163 | 1.70 |
| | 1967–70 | 82 | 144 | 1.76 |
| | 1971 | 74 | 133 | 1.80 |
| | 1972 | 70 | 127 | 1.81 |
| Suspended Sulfates (Arithmetic Mean) | 1959–62 | 5.2 | 10.4 | 2.00 |
| | 1963–66 | 8.1 | 10.5 | 1.30 |
| | 1967–70 | 9.4 | 12.2 | 1.30 |
| | 1971 | 9.6 | 11.8 | 1.23 |
| | 1972 | 9.7 | 13.3 | 1.37 |
| Suspended Nitrates (Arithmetic Mean) | 1959–62 | 1.8 | 2.2 | 1.22 |
| | 1963–66 | 2.2 | 2.5 | 1.14 |
| | 1967–70 | 1.4 | 2.8 | 2.00 |
| | 1971 | 1.7 | 2.5 | 1.47 |
| | 1972 | 1.0 | 1.7 | 1.70 |
| Sulfur Dioxide ($SO_2$) (Arithmetic Mean) | 1959–62 | 17 | <25x | <1.47 |
| | 1963–66 | 17 | <15 | <0.88 |
| | 1967–70 | 16 | <12 | 0.75 |
| | 1971 | 16 | 12 | 0.75 |
| | 1972 | 16 | 11 | 0.69 |

**Table 7**
**Average FEV Readings in Liters, in Charlotte and Birmingham in the 1971–1972 School Year, Distributed by Season, Age, Sex and Race**

| | Males | | | | Females | | | |
|---|---|---|---|---|---|---|---|---|
| | Fall | Winter | Spring | Whole Year | Fall | Winter | Spring | Whole Year |
| Blacks 5–8 Years | | | | | | | | |
| Charlotte | 1.224 | 1.224 | 1.232 | 1.215 | 1.137 | 1.168 | 1.128 | 1.169 |
| Birmingham | 1.185 | 1.207 | 1.140 | 1.181 | 1.140 | 1.131 | 1.058 | 1.113 |
| Significance level of difference | 0.05 | NS | 0.01 | 0.05 | NS | 0.05 | 0.01 | 0.01 |
| Blacks 9–13 Years | | | | | | | | |
| Charlotte | 1.678 | 1.675 | 1.695 | 1.665 | 1.711 | 1.691 | 1.723 | 1.702 |
| Birmingham | 1.692 | 1.669 | 1.560 | 1.642 | 1.639 | 1.646 | 1.534 | 1.603 |
| Significance level of difference | NS | NS | 0.01 | NS | NS | 0.01 | 0.01 | 0.01 |
| Whites 5–8 Years | | | | | | | | |
| Charlotte | 1.311 | 1.344 | 1.341 | 1.342 | 1.182 | 1.219 | 1.217 | 1.214 |
| Birmingham | 1.291 | 1.272 | 1.229 | 1.273 | 1.191 | 1.129 | 1.116 | 1.147 |
| Significance level of difference | NS | 0.01 | 0.01 | 0.01 | NS | 0.01 | 0.01 | 0.01 |
| Whites 9–13 Years | | | | | | | | |
| Charlotte | 1.822 | 1.836 | 1.873 | 1.839 | 1.747 | 1.750 | 1.801 | 1.763 |
| Birmingham | 1.823 | 1.797 | 1.773 | 1.776 | 1.695 | 1.740 | 1.688 | 1.710 |
| Significance level of difference | NS | 0.05 | 0.01 | 0.01 | 0.01 | NS | 0.01 | 0.01 |

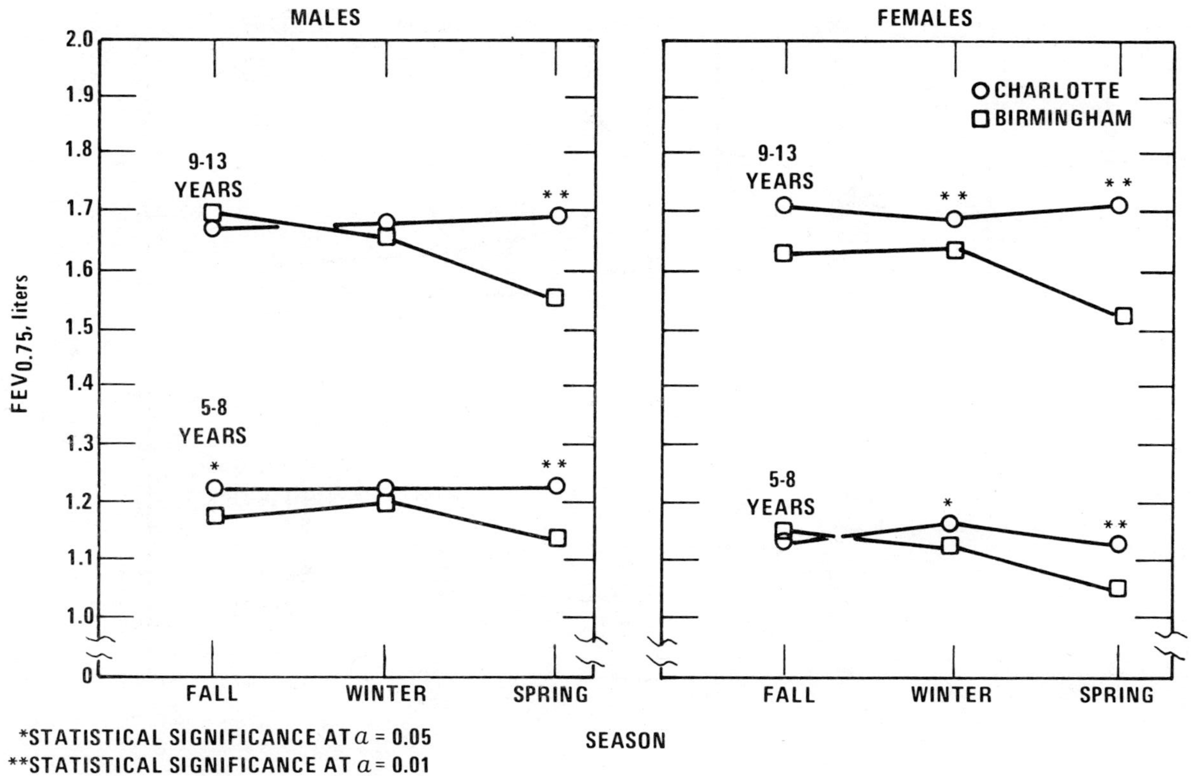


**Figure 1.** Average $FEV_{0.75}$ in black children in Charlotte and Birmingham, 1971–1972

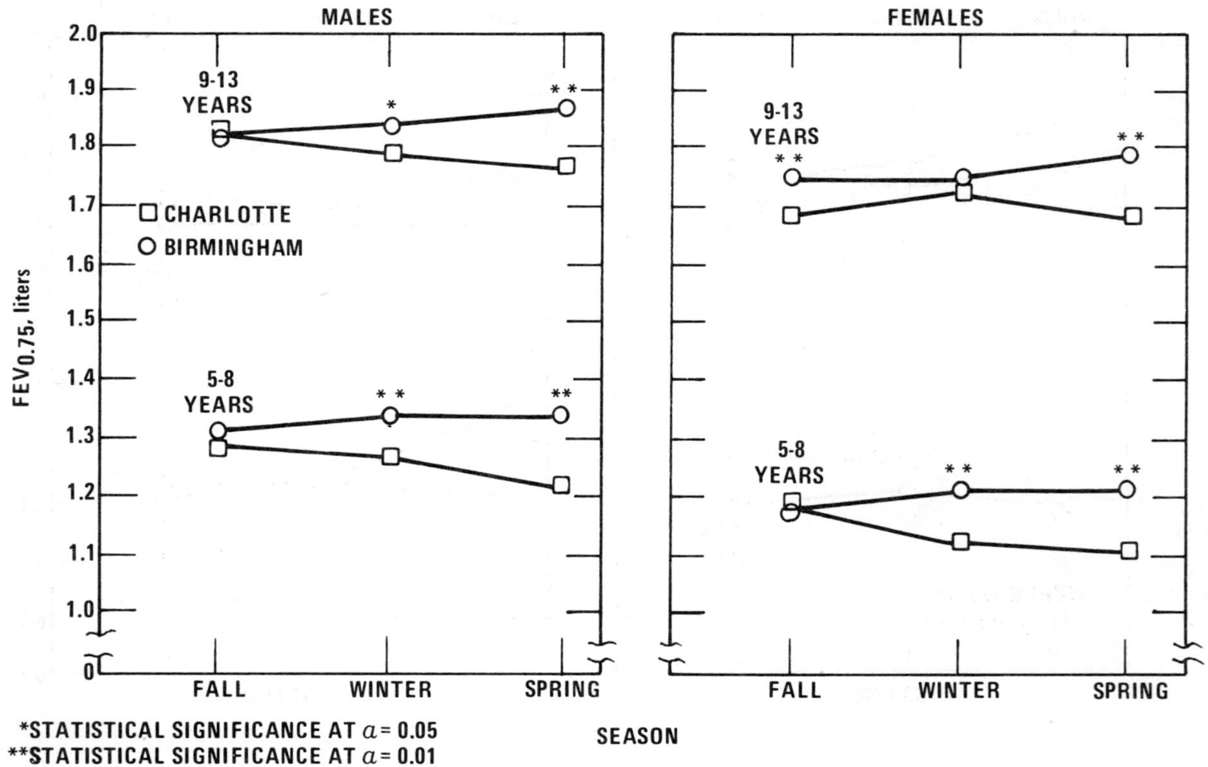


**Figure 2.** Average $FEV_{0.75}$ in white children in Charlotte and Birmingham, 1971–1972.

higher mean $FEV_{0.75}$ than blacks. For males the difference in mean $FEV_{0.75}$ between whites and blacks was 9% (133 cc.). For females, this mean difference was less than half as great, 4% (60 cc.). Percent differences in mean $FEV_{0.75}$ between races were fairly similar for younger children (7% or 84 cc.) and older ones (6% or 110 cc.). The greatest interracial difference in mean $FEV_{0.75}$ was observed in older males (10% or 161 cc.).

Perhaps the most striking single finding in Charlotte was that among older blacks, the mean $FEV_{0.75}$ of the females averaged 2% (26 cc.) higher than that of the males. In other age and race groups, the mean $FEV_{0.75}$ of the males averaged 7% (95 cc.) higher than that of the females. Generally, sex differences in $FEV_{0.75}$ were greater in whites than in blacks, and were greater in younger than in older children.

## Pulmonary Function Testing — Birmingham

Mean $FEV_{0.75}$ values in Birmingham showed a distinct tendency to decrease as the school year progressed. Decreases in $FEV_{0.75}$ from winter to spring were larger and more consistent than from fall to winter. Among males and females, seasonal decreases in $FEV_{0.75}$ were most marked in older blacks. As in Charlotte, white children consistently had higher mean $FEV_{0.75}$ values than black children of the same age and sex group. Average differences between races were larger in older children (8% or 129 cc.) than in younger ones (5% or 62. cc.), and considerably larger in males (9% or 122 cc.) than in females (5% or 69 cc.). As in Charlotte, the greatest interracial differences in mean $FEV_{0.75}$ were observed in older males (10% or 157 cc.). Within age and race groups, males consistently had higher mean $FEV_{0.75}$ values than females. On the average, differences between sexes were greater in whites (7% or 105 cc.) than in blacks (4% or 51 cc.), and were greater in younger children (8% or 94 cc.) than in older ones (4% or 62 cc.).

## Pulmonary Function Testing — Charlotte vs. Birmingham

Age- and height-adjusted mean $FEV_{0.75}$ values were higher in Charlotte than in Birmingham in 28 of 32 comparisons specific for age, sex, race, and season (Table 7). Of these 28 comparisons, 18 were statistically significant at $\alpha = 0.01$, and four more were significant at $\alpha = 0.05$. Mean $FEV_{0.75}$ values were higher in Birmingham than in Charlotte only in the fall, for younger females and older males of both races. In no group did Birmingham's mean $FEV_{0.75}$ exceed Charlotte's by as much as 1%. As the school year progressed, there was a distinct tendency for the percent difference between Charlotte and Birmingham $FEV_{0.75}$ values to increase. Charlotte's total mean $FEV_{0.75}$ was 1% (156 cc.) higher than Birmingham's in fall, 3% (316 cc.) higher in winter, and 8% (912 cc.) higher in spring. For whites, on the average, the difference between the two cities' $FEV_{0.75}$ values increased fairly steadily as the school year progressed. For blacks, on the average, this difference remained fairly constant at 2% until spring, when it increased to 9%.

For each age- , sex- , and race-specific group, the mean $FEV_{0.75}$ for the whole year in Charlotte was compared to the corresponding mean $FEV_{0.75}$ in Birmingham. In eight of eight possible comparisons, mean $FEV_{0.75}$ values in Charlotte were higher than those in Birmingham (Table 8). As shown in Table 8, the intercity percent difference in $FEV_{0.75}$ was distinctly larger in older black females than in older black males. In older white females, this difference was slightly smaller than in older white males. In younger children of both races, females exhibited larger intercity percent differences than males. Table 8 also reveals that, in younger females and in males of both age groups, the intercity percent difference in $FEV_{0.75}$ was slightly larger in whites than in blacks. The opposite was true in older females. Table 8 further reveals that, in both races, the intercity percent difference in $FEV_{0.75}$ was larger in younger than in older males. In white females, this percent difference was distinctly larger in the younger age group. In black females this percent difference was slightly larger in the older age group. On balance, intercity percent differences in mean $FEV_{0.75}$ were greater in females than in males, greater in whites than in blacks, and greater in younger than in older children.

## DISCUSSION

The findings of this study were consistent with the original study hypothesis, that exposure to ambient levels of particulate air pollution promotes impairments in children's pulmonary function. In all eight age-, sex-, and race-specific groups, children in the more heavily polluted city,

**Table 8**
**Margin by Which Mean $FEV_{0.75}$ in Charlotte Exceeded Mean $FEV_{0.75}$ in Birmingham in the 1971–1972 School Year, Distributed by Age, Sex, and Race**

| Age Group | Percent Difference Between Charlotte and Birmingham $FEV_{0.75}$ | | | |
|---|---|---|---|---|
| | Blacks | | Whites | |
| | Males | Females | Males | Females |
| 5–8 years | 3% (34cc)* | 5% (56cc) | 5% (69cc) | 6% (67cc) |
| 9–13 years | 1% (23cc) | 6% (99cc) | 4% (63cc) | 3% (53cc) |

*Absolute intercity differences, in cc, are in parentheses.

Birmingham, had lower mean $FEV_{0.75}$ values than those in the cleaner city, Charlotte. It is reasonable to ascribe pollutant effects on $FEV_{0.75}$ to particulates rather than other pollutants, since measured levels of $SO_2$ were low, and since there is no reason to suspect that continued high levels of any other gaseous pollutant would ever have been present in either city.

Mean $FEV_{0.75}$ values in Charlotte tended to stay fairly constant or to rise slightly with the passage of the seasons. In Birmingham mean $FEV_{0.75}$ values tended to fall slightly as the school year progressed. These results could have been partially explained by a greater growth trend in Charlotte children than in Birmingham children during the 1971–1972 school year. In fact, however, during the school year the average height of children in Birmingham increased 59% more than that in Charlotte.

Intercity differences in mean $FEV_{0.75}$ were smallest in fall, greater in winter, and greatest in spring. Intercity differences in TSP, of all pollutants measured, most dramatically paralleled this pattern. Generally similar seasonal patterns were also observed for suspended nitrates, RSP, and, to a lesser extent, for suspended sulfates. This study did not allow us formally to differentiate the relative short-term effect on $FEV_{0.75}$ of each of these four pollutants, since testing was performed only three times. It does seem unlikely, however, that suspended nitrates exerted a strong effect on $FEV_{0.75}$, since even in Birmingham, estimated annual average concentrations of the pollutant had never exceeded the low level of $3 \mu g/m^3$ during the lives of children under study.

Because intercity differences in $FEV_{0.75}$ were somewhat more marked in younger children than in older ones, it is tempting to hypothesize that these differences were at least partly attributable to differences in exposure to RSP from the earliest years of life. Of the pollutants measured, only RSP had been considerably more concentrated in Birmingham than in Charlotte for the earliest years of the younger, but not the older children's lives.

Three further considerations prohibit the development of precise dose-response estimates from these results. First, though Charlotte has consistently had lower particulate levels than Birmingham, Charlotte cannot strictly be called a control city. Estimates of annual average TSP levels in Charlotte for at least the last 15 years have exceeded the substantial figure of 70 $\mu g/m^3$, which is only $5 \mu g/m^3$ below the national annual particulate standard of 75 $\mu g/m^3$. Thus it may be that the $FEV_{0.75}$ values obtained in Charlotte themselves reflect some impairment due to particulate pollution exposure. Second, we cannot rule out the possibility of slight differences in the performance of testing personnel in the two cities, even though these personnel were trained and observed by the same EPA supervisors. Third, black and white parents in Charlotte had generally attained a somewhat higher educational level than their counterparts in Birmingham. (As mentioned, intercity differences in educational attainment were considerably greater for whites than for blacks.) In Charlotte, there was a slight tendency for children's $FEV_{0.75}$ values to vary inversely with parents' educational attainment. In Birmingham, to a small degree, the opposite tendency was observed. These opposing tendencies would have operated to drive mean $FEV_{0.75}$ values downward in each city relative to values from the other city. Thus the magnitude of intercity differences in $FEV_{0.75}$ was probably affected very little by any effect of parents' education or children's $FEV_{0.75}$. Still, we cannot eliminate the possibility that such an effect may have exerted a small influence on the observed results.

We doubt whether intercity differences in outdoor temperature biased the results of this study, since temperatures in both cities were very similar in all testing seasons. Any effect that temperature may have exerted on our results would probably have dampened observed intercity differences in $FEV_{0.75}$[12] since temperatures in Birmingham were slightly warmer than in Charlotte.

## SUMMARY

A survey of the ventilatory function of 7,997 black and white elementary school children was performed in Charlotte, North Carolina, and Birmingham, Alabama, during the 1971–1972 school year. These cities had been selected for study because they exhibited a gradient of exposure to suspended particulates and had low levels of other pollutants. The ventilatory function test employed was the three-quarter second forced expiratory volume ($FEV_{0.75}$).

In all eight age-, sex-, and race-specific subgroups, mean age- and height-adjusted $FEV_{0.75}$ readings were consistently lowest in the more polluted city, Birmingham. This finding strongly indicated that exposure to particulate pollution had exerted a deleterious effect on the $FEV_{0.75}$ of children in Birmingham. Our results were consistent with the alternate hypothesis that exposure to respirable suspended particulates (RSP), including suspended sulfates, from the beginning of life onward promotes impairments in $FEV_{0.75}$ in later childhood. Serial ventilatory function studies that will test this hypothesis are now under way.

Intercity differences in mean $FEV_{0.75}$ were smallest in fall, greater in winter, and greatest in spring. Intercity differences in total suspended particulates (TSP), RSP, suspended sulfates, and suspended nitrates paralleled this pattern. Because ventilatory function testing was performed on only three occasions, it was not possible formally to test the hypothesis that the $FEV_{0.75}$ is sensitive to seasonal changes in particulate concentrations.

Sulfur dioxide and suspended nitrates were present in low concentrations in both cities, both during the year of study and throughout the lives of the children under study. Thus it was unlikely that either of these pollutants had exerted important deleterious effects on $FEV_{0.75}$ in either city. Beyond this point, it was not possible to determine which specific particulate fraction or fractions had exerted the strongest effects on $FEV_{0.75}$.

## REFERENCES


1. Toyama T: Air Pollution and Its Health Effects in Japan. *Arch Environ Health* 8:53–173, 1964.

2. Watanabe H et al: Effects of Air Pollution on Health. Report No. 1: Peak Flow Rate and Vital Capacity of Primary School Children. *Reports of the Osaka City Institute of Hygiene*. 26:32–37, 1964.

3. Anderson DO and Kinnis C: An Epidemiologic Assessment of a Pediatric Peak Flowmeter. *Amer Rev Resp Dis* 95:73–80, 1967.

4. Holland WW et al: Factors Influencing the Onset of Chronic Respiratory Disease. *Brit Med J* 2:205–208, 1969.

5. Shy CM et al: Ventilatory Function in School Children: 1970–71 Testing in New York Communities. In "Health Consequences of Sulfur Oxides: A Report From CHESS." U.S. Government Document EPA–650/1–74–004. U.S. Government Printing Office, May 1974.

6. Shy CM et al: Ventilatory Function in School Children: 1967–68 Testing in Cincinnati Neighborhoods. In "Health Consequences of Sulfur Oxides: A Report From CHESS." U.S. Government Document EPA 650/1–74–004. U.S. Government Printing Office, May 1974.

7. Hinton DO et al: Human Exposure to Fine Particulates in the Southeastern CHESS Area. EPA Intramural Technical Report, February 21, 1974.

8. Chapman RS et al: CHESS Pulmonary Function Testing Methods: Optimizing the Number of Readings in the Field. EPA Intramural Technical Report, December 23, 1971.

9. Oscherwitz M et al: Differences in Pulmonary Functions in Various Racial Groups. *Am J Epidemiology* 96:319–327, November 1972.

10. Higgins MW and Keller JB: Seven Measures of Ventilatory Lung Function. *Am Rev Resp Dis* 108:258–272, 1973.

11. Dunnill MS: Postnatal Growth of the Lung. *Thorax* 17:329, 1962.

12. McKerrow CB and Rossiter CE: An Annual Cycle in the Ventilatory Capacity of Men with Pneumoconiosis and of Normal Subjects. *Thorax* 23:340–349, July 1968.

# 23 Pulmonary Function Test in Children Exposed to Air Pollution

Jun Kagawa, M.D.
Toshio Toyama, M.D.
Masahiro Nakaza

## INTRODUCTION

Photochemical air pollution consists of mixtures of gaseous and particulate products (aerosols) resulting from atmospheric photochemical reactions of gases evolved primarily from the combustion of organic fuels. It is characterized by the presence of high atmospheric oxidant levels. Hammer et al.[1] reported that eye discomfort, cough, and chest discomfort all increased with maximum hourly oxidant levels. Wayne et al.[2] reported decreased performance of cross-country runners associated with increasing ambient photochemical oxidant concentrations in Los Angeles. The major pollutants that cause these adverse effects are oxidants, nitrogen oxides, and photochemically reactive organic substances. These compounds may irritate mucous membranes.

In this study we tried to examine how the respiratory airway might be irritated as the concentrations of environmental pollutants changed. More specifically, we sought to find out if pulmonary function would be significantly influenced by a group of specific environmental factors. The study period was from June 1972 to December 1973. The results from June 1972 to December 1972 have been reported in a previous study.[3] These results were as follows: The maximum expiratory flow rate ($\dot{V}$ max) at 50% and 25% forced vital capacity (FVC) among the various respiratory function tests indicated high correlation with the greatest number of environmental factors. Among environmental factors, temperature highly affected various pulmonary function tests. $O_3$ was significantly associated with airway resistance (Raw) or specific airway conductance (Gaw/Vtg), NO or $No_2$ with $\dot{V}$ max, and temperature with Raw, Gaw/Vtg and $\dot{V}$ max. There were some reactors to the environmental factors.

## STUDY METHODS

The study methods will be stated briefly as they were described in details in the previous study.

### Selection of a Place for Investigation and Study Subject

Arai elementary school in the residential area in Nakano-ku in Tokyo was chosen for the study because it was likely to provide the most cooperation in order to carry out the study smoothly for 18 months. For the purpose of tracing the same set of individuals, 20 fourth graders (11 years old) were selected and followed, and consisted of 10 males and 10 females who were considered healthy on the basis of present physical status.

### Environmental Factors

Air pollutant monitoring station set up in the school provided the concentrations of 5 types of air-pollutants and the degree of temperature (Table 1).

**Table 1**
**Measurement of Environmental Variables**

| Environmental Variables | Units | Measurement Methods | Measurement Interval |
|---|---|---|---|
| 1. Ozone | pphm | Ethylene-chemiluminescence method | Continuous & 1-hour average |
| 2. NO | pphm | Colorimeter with Saltzman method | 1-hour average |
| 3. $NO_2$ | pphm | Colorimeter with Saltzman method | 1-hour average |
| 4. $SO_2$ | pphm | Conductimetric method | 1-hour average |
| 5. Suspended particulate matter (SPM) | $mg/m^3$ | Light scattering method | 1-hour average |
| 6. Temperature | °C | Platinum resister | Continuous |

## Pulmonary Function Tests

With the object of detecting the variation in the lower airway, $\dot{V}$ max determinations at 50% and 25% FVC were obtained from a flow-volume curve. In order to study the variation in the upper airway, airway resistance (Raw) near functional residual capacity (FRC) was first obtained by the body plethysmograph,[4] then specific airway conductance (Gaw/Vtg, where Vtg was the lung volume at which Raw was measured) was calculated to allow for possible changes in lung volume during the study.

A test consisted of three stages in order to allow for possible effects in volume history: First, 6 readings of Raw and 3 measurements of VTg were made in order to obtain FRC. As a next step, a flow-volume curve was recorded 3 times on X-Y recorder. Six measurements of Raw and Gaw/Vtg were averaged. For FVC, $\dot{V}$ max at 50% or 25% FVC, each of the maximum values was used as a representative value.

A pulmonary function test was performed on each of the 20 subjects on every Wednesday from 1 PM to 3 PM as a general rule, and especially on every Wednesday and Friday from May to July.

## Data Analysis

Since the previous study showed that temperature highly affected various pulmonary function tests, the results of the present study were analyzed by dividing one year into two seasons in order to minimize the effect of seasonal temperature variation on pulmonary function. Two study periods were chosen: one period from November 1972 to March 1973 as a low temperature season, the other from April 1973 to October 1973 as a high temperature season.

For each of the environmental factors, the mean of the hourly means was calculated for each of the four phases; during the measurement of pulmonary function (although a pulmonary function test was performed from 1 PM to 3 PM as a general rule, the phase during the measurement of pulmonary function in this analysis was an average of two phases, the hourly average from 12 noon to 1 PM and the hourly average from 1 PM to 2 PM, since the first subject has been affected by the environmental factors before 1 PM, not after 1 PM) an hour prior to the measurement (12 noon to 1 PM), two-hours prior to the measurement (11 AM to 12 noon), and three-hours prior to the measurement (10 AM to 11 AM). Finally, the correlation coefficients between pulmonary function measurement and each of the environmental factors were calculated for each of the above four phases.

The number of study days for pulmonary function measurement was 16 days during the low temperature season, and 30 days during the high temperature season. The data were analyzed for 19 subjects (10 males and 9 females) because one person had changed schools after the study was initiated.

## RESULTS

### Variations of Environmental Factors

The previous study showed that among environmental factors the ones that were highly associated with various pulmonary function tests were temperature, NO, $O_3$, hydrocarbon, $SO_2$, relative humidity, SPM (suspended particulate matter), and oxidant in that order. Figure 1 shows the variation in hourly average of main factors among these environmental factors during the measurement of pulmonary function. Table 2 shows the correlation coefficients between these environmental variables during the measurement of pulmonary function. NOx is the sum total of NO and

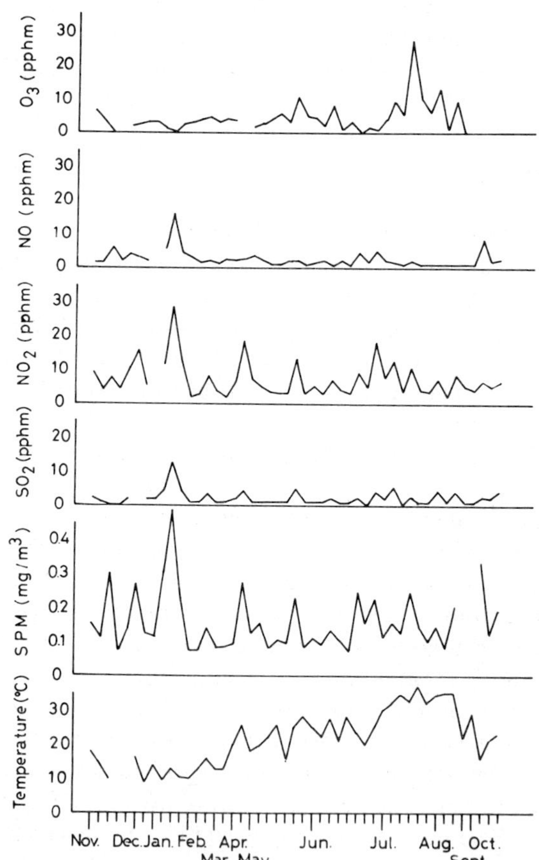


**Figure 1.** Variation of an hourly average of the environmental factors during the measurement of pulmonary function.

**Table 2**
**Correlation Coefficients between Environmental Variables during the Measurements of Pulmonary Function**

| Low-Temperature Season (upper right) / High-Temperature Season (lower left) | $O_3$ | NO | $NO_2$ | NOx | $SO_2$ | SPM | Temperature |
|---|---|---|---|---|---|---|---|
| $O_3$ | | −0.713† | −0.516 | −0.606* | −0.405 | −0.695† | 0.594* |
| NO | −0.259 | | 0.858† | 0.938† | 0.868† | 0.882† | −0.192 |
| $NO_2$ | 0.290 | 0.370* | | 0.982† | 0.942† | 0.907† | 0.148 |
| NOx | 0.183 | 0.626† | 0.956† | | 0.934† | 0.929† | 0.029 |
| $SO_2$ | 0.421* | 0.213 | 0.784† | 0.726† | | 0.821† | 0.046 |
| SPM | 0.428* | 0.690† | 0.687† | 0.795† | 0.572† | | −0.088 |
| Temperature | 0.529† | −0.500 | 0.051 | −0.114 | 0.189 | −0.118 | |

Level of Significance:
*$P<0.05$
†$P<0.01$.

$NO_2$. In both seasons, significant positive correlations were observed between temperature and $O_3$, SPM and NO, SPM and $NO_2$, SPM and NOx, SPM and $SO_2$, $SO_2$ and $NO_2$, $SO_2$ and NOx, NO and NOx, $NO_2$ and NOx. Significant negative correlations were observed between SPM and $O_3$, and between NOx and $O_3$, only during the high temperature season.

## Correlation between Pulmonary Function Measurement and Environmental Factor

Selected measurements of respiratory function, with the arithmetic means and standard deviations, and the number of study days are listed in Table 3. Table 4 shows the percentage of subjects who showed positive or negative correlation coefficients from calculation of simple correlation coefficients between pulmonary function measurements and environmental factors, and the numbers of subjects who showed a significant positive or negative correlation at the $P<0.05$ level. (In this study, significance means significant at the $P<0.05$ level.) Table 5 shows subject numbers and their correlation co-efficients between environmental factors and pulmonary function measurements that could be explained as a physiological response, when $O_3$, NO, $NO_2$, NOx, $SO_2$ and SPM were respiratory irritants.

## COMMENT

In the previous study, the environmental factor that highly affected various pulmonary function tests was temperature. Therefore, in order to

**Table 3**
**Values for Selected Pulmonary Function Measurements with Arithmetic Mean and Standard Deviation**

| Subject No. | Sex | Standing Height (cm) April 1973 | Study Days | | Raw ($cmH_2O/L/sec$) | Gaw/Vtg ($1/cmH_2O \cdot sec$) | $\dot{V}$ max at 50% FVC (L/min) | $\dot{V}$ max at 25% FVC (L/min) | FVC (L) |
|---|---|---|---|---|---|---|---|---|---|
| | | | Low-Temp.* | High-Temp.† | | | | | |
| 1 | M | 145.2 | 15 | 25 | 2.01±0.24 | 0.237±0.024 | 140.8±17.4 | 91.1±10.3 | 2.2±0.1 |
| 2 | M | 133.8 | 15 | 27 | 2.17±0.25 | 0.278±0.027 | 178.6±22.3 | 88.7±13.0 | 1.8±0.0 |
| 3 | M | 146.5 | 16 | 26 | 1.96±0.26 | 0.246±0.030 | 219.4±29.8 | 104.7±19.3 | 2.2±0.1 |
| 4 | M | 148.4 | 16 | 25 | 2.83±0.39 | 0.187±0.021 | 128.8±17.5 | 61.5±10.7 | 2.1±0.1 |
| 5 | M | 130.4 | 15 | 25 | 2.45±0.29 | 0.261±0.022 | 117.6±13.0 | 52.4±8.7 | 1.4±0.0 |
| 6 | F | 146.2 | 16 | 30 | 1.97±0.22 | 0.284±0.028 | 158.6±12.7 | 78.5±8.4 | 1.9±0.1 |
| 7 | F | 142.3 | 16 | 27 | 2.54±0.26 | 0.237±0.019 | 124.3±12.5 | 78.3±13.2 | 1.7±0.1 |
| 8 | F | 141.8 | 16 | 28 | 2.49±0.31 | 0.268±0.029 | 158.4±26.2 | 72.1±17.0 | 1.6±0.1 |
| 9 | F | 124.8 | 16 | 26 | 2.34±0.32 | 0.294±0.026 | 129.3±13.4 | 70.9±9.6 | 1.3±0.1 |
| 10 | M | 134.8 | 14 | 26 | 2.48±0.29 | 0.237±0.023 | 145.0±14.2 | 67.5±7.6 | 1.9±0.1 |
| 11 | M | 139.5 | 9 | 25 | 2.63±0.39 | 0.221±0.025 | 172.8±14.0 | 75.4±7.4 | 2.1±0.1 |
| 12 | M | 141.6 | 13 | 20 | 2.16±0.21 | 0.263±0.025 | 140.4±13.0 | 65.2±8.0 | 2.0±0.1 |
| 13 | M | 129.8 | 13 | 22 | 2.52±0.38 | 0.260±0.027 | 117.0±12.2 | 49.9±6.4 | 1.5±0.1 |
| 14 | M | 131.4 | 16 | 27 | 2.09±0.23 | 0.274±0.027 | 165.0±15.1 | 89.8±10.8 | 1.6±0.1 |
| 15 | F | 134.4 | 14 | 28 | 2.67±0.42 | 0.260±0.028 | 135.4±16.2 | 53.8±7.7 | 1.5±0.1 |
| 16 | F | 130.5 | 14 | 25 | 2.53±0.37 | 0.275±0.027 | 107.3±17.0 | 48.5±8.8 | 1.5±0.1 |
| 17 | F | 135.2 | 15 | 26 | 2.17±0.37 | 0.269±0.028 | 157.1±21.4 | 93.3±17.4 | 1.5±0.1 |
| 18 | F | 141.5 | 14 | 23 | 2.27±0.34 | 0.282±0.027 | 172.2±25.5 | 95.0±14.5 | 1.5±0.1 |
| 19 | F | 139.9 | 16 | 25 | 2.31±0.22 | 0.241±0.021 | 129.2±14.8 | 58.8±7.9 | 1.8±0.2 |

*Low temperature season: November, 1972–March, 1973.
†High temperature season: April, 1973–October, 1973.

minimize the effect of temperature, the present study was analyzed by dividing one year into two seasons. But, even with the study period divided into two seasons, 68.4% of subjects showed that their Raw increased with increasing temperature as the previous study showed. However, when the data were analyzed for one full year, 89.4% of subjects showed their Raw decreased when temperature increased and when temperature decreased Raw increased, among whom 9 subjects showed significant negative correlation. As Raw will change with Vtg, Gaw/Vtg, which was calculated to allow for possible changes in lung volume, decreased with increasing temperature in 57.8% of subjects, among whom 3 subjects showed significant negative correlation and the other 2 subjects significant positive correlation. Therefore, when the effect of temperature on Raw was evaluated without considering the change of Vtg, Raw decreased when temperature increased, and when temperature decreased Raw increased as other researchers have noted.[5,6] Then how should this opposite result to the previous study be explained? In the previous study Raw increased with increasing temperature in all subjects. When the data were analyzed for the same period as the previous study from June 1973 to December 1973, Raw increased with increasing temperature in 68.4% of subjects and Gaw/Vtg decreased in 84.2% of subjects. Therefore it seemed that the result might be influenced by the selection of the study period.

$\dot{V}$ max increased with increasing temperature as the previous study showed. As for $\dot{V}$ max at 50% FVC for one full year's data, all subjects showed the significant positive correlations at the $P<0.01$ level and no remarkable individual differences were observed. Almost the same results were obtained between Raw, Gaw/Vtg or $\dot{V}$ max and temperature at one, two, and three hours prior to the measurement of pulmonary function. Therefore, if Raw is considered as a main reflector of change in the upper airway, and $\dot{V}$ max as a reflector of change in the lower airway, it may be said that the upper and lower airways dilate with increasing temperature, while at low temperature the opposite will be observed.

As for $O_3$, Raw and Gaw/Vtg seem to react well with $O_3$ concentration during the low temperature season, from the significance of correlation coefficients. Raw increased with increasing $O_3$ concentration in 94.4% of subjects, among whom 8 subjects showed significant correlation. During the high temperature season, such a tendency was observed in 63.1% of subjects in Raw and 73.6% of subjects in Gaw/Vtg, among whom one subject showed significant correlation with Gaw/Vtg. Since a significant correlation between $O_3$ and Raw during the low temperature season might have been observed because of association with temperature, partial correlation coefficients were calculated in order to eliminate the effect of temperature. The results showed that 5 subjects showed the significant partial correlation between $O_3$ and Raw. Therefore, regardless of temperature, $O_3$ alone or some unknown factor coexisting with $O_3$ may play an important

**Table 4**
**Percentage of Subjects Who Showed Positive or Negative Correlation Coefficient from Calculation of Simple Correlation Coefficients between Environmental and Respiratory Function Measurements**

| Pulmonary Function Tests / Season / Environmental Variables | Raw | | Gaw/Vtg | | $\dot{V}$ max (50% FVC) | | $\dot{V}$ max (25% FVC) | |
|---|---|---|---|---|---|---|---|---|
| | Low-Temp.* | High-Temp.† | Low-Temp. | High-Temp. | Low-Temp. | High-Temp. | Low-Temp. | High-Temp. |
| Temperature | ↗ 68.4%<br>+ − <br>2 0 | ↗ 52.6%<br>+ −<br>0 1 | ↘ 68.4%<br>+ −<br>0 1 | ↘ 68.4%<br>+ −<br>0 2 | ↗ 100%<br>+ −<br>2 0 | ↗ 94.7%<br>+ −<br>7 0 | ↗ 78.9%<br>+ −<br>3 0 | ↗ 94.7%<br>+ −<br>6 0 |
| $O_3$ | ↗ 94.7%<br>+ −<br>8 0 | ↗ 63.1%<br>+ −<br>0 0 | ↘ 100%<br>+ −<br>0 7 | ↘ 73.6%<br>+ −<br>1 1 | ↗ 78.9%<br>+ −<br>2 0 | ↗ 57.8%<br>+ −<br>0 1 | ↗ 57.8%<br>+ −<br>4 0 | ↗ 52.6%<br>+ −<br>2 1 |
| NO | ↘ 73.6%<br>+ −<br>0 1 | ↘ 63.1%<br>+ −<br>1 0 | ↗ 89.4%<br>+ −<br>3 0 | ↗ 57.8%<br>+ −<br>0 0 | ↘ 52.6%<br>+ −<br>0 2 | ↘ 78.9%<br>+ −<br>0 7 | ↘ 57.8%<br>+ −<br>0 2 | ↘ 73.6%<br>+ −<br>0 2 |

**Table 4** *(continued)*

**Percentage of Subjects Who Showed Positive or Negative Correlation Coefficient from Calculation of Simple Correlation Coefficients between Environmental and Respiratory Function Measurments**

| Environmental Variables \ Season \ Pulmonary Function Tests | Raw | | Gaw/Vtg | | $\dot{V}$ max (50% FVC) | | $\dot{V}$ max (25% FVC) | |
|---|---|---|---|---|---|---|---|---|
| | Low-Temp.* | High-Temp.† | Low-Temp. | High-Temp. | Low-Temp. | High-Temp. | Low-Temp. | High-Temp. |
| $NO_2$ | ⟍ 78.9% | ⟋ 78.9% | ⟋ 84.2% | ⟍ 84.2% | ⟍ 68.4% | ⟍ 78.9% | ⟍ 57.8% | ⟍ 68.4% |
| | + − | + − | + − | + − | + − | + − | + − | + − |
| | 0 1 | 6 0 | 0 0 | 0 8 | 1 2 | 0 1 | 0 1 | 0 2 |
| NOx | ⟍ 78.9% | ⟋ 84.2% | ⟋ 89.4% | ⟍ 78.9% | ⟍ 57.8% | ⟍ 89.4% | ⟍ 57.8% | ⟍ 73.6% |
| | + − | + − | + − | + − | + − | + − | + − | + − |
| | 0 2 | 3 0 | 0 0 | 0 3 | 0 2 | 0 4 | 0 2 | 0 2 |
| $SO_2$ | ⟍ 68.4% | ⟋ 78.9% | ⟋ 78.9% | ⟍ 78.9% | ⟍ 63.1% | ⟍ 78.9% | ⟍ 57.8% | ⟍ 68.4% |
| | + − | + − | + − | + − | + − | + − | + − | + − |
| | 0 0 | 1 0 | 0 0 | 0 0 | 0 2 | 0 1 | 0 1 | 0 1 |
| S.P.M. | ⟍ 94.7% | ⟋ 52.6% | ⟋ 94.7% | ⟍ 68.4% | ⟍ 57.8% | ⟍ 68.4% | ⟍ 52.6% | ⟍ 73.6% |
| | + − | + − | + − | + − | + − | + − | + − | + − |
| | 0 4 | 2 0 | 4 0 | 0 2 | 0 1 | 0 3 | 0 2 | 0 1 |

*Low temperature season, November, 1972–March, 1973
†High temperature season, April, 1973–October, 1973
‡Numbers of subject who showed a significant positive or negative correlation coefficient at $P < 0.05$ among 19 subjects.

**Table 5**
**Subject Numbers and Their Correlation Coefficients between Environmental Variables and Pulmonary Function Tests**

| Environmental Variables (Season / Pulmonary Function) | Raw | | Gaw/Vtg | | $\dot{V}$ max (50% FVC) | | $\dot{V}$ max (25% FVC) | |
|---|---|---|---|---|---|---|---|---|
| | Low-Temp.* | High-Temp.† | Low-Temp. | High-Temp. | Low-Temp. | High-Temp. | Low-Temp. | High-Temp. |
| Temperature | S14(0.64)‡<br>S9 (0.49) | S5 (−0.43) | S14(−0.50) | S15(−0.43)<br>S19(−0.42) | S13(0.69)<br>S4 (0.53) | S1 (0.71)<br>S4 (0.65)<br>S3 (0.56)<br>S12(0.55)<br>S18(0.46)<br>S14(0.43)<br>S2 (0.40) | S13(0.81)<br>S4 (0.61)<br>S5 (0.56) | S3 (0.63)<br>S15(0.52)<br>S1 (0.51)<br>S11(0.49)<br>S12(0.49)<br>S16(0.45) |
| $O_3$ | S2 (0.64)<br>S15(0.62)<br>S7 (0.60)<br>S6 (0.59)<br>S18(0.58)<br>S17(0.57)<br>S12(0.56)<br>S14(0.50) | | S15(−0.72)<br>S6 (−0.69)<br>S12(−0.65)<br>S19(−0.65)<br>S18(−0.61)<br>S7 (−0.58)<br>S11(−0.56) | S15(−0.52) | | S19(−0.72) | | S19(−0.41) |
| NO | | S5 (0.50) | | | S3 (−0.81)<br>S9 (−0.63) | S1 (−0.60)<br>S18(−0.51)<br>S11(−0.50)<br>S4 (−0.49)<br>S2 (−0.44)<br>S16(−0.43)<br>S17(−0.41) | S3 (−0.55)<br>S14(−0.53) | S11(−0.52)<br>S15(−0.42) |

**Table 5** *(continued)*
**Subject Numbers and Their Correlation Coefficients between Environmental Variables and Pulmonary Function Tests**

| Environmental Variables \ Season \ Pulmonary Function | Raw | | Gaw/Vtg | | $\dot{V}$ max (50% FVC) | | $\dot{V}$ max (25% FVC) | |
|---|---|---|---|---|---|---|---|---|
| | Low-Temp.* | High-Temp.† | Low-Temp. | High-Temp. | Low-Temp. | High-Temp. | Low-Temp. | High-Temp. |
| $NO_2$ | | S8 (0.65)<br>S11(0.50)<br>S16(0.45)<br>S1 (0.44)<br>S14(0.40)<br>S15(0.38) | | S1 (−0.52)<br>S2 (−0.49)<br>S15(−0.49)<br>S8 (−0.46)<br>S9 (−0.46)<br>S6 (−0.41)<br>S4 (−0.40)<br>S11(−0.39) | S3 (−0.75)<br>S9 (−0.55) | S16(−0.45) | S3 (−0.69) | S4 (−0.47)<br>S16(−0.42) |
| NOx | | S8 (0.48)<br>S16(0.42)<br>S11(0.41) | | S2 (−0.51)<br>S16(−0.39)<br>S1 (−0.38) | S3 (−0.79)<br>S9 (−0.59) | S16(−0.52)<br>S18(−0.51)<br>S2 (−0.44)<br>S4 (−0.41) | S3 (−0.66)<br>S9 (−0.48) | S16(−0.44)<br>S4 (−0.42) |
| $SO_2$ | | S8 (0.55) | | | S3 (−0.72)<br>S9 (−0.55) | S16(−0.40) | S3 (−0.55) | S4 (−0.45) |
| S.P.M. | | S9 (0.43)<br>S2 (0.38) | | S2 (−0.51)<br>S16(−0.39) | S3 (−0.67) | S16(−0.54)<br>S18(−0.50)<br>S2 (−0.48) | S14(−0.61)<br>S3 (−0.51) | S11(−0.38) |

*Low-temperature season, November, 1972–March, 1973. Number of study days for pulmonary function tests: 16.
†High-temperature season, April, 1973–October, 1973. Number of study days for pulmonary function tests: 30.
‡Subjects number. Number in parentheses is a correlation coefficient.

role in affecting pulmonary function among some individuals. However, if $O_3$ alone could affect Raw, it may be expected that the number of subjects who show the significant correlation coefficients may increase during the high temperature season in which $O_3$ concentration will become higher. However, this was not the case. During the high temperature season, only one subject showed the significant correlation with Gaw/Vtg and his partial correlation coefficient between $O_3$ and Gaw/Vtg was significant when the effect of temperature variable was removed. The significant correlation between $O_3$ and Raw during the low temperature season may have been observed on the ground of $O_3$ and/or some unknown factor coexisting with $O_3$. Negative correlation between $O_3$ and Gaw/Vtg was exhibited in 84.2% of the subjects, among whom 4 subjects showed the significant correlation when the correlation coefficients were calculated for one full year's data.

As for $\dot{V}$ max, it may not result from the effect of $O_3$ that $\dot{V}$ max increased with increasing $O_3$ concentration in 52.6% to 78.9% of subjects. During the low temperature season, 4 subjects showed significant positive correlation between $\dot{V}$ max at 25% FVC and $O_3$, among whom none showed any significant partial correlation coefficient when the effect of temperature variable was taken away. But the partial correlation coefficients were still positive. This phenomena could not be explained from the view point of ozone toxicology, and the real cause is not known, but might be temperature. During the high temperature season, only 1 subject showed a significant negative correlation. However, the number of subjects who showed significant negative partial correlation coefficients between $\dot{V}$ max at 50% FVC and $O_3$ when the effect of temperature variable was taken away, increased by 4, although only 1 subject showed significant simple correlation between $\dot{V}$ max at 50% FVC and $O_3$. As was noted above, it might be said that $O_3$ during the low temperature season would affect the upper airway by some unknown factors, and $O_3$ during the high temperature season would affect the upper and lower airways in some subjects.

As for NO, Raw and $\dot{V}$ max decreased when NO concentration increased, as the previous study showed. Especially in the high temperature season, 7 subjects showed significant negative correlation between $\dot{V}$ max at 50% FVC and NO, among whom 2 subjects showed significant partial correlation coefficients regardless of the effect of temperature. But it is difficult to discuss the meaning of the correlation since the effect of NO on the pulmonary mechanism is not known.

As for $NO_2$, during the low temperature season Raw decreased with increasing $NO_2$ concentration. On the other hand in the high temperature season, Raw increased with increasing $NO_2$ concentration. Such a tendency was observed for the $NO_2$ concentration during the measurements of pulmonary function tests, and one and two hours prior to the measurement.

However, for the concentration of $NO_2$ three hours prior to the measurement of pulmonary function tests, the number of subjects who showed negative correlation with Raw increased by more than half. In the high temperature season 84.2% of subjects showed negative correlation with Gaw/Vtg for the concentration during the measurement of pulmonary function tests, among whom 8 subjects showed the significant correlation, and these same 8 subjects showed negative partial correlation coefficients regardless of the effect of temperature.

$\dot{V}$ max showed a decreasing tendency with increasing $NO_2$ concentration. Significant negative partial correlation coefficients when the effect of temperature was taken away were shown by 2 subjects during the low temperature season and 4 subjects during the high temperature season. When the correlation coefficients were calculated for one full year's data, 94.7% of subjects showed negative correlation with $\dot{V}$ max at 50% FVC, among whom 7 subjects showed significant correlation. A similar tendency was noted as for NOx.

From the above results, it may be said that $NO_2$ during the low temperature season will affect the lower airway and $NO_2$ during the high temperature season will affect the upper and lower airways. It can not be explained why there are many subjects who showed negative correlation between $NO_2$ and Raw during the low temperature season. From the relationship between $NO_2$ and $\dot{V}$ max at 50% FVC in subject No. 16, who showed the highest correlation, it seems that $NO_2$ started having a significant effect above a concentration of 4 pphm (Figure 2).

As for $SO_2$, 78.9% of subjects showed positive correlation with Gaw/Vtg during the high temperature season. Over 63.1% of subjects showed negative correlation with $\dot{V}$ max at 50% FVC. From the above results, it may be said that $SO_2$ during the low temperature season will affect the lower airway and $SO_2$ during the high temperature season the upper and lower airways. High correlation coefficients and an increased number of subjects who showed negative correlation with $\dot{V}$ max, were observed with the $SO_2$ concentration at 2 and 3 hours prior to the pulmonary function measurement during the low temperature season. However, in the high temperature season, such a tendency was observed between Raw or $\dot{V}$ max and the $SO_2$ concentration only during the pulmonary function measurement.

As for SPM, significant positive correlation with Raw was seen only in the high temperature season. The negative correlation between SPM and Raw during the low temperature season may have resulted from other factors than SPM. More than half of the subjects showed a negative correlation with $\dot{V}$ max, among whom were 3 subjects who showed significant negative correlation with $\dot{V}$ max at 50% FVC during the high temperature season. From the above results, it may be said that SPM during the low temperature season will affect the lower airway and SPM during the high temperature season the upper and lower airways.

The results presented here show that the relationships between environmental factors and pulmonary function tests that could be explained as a physiological response in more than 72.5% of subjects regardless of the season, were the one between temperature and $\dot{V}$ max at 50 or 25% FVC and the one between $O_3$ and Gaw/Vtg. A physiological response in more than 52.6% of subjects regardless of the season was seen in the relationships between $NO_2$ and $\dot{V}$ max at 50 or 25% FVC, between NOx and $\dot{V}$ max at 50 or 25% FVC, between $SO_2$ and $\dot{V}$ max at 50 or 25% FVC, and between SPM and $\dot{V}$ max at 50 or 25% FVC. Therefore, it may be said that $O_3$ will affect the upper airway, and $NO_2$, NOx, $SO_2$ and SPM will affect the lower airway. During the high temperature season, $NO_2$, NOx, $SO_2$, and SPM may affect the upper and lower airways, and $O_3$ may affect the lower airway in some subjects.

The above results were obtained by evaluating only the simple factors among environmental variables. In order to see the effect of complex pollution involving more than one simple factor, correlation coefficients were calculated between pulmonary function measurements and complex environmental factors: temperature x $O_3$ (multiplying the degree of temperature by the concentration of $O_3$), temperature x $SO_2$, temperature x NOx, SPM x $O_3$, SPM x $SO_2$, and SPM x NOx. The correlation coefficients did

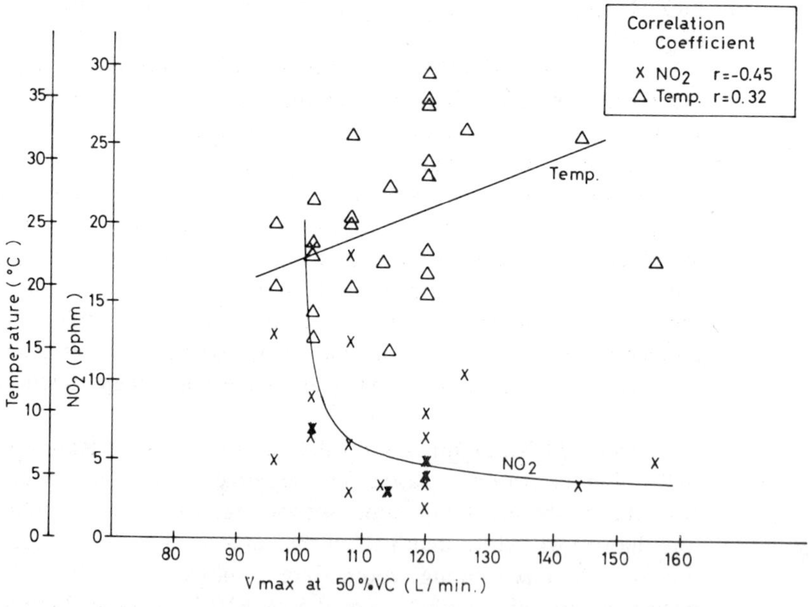


**Figure 2.** Relationship between $\dot{V}$ max and environmental variables in subject 16 during the high-temperature season (study days: 25).

not show a tendency such as this when the effect of the complex pollution was strengthened. Subjects No. 2,3,4,11,15,16, and 18 showed significant correlation with more than three factors among the environmental factors of temperature, $O_3$, $SO_2$, oxides of nitrogen and SPM. Subjects No. 1,2,4,9,11,14,15,16, and 19 showed significant correlation between environmental factors and both Raw or Gaw/Vtg and $\dot{V}$ max during any of the seasons — namely, their upper and lower airways were affected. Subjects No. 2,4,11,15, and 16 showed significant correlation with more than three factors among the environmental factors and their upper and lower airways were affected during any of the seasons. They were considered to be reactors to the environmental factors.

Although it may be possible to analyze the raw data further in detail, it may be difficult to evaluate the analyzed data. This situation exists because it may be difficult to evaluate the real factor that affects pulmonary function, since the subject has been already affected by NO, $NO_2$ or reactive organic substances before a high $O_3$ concentration is recorded, even if his pulmonary function shows significant correlation with $O_3$. The effect of photochemical air pollution on pulmonary function should be evaluated from multiple viewpoints. Controlled exposure experiments are needed to confirm and quantify the relationships found in this study.

## CONCLUSIONS

1. Since the previous study showed that temperature highly affected various pulmonary function tests among environmental factors, the study period was divided into two seasons in order to eliminate the effect of temperature; the period from November 1972 to March 1973 was chosen as the low temperature season and the period from April 1973 to October 1973 as the high temperature season.

In both seasons Raw increased with increasing $O_3$ concentration and temperature, and $\dot{V}$ max decreased with increasing oxides of nitrogen and decreasing temperature in more than half subjects.

However, the analysis of one full year's data showed that Raw increased with increasing temperature in all subjects. Therefore, the upper and lower airways seem to dilate with increasing temperature while for low temperatures the opposite was observed.

2. $O_3$ may affect the upper airway, and $NO_2$, NOx, $SO_2$, and SPM affect the lower airway. During the high temperature season, $NO_2$, NOx, $SO_2$ and SPM may affect the upper and lower airways, and $O_3$ may affect the lower airway in some individuals.

3. Five subjects among 19 showed significant correlation of their pulmonary function tests with more than three environmental factors, and both their upper and lower airways seemed to be affected during any of the

seasons. Although they did not show any respiratory symptoms, they were considered as reactors to the environmental factors. They might be expected to show the stronger clinical symptoms if their respiratory airway were infected. We have not thought it advisable to expose subjects known to have asthma or suffering from chronic bronchitis and emphysema to the maximum concentrations of $O_3$, $NO_2$, and SPM, which have been observed in this study, although we do not know whether these subjects would be more sensitive or less sensitive than these school children.

## ACKNOWLEDGEMENT

This study was supported by Tokyo Photochemical Air Pollution Research Grant from the Department of Hygiene, Tokyo Metropolis. We are indebted to Dr. M. Nagasaki and Mr. T. Hitomi for their cooperation, and to Miss I. Tominaga and Y. Ushiki for their assistance in preparing this paper.

## REFERENCES

1. Hammer DI, Hasselblad V, Portnoy B, et al: Los Angeles Student Nurse Study. *Arch Environ Health* 28:255–258, 1974.
2. Wayne WS, Wehrle PF, Carroll RE, et al: Oxidant Air Pollution and Athletic Performance. *JAMA* 199:901–904, 1967.
3. Kagawa J, Toyama T: Photochemical Air Pollution and Its Effects on Respiratory Function of Elementary School Children. Accepted by *Arch Environ Health*.
4. DuBois AB, et al: A New Method for Measuring Airway Resistance in Man Using a Body Plethysmograph: Values in Normal Subjects and in Patients With Respiratory Diseases. *J Clin Invest* 35:327–335, 1956.
5. Spodnik MJ Jr, Chushman GD, Kerr DH, et al: Effects of Environment on Respiratory Function: Weekly Studies on Young Male Adults. *Arch Environ Health* 13:243–254, 1966.
6. Wells RE Jr, Walker JE, Hicklea RB: Effects of Cold Air on Respiratory Air Flow Resistance in Patients With Respiratory-tract Disease. *New Engl J Med* 263:268–273, 1960.

# 24 Air Pollution and Childhood Lower Respiratory Disease I. Exposure to Sulfur Oxides and Particulate Matter in New York, 1972

D.I. Hammer, M.D.
F.J. Miller, M.S.
A.G. Stead, M.S.
C.G. Hayes, Ph.D.

## INTRODUCTION

Illnesses of the respiratory tract are the most common cause of morbidity in the United States and account for 50% to 80% of all childhood illnesses.[1–4] Repeated lower respiratory tract illnesses in childhood may well contribute to the development of chronic respiratory disease in adult life.[5] Studies in England, Japan, and Russia have implicated ambient sulfur dioxide and particulate matter as causes of increased respiratory morbidity in children.[6–10] Prospective studies in New York and Chicago found increased acute respiratory disease morbidity associated with exposure to sulfur oxide and particulate air pollution among parents and their children[11,12] Recent retrospective surveys of acute lower respiratory disease in children living in U.S. western smelter communities found excess illness associated with sulfur oxide air pollution.[13,14] Thus, there is epidemiologic evidence from several populations from different countries associating measured childhood respiratory morbidity with exposure to sulfur oxide and particulate air pollution.

Although New York City has historically experienced high levels of air pollutants resulting from fossil fuel combustion, ambient air quality has improved markedly during the past few years.[15] However, in May 1972, when this survey was done, children living in the higher exposure communities had been exposed previously to high levels of sulfur oxide and particulate air pollution. We describe here a restropective survey of acute lower respiratory illnesses among children in four New York communities representing an exposure gradient to sulfur oxide and particulate air pollution. The primary study hypothesis was that reported respiratory illness rates in the three high exposure communities would be higher than in the low exposure community.

## MATERIALS AND METHODS

### Community Selection

Communities that were socioeconomically similar but different in air pollution exposure were selected on the basis of historical air quality data (1944–1970) obtained from the New York City Department of Air Resources (NYC-DAR) and the Suffolk County (New York) Department of Health, in conjunction with census data and current information on house and property values. Communities were ranked on the basis of their exposure to sulfur dioxide and total suspended particulate matter. The rankings were relative in relation to the pertinent U.S. national primary air quality standards. Riverhead, Long Island, was designated the Low exposure community. The Howard Beach section of Queens and the Westchester section of Bronx were ranked as Intermediate exposure communities.[16] Howard Beach is located close to the John F. Kennedy Airport, and its residents may well have been exposed to pollutants such as organic air pollutants and noise that were not monitored during this study and could not be accurately estimated during the previous years. Thus, Sheepshead Bay in Brooklyn, located geographically close to Howard Beach, but farther from the airport, was chosen as a third Intermediate exposure community. Detailed descriptions of the first three communities and the reasons for their selection have been published.[15]

### Assessing Air Pollution Exposure

Community Health and Environmental Surveillance System (CHESS) air monitoring stations were established in Riverhead, Queens, and Bronx during 1970 and were located about 35 to 40 feet above the ground on tops of buildings. At each station, 24-hour integrated samples of sulfur dioxide (modified West-Gaeke method), total suspended particulates (high volume sampler), respirable particulates (cyclone separator, from 1972 on), suspended sulfates, and suspended nitrates (chemical analysis of portion of total suspended particulate filter) were monitored on a daily basis. Dustfall was determined from monthly samples. A monitoring site was not established in Sheepshead Bay before this study was completed. Since we were primarily concerned with the effects of long-term exposure and since Sheepshead Bay is located near Howard Beach, current and historical aerometric data from the latter station were also used for Sheepshead Bay. Estimates of past exposures were derived from historical information supplied by NYC-DAR and the Suffolk County (New York) Department of Health. A complete description of the location of monitoring stations, the methodology for current monitoring, and the estimation of past exposures have been published previously.[15]

## Collection of Health and Demographic Data

In May 1972, elementary school children (grades 1 through 6) in each of the selected study communities were asked to take explanatory letters and questionnaires home to their parents. Socioeconomically similar school districts in each area were selected after consultation with local school officials and school principals. Children in three schools in each area were surveyed. The frequency of acute lower respiratory disease was ascertained by means of a School and Family Health Questionnaire that was completed by the mother or female guardian in each family when possible and returned to the school. Mothers were asked to answer the questions about lower respiratory disease for all children twelve years of age or younger who lived in the household.

The questionnaire inquired about the frequency of treatment by a physician for pneumonia, croup, bronchitis, or any chest infection other than these three during the three years prior to the time of questionnaire completion. Hence, all morbidity rates presented encompass the three-year period from June 1969 through May 1972. Other information ascertained included hospitalizations for lower respiratory illnesses, name of children's current physician, history of asthma diagnosed by a doctor, length of residence in the community, education of head of household as an index of socioeconomic status, parents' smoking habits, race, family census, number of rooms per household, and use of a gas stove for cooking or a gas space heater in the home.

## Data Analysis and Hypothesis Testing

As planned in the protocol, children with a history of asthma and all children with less than three years' residence in the community were excluded from the analysis. Children with asthma are known to have a higher risk of respiratory illnesses than nonasthmatic children, as they did also in this study, and these results will be reported separately. Children with less than three years residence were excluded because they had been exposed to the community air for only a short time and because no knowledge of their previous residence and, hence, of their previous pollutant exposure was available. Moreover, illness itself has been associated with migration and recently migrated families would be least likely to have established patterns of medical care within a community, a situation that could confound the effect of pollution with the effect of migration on reported doctor-diagnosed illnesses within a community. Black and white children were analyzed separately because a previous study had shown differences in reported acute lower respiratory disease morbidity patterns by race.[17] Children under one year of age had to be excluded due to an error in coding instructions that made children under age one indistinguishable from children with missing information on age.

The major test hypothesis was that the frequency of reported physician treatment for ''any lower respiratory illness,'' pneumonia, croup, bronchitis, and ''other'' chest infections, appropriately adjusted would be higher in the Intermediate exposure communities than in the Low exposure community. A combined disease category, ''any lower respiratory disease,'' was constructed to include a child if he had either pneumonia, croup, bronchitis, or any deep chest infections other than the previous three (''other''). For each morbidity condition, two sets of analyses were done. In the first, the dependent variable was the percent of children reporting one or more episodes of each morbidity condition. In the second, the dependent variable was the percent of children reporting two or more episodes of each condition.

Each specific hypothesis was tested statistically in a general linear model for categorical data.[18] This technique utilizes weighted regression on categorical data and allows estimation of each individual factor adjusted for all other factors in the model. Independent variables used in the model were age (three categories: 1 to 4, 5 to 8, 9 to 12), sex, and, as an index of socioeconomic status, education of the head of household (2 categories: less than high school, high school or more). Therefore, there were 48 subpopulations in the model for each morbidity condition, viz. community exposure (4), age (3), sex (2), and education of the head of household (2). With approximately 48 subpopulations, the variance matrix used in the linear categorical model becomes unstable under two conditions: first, if more than a few cells have zero rates, and second, if less than 30–40 observations occur in any subpopulation. For the few times when these constraints were violated, a simpler model was fitted if possible. Descriptive rates were direct-adjusted according to standard methods that used all white nonasthmatic children with three or more years residence duration as the reference population.[19]

During preliminary statistical analyses, uncommonly high rates of pneumonia were observed in Riverhead. These rates were found not to be due to a systematic error. Subsequent study indicated that about 40% of the children in Riverhead were attended by one of two doctors in the same office for their lower respiratory illnesses, whereas all other children in Riverhead were seen by over 50 different physicians. Furthermore, lower respiratory disease rates in Riverhead varied significantly when similar children from the first group were compared to those in the second group. In Queens, three physicians saw enough children (13% to 15%) to be compared to all other physicians, and no differences in lower respiratory disease rates were found. For Bronx and Sheepshead Bay no one physician saw enough children to permit such comparisons. Consequently, an additional independent variable was added to the model for white children, viz. "doctor in Riverhead," and consisted of the contrast of each morbidity rate between children seen by the first two doctors and those children seen by all other doctors in Riverhead. Rates of pneumonia and one or more episodes of "any lower respiratory disease" and hospitalization were significantly higher among the children seen by the two previously mentioned doctors whereas rate of the remaining three morbidity conditions were similar. Because of smaller sample size, black children in Riverhead seen by the two doctors were excluded from all contrasts.

## RESULTS

### Environmental Exposure

Estimated exposure to ambient pollutants for all study communities from 1960 to 1972 are summarized in Table 1. For the reasons previously mentioned, exposure in Sheepshead Bay was assumed to be similar to that of Queens. In Queens and Bronx, estimated annual sulfur dioxide levels were similar, ranging from about 175 to 425 $\mu g/m^3$ from 1960 through 1970 and about 40 to 50 $\mu g/m^3$ during 1971 and 1972. Average annual sulfur dioxide levels in Riverhead were about 25 $\mu g/m^3$ in 1971 and 1972 and were assumed to be quite low during 1960–1970, although measured data are not available. Estimated average annual total suspended particulate levels in Queens and Bronx ranged from about 85 to 195 $\mu g/m^3$ during 1960 to 1970 and from about 60 to 90 $\mu g/m^3$ in 1971 and 1972. Comparable total suspended particulate levels in Riverhead during 1971 and 1972 were about 35 $\mu g/m^3$, and no data were available for the period 1960 to 1970. Respirable suspended particulates measured only in 1972 were only slightly higher in Queens and Bronx than in Riverhead.

Suspended sulfate levels in Queens and Bronx ranged from about 9 to 20 $\mu g/m^3$ during 1960 to 1970 and were about 14 $\mu g/m^3$ during 1971 and 1972. Data from Riverhead indicated levels of about 10 $\mu g/m^3$ during 1971 and 1972. Annual average suspended nitrate levels in all three communities were low during 1971 and 1972, being lowest in Riverhead. When compared with children in Riverhead, children in Queens and Bronx were exposed to higher levels of sulfur dioxide and total suspended particulate matter until about a year and one-half before the survey. Suspended sulfate levels, which were also higher in Queens and Bronx during 1960 to 1970, remained moderately higher than those in Riverhead during 1971 and 1972.

## Characteristics of Family Households

Response rates were good, about 80% in all four communities. Characteristics of study families by community and race are given in Table 2. Except for whites in Riverhead (89%), 95% or more of all families lived in their communities for three or more years. Although the median age of

**Table 1**
**Estimated Average Annual Pollutant Exposure Levels for Three* Study Communities in the New York Metropolitan Area, 1960–1972**

| | | Estimated Average Annual Concentration, $\mu g/m^3$† | | | | National Primary Air Quality Standard, |
|---|---|---|---|---|---|---|
| Pollutant | Community | 1960–67 | 1968–70, | 1971‡ | 1972‡ | Annual Average |
| Sulfur dioxide | Riverhead | Not Available | | 23 | 22 | 80 $\mu g/m^3$ |
| | Queens | ~425 | ~175 | 51 | 50 | (0.03 ppm) |
| | Bronx | ~400 | ~250 | 51 | 38 | |
| Total suspended particulates | Riverhead | Not Available | | 34 | 36(32)§ | 75 $\mu g/m^3$ |
| | Queens | ~185 | ~85 | 63 | 89(79) | (Geometric |
| | Bronx | ~175 | ~110 | 86 | 60(55) | Mean) |
| Respirable suspended particulates | Riverhead | | | | 28 | |
| | Queens | Not Available | | | 34 | None |
| | Bronx | | | | 43 | |
| Suspended sulfates | Riverhead | Not Available | | 10 | 11 | |
| | Queens | ~20 | ~9 | 13 | 14 | None |
| | Bronx | ~18 | ~15 | 14 | 14 | |
| Suspended nitrates | Riverhead | | | 2 | 2 | |
| | Queens | Not Available | | 4 | 3 | None |
| | Bronx | | | 4 | 3 | |

*Although aerometric data were not available, it was assumed that exposure in Sheepshead Bay, the fourth study community, would be most similar to the Queens community of Howard Beach since it is geographically contiguous to it..
†Adapted from Table 5.14, reference 15.
‡Measured data.
§Geometric mean in parentheses.

**Table 2**
**Characteristics of Study Families, by Community and Race**

| Community (Number of Households) | Race | Three or More Years Residence Duration (%) | Median Age of Mothers (years) | Both Parents or Guardians Present (%) | Fathers Completing High School (%) | | Current Cigarette Smokers (%)† | |
|---|---|---|---|---|---|---|---|---|
| | | | | | ≤39 yrs | ≤40 yrs | Mothers | Fathers |
| Riverhead (275) | Black | 95 | 33 | 59 | 35 | 27 | 41 | 49 |
| Q,B,S* Combined (302) | | 96 | 34 | 53 | 79 | 67 | 45 | 49 |
| Riverhead (620) | White | 89 | 34 | 85 | 78 | 46 | 45 | 51 |
| Queens (991) | | 95 | 35 | 93 | 82 | 73 | 49 | 48 |
| Bronx (957) | | 96 | 35 | 87 | 72 | 58 | 46 | 53 |
| Sheepshead Bay (1066) | | 97 | 35 | 88 | 85 | 83 | 50 | 50 |

*Queens, Bronx, and Sheepshead Bay combined due to small sample size.
†Direct adjusted for age and education.

mothers or maternal guardians was similar in all communities, significantly fewer black families had both parents or guardians present. Black families in Riverhead had a lower proportion of heads of household with a high school or better education than any of the others. Parental cigarette smoking habits were comparable by race and community, with females smoking somewhat less than males. Black families tended to have more persons per household and fewer rooms per household when compared to white families (Table 3). Riverhead had a lower proportion of households using either gas stoves or gas space heaters.

## Lower Respiratory Illnesses in Children

As expected, the age distribution of children used in the analyses was comparable for both rates in all communities (Table 4). Morbidity rates generally decreased with age for both races, as expected. Among black children, bronchitis, croup, and "other" chest infections were reported in decreasing frequency and all conditions except pneumonia and hospitalization were reported more frequently in the more polluted communities than in Riverhead (Table 5). Among white children, bronchitis was reported a bit more frequently than croup and "other" chest infections (Table 6). Again, all conditions except pneumonia and hospitalization were reported more frequently in the more polluted communities, when compared to Riverhead. When compared to white children, black children had higher rates of pneumonia and lower rates of croup, bronchitis, and "other" chest infections. Hospitalization rates were similar among blacks and whites. Patterns for two or more episodes were similar to those for one or more episodes for each condition.

**Table 3**
**Characteristics of Households, by Community and Race**

| | | Number of Persons per Household† | | Number of Rooms per Household | | Presence of Gas Stove or Gas Space Heater | |
|---|---|---|---|---|---|---|---|
| Community | Race | Median | Four or More (%) | Median | Six or More (%) | Stove Only (%) | Heater Only or Both‡ (%) |
| Riverhead | Black | 4 | 51 | 5 | 45 | 77 | 12 |
| Q,B,S* Combined | | 3 | 42 | 5 | 20 | 93 | 4 |
| Riverhead | White | 3 | 32 | 6 | 64 | 54 | 5 |
| Queens | | 3 | 22 | 7 | 87 | 71 | 25 |
| Bronx | | 3 | 23 | 5 | 42 | 87 | 11 |
| Sheepshead Bay | | 2 | 13 | 5 | 44 | 90 | 8 |

*Queens, Bronx, and Sheepshead Bay combined due to small sample size.
†Excluding parents or guardians.
‡Less than 1% of households had a gas space heater only.

Model adjusted morbidity rates for blacks and whites are shown in Table 7 and Table 8. Among black children, rates of "any lower respiratory disease," croup, two or more episodes of bronchitis, and one or more episode of "other" chest infections were significantly higher in the more polluted communities with relative increases ranging from about 1.7 for

**Table 4**
**Age Distribution of Nonasthmatic Children with Three or More Years of Residence, by Community and Race†**

| | | Number of Children | | | |
|---|---|---|---|---|---|
| Community | Race | 1–4 Years | 5–8 Years | 9–12 Years | Total |
| Riverhead | Black | 102 | 218 | 224 | 544 |
| Q,B,S* Combined | | 85 | 218 | 287 | 590 |
| Riverhead | White | 220 | 495 | 481 | 1,196 |
| Queens | | 328 | 792 | 720 | 1,840 |
| Bronx | | 273 | 769 | 718 | 1,760 |
| Sheepshead Bay | | 310 | 793 | 726 | 1,829 |

*Queens, Bronx, and Sheepshead Bay combined due to small sample size.
‡Excludes children with missing information on morbidity, age, race, sex, education of head of household, or duration of residence (3% of sample for whites and 11% of sample for blacks).

**Table 5**
**Three Year Frequency of One or More Episodes of Morbidity Among Black Children in New York by Age and Community**

| | | Age-Specific Rates* | | |
|---|---|---|---|---|
| Community† | Morbidity Condition | 1–4 Years (%) | 5–8 Years (%) | 9–12 Years (%) |
| Riverhead | "Any lower respiratory disease" | 20.3 | 16.4 | 6.9 |
| Q,B,S Combined† | | 32.9 | 18.8 | 18.1 |
| Riverhead | Pneumonia | 10.1 | 10.3 | 2.9 |
| Q,B,S Combined | | 16.5 | 5.8 | 4.5 |
| Riverhead | Croup | 3.8 | 2.4 | 1.7 |
| Q,B,S Combined | | 11.8 | 5.5 | 6.6 |
| Riverhead | Bronchitis | 8.9 | 6.7 | 2.9 |
| Q,B,S Combined | | 12.9 | 8.2 | 6.6 |
| Riverhead | "Other" chest infections | 2.5 | 1.2 | 0.6 |
| Q,B,S Combined | | 4.7 | 3.7 | 3.8 |
| Riverhead | Hospitalization | 7.6 | 6.1 | 1.1 |
| Q,B,S Combined | | 1.2 | 0.9 | 1.3 |

*Rates of "any lower respiratory disease," bronchitis, and pneumonia declined significantly with sex or education of head of the household.
†Queens, Bronx, and Sheepshead Bay combined due to small sample size.

one or more episodes of "any lower respiratory disease" to about 3.2 for one or more episodes of "other" chest infections. Among white children, all morbidity conditions except pneumonia and hospitalization were significantly higher in the more polluted communities with relative increases ranging from about 1.1 for one or more episodes of croup to about 1.5 for one or more episodes of "other" chest infections. Bronchitis and one or more episodes of "any lower respiratory disease" were significant only among children aged 5 to 12 years. When rates in Queens, Sheepshead Bay, and Bronx were contrasted, Bronx was lowest the few times that differences were significant (Table 8). Pneumonia and hospitalization rates were significantly higher in Riverhead, although the absolute rates for both

**Table 6**
**Three-Year Frequency of One or More Episodes of Morbidity Among White Children in New York by Age and Community**

| | | Sex and Education of Head of Household Adjusted Rates* | | |
|---|---|---|---|---|
| Community | Morbidity Condition | 1–4 Years (%) | 5–8 Years (%) | 9–12 Years (%) |
| Riverhead | "Any lower respiratory disease" | 30.8 | 27.6 | 11.2 |
| Queens | | 38.6 | 36.7 | 28.4 |
| Bronx | | 34.2 | 31.0 | 22.9 |
| Sheepshead Bay | | 35.6 | 39.9 | 24.5 |
| Riverhead | Pneumonia | 9.5 | 3.8 | 1.5 |
| Queens | | 2.1 | 2.0 | 2.2 |
| Bronx | | 2.8 | 4.9 | 2.8 |
| Sheepshead Bay | | 1.7 | 2.4 | 2.6 |
| Riverhead | Croup | 10.6 | 11.4 | 1.9 |
| Queens | | 16.6 | 13.9 | 6.4 |
| Bronx | | 12.9 | 9.9 | 6.2 |
| Sheepshead Bay | | 15.1 | 13.7 | 5.7 |
| Riverhead | Bronchitis | 15.8 | 15.4 | 7.2 |
| Queens | | 14.7 | 15.4 | 13.1 |
| Bronx | | 17.2 | 17.6 | 10.7 |
| Sheepshead Bay | | 12.5 | 18.1 | 11.8 |
| Riverhead | "Other" chest infections | 5.0 | 4.5 | 4.4 |
| Queens | | 15.0 | 14.4 | 12.4 |
| Bronx | | 11.0 | 8.4 | 9.2 |
| Sheepshead Bay | | 16.7 | 15.9 | 11.4 |
| Riverhead | Hospitalization | 7.3 | 2.8 | 0.8 |
| Queens | | 1.8 | 1.0 | 0.4 |
| Bronx | | 2.3 | 1.5 | 0.6 |
| Sheepshead Bay | | 1.6 | 1.5 | 0.6 |

*All conditions decreased significantly with age, all but pneumonia and hospitalization were reported more frequently among households with a high school or greater education; "any lower respiratory disease," croup, and pneumonia were reported more frequently among males.

conditions were low. When significant, morbidity rates were higher among males and among children from households with a high school or greater education.

## DISCUSSION

Reported lower respiratory disease morbidity was significantly higher among children exposed to ambient sulfur oxide and particulate air pollution than comparable children not so exposed. Rates of "any lower respiratory disease," croup, bronchitis, and "other" chest infections were increased among black and white children residing in the more polluted communities (Tables 7, 8). Rates of pneumonia and hospitalization were significantly increased in the low pollution community only among white children and the absolute rates for both conditions in all communities were low. As expected, morbidity rates declined with age and were generally higher in males and among children from households with a high school or greater education.

**Table 7**
**Three-Year Frequency of Morbidity: Model Predicted Rates for Black Children Aged 1 to 12 Years, by Morbidity Condition and Number of Episodes**

| Morbidity Condition | Number of Episodes | Model Predicted Rates* Riverhead (%) | Model Predicted Rates* Queens, Bronx, and Sheepshead Bay Combined† (%) | High/Low Polluted Community ($p < 0.05$) |
|---|---|---|---|---|
| "Any lower respiratory disease" | ≥1 | 14.0 | 23.2 | 1.66 |
| | ≥2 | 4.9 | 9.4 | 1.92 |
| Pneumonia | ≥1 | No significant difference between communities | | |
| | ≥2 | | | |
| Croup | ≥1 | 2.8 | 7.3 | 2.61 |
| | ≥2 | 0.1 | 2.8 | 28.0 |
| Bronchitis | ≥1 | No significant difference between communities | | |
| | ≥2 | 1.0 | 3.0 | 3.0 |
| "Other" chest infections | ≥1 | 1.3 | 4.2 | 3.23 |
| | ≥2 | No significant difference between communities | | |
| Hospitalization | ≥1 | Rates too low to fit model | | |
| | ≥2 | | | |

*Restricted to nonasthmatic children with three or more years residence duration; adjusted for age, sex, and education of head of household for "any lower respiratory disease," pneumonia, and bronchitis; adjusted for age for croup and "other" chest infections.
†Queens, Bronx, and Sheepshead Bay combined due to small sample size.

Several other known determinants of childhood respiratory disease morbidity besides ambient air pollution were comparable among the communities for each race. Exposure to sidestream smoke has been associated with increased acute respiratory disease morbidity in children.[20] But parental smoking habits were similar in all four communities and lower respiratory disease morbidity did not differ among children of smoking and nonsmoking parents. Family size and composition and household size did not differ among the communities. Age, sex, and education of the head of the household were accounted for in the analysis.

Indoor air pollution such as oxides of nitrogen and carbon monoxide emanating from gas stoves or gas space heaters has been mentioned as a possible risk factor for acute respiratory disease. Although oxides of nitrogen have been linked to excess acute lower respiratory disease morbidity, evidence for such an association with carbon monoxide is currently lacking.[22–24] Except for whites in Riverhead, from 90% to 98% of all households used a gas stove for cooking and/or a gas space heater (Table 3). Due

**Table 8**
**Three-Year Frequency of Morbidity: Model Predicted Rates for White Children Aged 1 to 12 Years by Morbidity Condition and Number of Episodes**

| Morbidity Condition | Number of Episodes | Model Predicted Rates* Riverhead (%) | Queens, Bronx, and Sheepshead Bay Combined* (%) | High/Low Polluted Community (P<0.05) |
|---|---|---|---|---|
| "Any lower respiratory disease" | ⩾1‡ | 27.2 | 32.0 | 1.18 |
| | ⩾2§ | 13.4 | 16.3 | 1.22 |
| Pneumonia | ⩾1 | 5.7 | 2.6 | 0.46 |
| | ⩾2 | 1.7 | 0.7 | 0.41 |
| Croup | ⩾1 | 8.4 | 9.1 | 1.08 |
| | ⩾2 | No significant difference between communities | | |
| Bronchitis | ⩾1‡ | 11.6 | 14.2 | 1.22 |
| | ⩾2‡ | 4.2 | 5.2 | 1.24 |
| "Other" chest infections | ⩾1 | 7.7 | 11.6 | 1.51 |
| | ⩾2 | 3.9 | 5.7 | 1.46 |
| Hospitalization | ⩾1 | 2.8 | 1.5 | 0.54 |
| | ⩾2 | Rates too low to fit model | | |

*Restricted to nonasthmatic children with three or more years residence duration; adjusted for age, sex, education of head of household, and doctor in Riverhead.
†No significant difference among three communities for pneumonia, croup, bronchitis, and hospitalization: Queens and Sheepshead Bay equal to each other and higher than Bronx for "other" chest infections; Sheepshead Bay > Queens > Bronx for "any lower respiratory disease" among 5–8 year olds only.
‡Significant among children aged 9–12 years only.
§Significant among children aged 5–12 years only.

to small sample size, only white, non-asthmatic, residentially stable children from households with a high school or greater education could be analyzed in a model testing the relationship of all respiratory morbidity categories as a function of the presence of gas stoves or space heaters in the household, age, and sex. No significant differences or consistent trends were found with regard to respiratory morbidity and the presence of gas stoves and/or space heaters in the household. Thus, our data do not associate excess acute lower respiratory disease in children with such exposures. Perhaps if studies were restricted to persons who would be most likely to be exposed — viz. mothers or the person who cooks — an association might be found. Nevertheless, the difference in the presence of gas stoves or space heaters between Riverhead and the other three communities surely was not the cause of the excess morbidity observed in this study.

This study corroborates similar earlier observations of differences in morbidity reporting between blacks and whites and the possible explanations for such differences were discussed in detail in the previous report.[17] In this study, when compared to white families, black families were larger, had only slightly less average living space within their respective community, were less educated only in Riverhead, but did have considerably fewer households with both parents or guardians present — viz. about 56% for blacks versus 85% to 93% for whites. Patterns of reporting for blacks and whites were similar with regard to age and history of childhood asthma suggesting the difference was not a statistical artifact. Differences in medical care utilization could partially explain our results. In a recent U.S. national survey, nonwhites reported fewer physician visits than whites did for each family size category.[25] Black-white differences in pulmonary function and susceptibility to cigarette smoke have also been reported.[26–28] Differences in recall, access to care, parental perception of illness, as well as other cultural or genetic differences, all may be associated with the observed black-white disparity in reported morbidity. Although answers to these questions are of clinical and public health interest, this study was not designed to address them.

In this study, excess acute lower respiratory disease morbidity in children was associated with estimated exposure to average annual sulfur dioxide levels of about 38 to 425 $\mu g/m^3$, total suspended particulate levels of about 60 to 185 $\mu g/m^3$, and suspended sulfate, levels of about 9 to 20 $\mu g/m^3$. Respirable particulates were 34 to 43 $\mu g/m^3$ in Queens and Bronx compared to 28 $\mu g/m^3$ in Riverhead, but data were available only for the five months of 1972 that the study encompassed. Since the study encompassed up to twelve years of exposure for the oldest children and respirable particulates are not related to total suspended particulate matter in a simple linear fashion, it is not possible to estimate a threshold effects level for respirable particulates per se. Recent work has linked excess morbidity with exposure to annual total suspended particulate concentrations of about

100 to 140 $\mu g/m^3$ and suspended sulfate levels of about 13 to 14 $\mu g/m^3$ compared to respective levels of about 75 $\mu g/m^3$ and 10 $\mu g/m^3$ in the lower exposure community, and average annual sulfur dioxide levels were only 25 $\mu g/m^3$ in both communities.[17] However, information regarding lower respiratory disease and exposure to average annual particulate levels of about 60 to 89 $\mu g/m^3$ compared to about 35 $\mu g/m^3$ is not currently available. Therefore, a ''best judgment'' estimate of the levels required to produce effects would be exposure to average annual levels of about 175 to 250 $\mu g/m^3$ of sulfur dioxide, 85 to 110 $\mu g/m^3$ of total suspended particulates, and about 13 to 14 $\mu g/m^3$ of suspended sulfates. Such concentrations are certainly within the range of those estimated from earlier studies.[6,7,13,14] Since sulfur dioxide levels have decreased considerably in Queens and Bronx while total suspended particulate levels remain near the Federal standard, a replication of this survey would provide an opportunity to further distinguish between the effects of these two pollutants.

## SUMMARY

Acute lower respiratory disease in children aged 1 to 12 years was surveyed retrospectively by questionnaire among parents in four New York metropolitan communities representing different exposures to sulfur dioxide, particulate matter, and suspended sulfates. Morbidity patterns were similar with regard to age for blacks and whites, but pneumonia was more frequent and bronchitis and ''other'' chest infections less frequent among blacks when compared to whites in each community. Rates of ''any lower respiratory disease'' (a combined category), croup, bronchitis, and ''other'' chest infections were significantly higher among black and white children residing in the higher pollution exposure communities. Pneumonia and hospitalization were significantly higher only among white children in the low exposure community but the absolute rates were low for both conditions in all communities. Differences in family size and composition, crowding, parental cigarette smoking, or indoor air pollution due to gas stoves or gas space heaters could not explain the morbidity excesses in the high exposure communities. The ''best judgment'' estimate of the average annual pollutant concentrations associated with excess childhood respiratory morbidity in this study was 175 to 250 $\mu g/m^3$ of sulfur dioxide, about 85 to 110 $\mu g/m^3$ of total suspended particulates, and about 13 to 14 $\mu g/m^3$ of suspended sulfates.

## REFERENCES

1. Acute Conditions (Incidence and Associated Disability, United States—June 1965-June 1966). National Center for Health Statistics, Public

Health Service, US Department of Health, Education and Welfare, Washington, DC. PHS Publication No. 1000, Series 10, No. 38, June 1967, 61 p.

2. Dingle JH, Badger GF, Jordan WS: Patterns of Illness. In: *Illness in the Home*. Cleveland, The Press of Western Reserve University, 1964, pp 33–37.

3. Valadian I, Stuart HC, Reed RB: Contribution of Respiratory Infections to the Total Illness Experiences of Healthy Children from Birth to 18 years. *Am J Public Health* 51:1320–1328, 1961.

4. Annotation: *Lancer* 1:865, 1960.

5. Reid DD: The Beginnings of Bronchitis. *Proc Roy Soc Med* 62:311–316, 1969.

6. Douglas JWB, Waller RE: Air Pollution and Respiratory Infection in Children. *Br J Prev Soc Med* 20:1–8, 1966.

7. Lunn JE, Knowelden J, Handyside HA: Patterns of Respiratory Illness in Sheffield Infant School Children. *Br J Prev Soc Med* 21:7–16, 1967.

8. Holland WW, Halil T, Bennett AE, et al: Factors Influencing The Onset of Chronic Respiratory Disease. *Br Med J* 2:205–208, 1969.

9. Toyama T: Air Pollution and Its Health Effects in Japan. *Arch Environ Health* 8:153–173, 1964.

10. Manzhenko EG: The Effect of Atmospheric Pollution on the Health of Children. *Hyg and Sanitation* (Moscow) 31:126–128, 1966.

11. Love GJ, Cohen AA, Finklea JF, et al: 5.3 Prospective Surveys of Acute Respiratory Disease in Volunteer Families: 1970–1971 New York Studies. In: Health Consequences of Sulfur Oxides: A Report from CHESS, 1970–71. US Environmental Protection Agency, ORD, NERC, Research Triangle Park, North Carolina, May 1974, EPA–650/1–74–004, pp 5–49–69.

12. Finklea JF, French JG, Lowrimore GR, et al: 4.3 Prospective Surveys of Acute Respiratory Disease in Volunteer Families: Chicago Nursery School Study, 1969–1970. In: Health Consequences of Sulfur Oxides: A Report from CHESS, 1970–71. US Environmental Protection Agency, ORD, NERC, Research Triangle Park, North Carolina, May 1974, EPA–650/1–74–004, pp 4–37–55.

13. Nelson WC, Finklea JF, House DE, et al: 2.3 Frequency of Acute Lower Respiratory Disease in Children: Retrospective Survey of Salt Lake Basin Communities, 1967–1970, In: Health Consequences of Sulfur Oxides: A Report from CHESS, 1970–71, US Environmental Protection Agency, ORD, NERC, Research Triangle Park, North Carolina, May 1974, EPA/650/1–74–004, pp 2–55–73.

14. Finklea JF, Hammer DI, House DE, et al: 3.3 Frequency of Acute Lower Respiratory Disease in Children: Retrospective Survey of Five Rocky Mountain Communities, 1967–1970, In: Health Conse-

quences of Sulfur Oxides: A Report from CHESS, 1970–71. US Environmental Protection Agency, ORD, NERC, Research Triangle Park, North Carolina, May 1974, EPA–650/1–74–004, pp 3–35–56.

15. English TD, Steen WB, Ireson RB, et al: 5.1 Human Exposure to Air Pollution in Selected New York Metropolitan Communities, 1944–1971, In: Health Consequences of Sulfur Oxides: A Report from CHESS, 1970–71. US Environmental Protection Agency, ORD, NERC, Research Triangle Park, North Carolina, May 1974, EPA–650/1–74–004, pp 5–3–32.

16. Goldberg HE, Finklea JF, Nelson CJ, et al: 5.2 Prevalence of Chronic Respiratory Disease Symptoms in Adults: 1970 Survey of New York Communities, In: Health Consequences of Sulfur Oxides: A Report from CHESS, 1970–71. US Environmental Protection Agency, ORD, NERC, Research Triangle Park, North Carolina, May 1974, EPA–650/1–74–004, pp 5–33–47.

17. Hammer DI, Miller FJ, House DE, et al: Frequency of Acute Lower Respiratory Diseases in Children: Retrospective Survey of Two Southeastern Communities, 1968–1971. EPA In-House Technical Report, May 8, 1974.

18. Grizzle JE, Starmer CF, Koch GG: Analysis of Categorical Data by Linear Models. *Biometrics* 24:489–504, 1969.

19. Hill AB: *Principles of Medical Statistics,* ed 7. Oxford University Press, New York, 1961, 367 pp.

20. Cameron P, Kostin JS, Zaks JM, et al: The health of smokers' and nonsmokers' children. *Journal of Allergy* 43(6):336–341, 1969.

21. Air Quality Criteria for Particulate Matter. USDHEW, PHS, EHS, Washington, DC, January 1969, p 149.

22. Shy CM, Creason JP, Pearlman ME, et al: The Chattanooga School Children Study: Effects of Community Exposure to Nitrogen Dioxide. I. Methods, description of pollutant exposure and results of ventilatory function testing. *J Air Pollut Control Assoc* 20:No. 8, 539, 1970.

23. Shy CM, Creason JP, Pearlman ME, et al: The Chattanooga School Children Study: Effects of Community Exposure to Nitrogen Dioxide. II. Incidence of Acute Respiratory Illness. *J Air Pollut Control Assoc* 20:No. 9, 582, 1970.

24. Pearlman ME, Finklea JF, Creason JP, et al: Nitrogen Dioxide and Lower Respiratory Illness. *Pediatrics* 47:391–398, 1971.

25. Acute Conditions, Incidence and Associated Disability, United States, July 1964–June 1965. National Health Survey, National Center for Health Statistics, Public Health Service Publication No. 1000, Series 10, No. 26, 1965, p 18.

26. Damon A: Negro-White Differences in Pulmonary Function (vital capacity, timed vital capacity, and expiratory flow rate). *Hum Biol* 38:380–393, 1966.

27. Prevalence of Selected Chronic Respiratory Conditions, United States-1970. National Health Survey, National Center for Health Statistics, Vital and Health Statistics—Series 10–No. 84. DHEW Publication No. (HRS) 74–1511, Rockville, Maryland, 1973.

28. Stebbings JH Jr: A Survey of Respiratory Diseases Among New York City Postal and Transit Workers. IV. Racial Differences in the $FEV_1$. *Environmental Research* 6:147–158, 1973.

# 25 Surveillance Techniques for Respiratory Illness

I.B. Tager, M.D., M.P.H.
F.E. Speizer, M.D.

## INTRODUCTION

The slowly progressive character of the obstructive component of chronic bronchitis (CB) and the lack of reversibility of the more severe functional abnormalities have directed many investigators' attention to study more sensitive methods of detecting obstructive airway disease (OAD) and to investigate younger individuals to determine which groups are at highest risk of developing OAD.[1] Several groups[2–7] have shown that air pollutants, cigarette smoking, and lower respiratory infections acting in childhood produce measurable effects on pulmonary symptomatology and function. These data have been interpreted to suggest that the origins of adult obstructive bronchitis might actually begin early in childhood.

In the above context, lower respiratory infections experienced in childhood are of particular interest. Studies of the impact of ambient pollutants on respiratory illness may have important consequences for the subsequent development of functional abnormalities, since a substantial proportion of these illnesses in very young children involve the lower respiratory tract[8,10,11] and particularly the small airways[12] thought to be important in the pathogenesis of the obstructive forms of CB in adults. Of particular interest in this regard is whether children who experience frequent or severe respiratory illness are at greater risk for the untoward effects of environmental pollutants and cigarette smoking on the development of lowered pulmonary function.

Studies on the etiology of the various agents responsible for respiratory illness have generally employed prospective methods with regard to the illnesses under study.[9–11] On the other hand, many studies concerned with the relation of respiratory illness to pulmonary function have been based on retrospective assessments of respiratory illness history that use parental histories with or without attempts to validate these histories with physician and hospital records.[2–5] Other workers have used prospective assessments that rely on a single method of data collection (e.g., questionnaire alone) to categorize illness histories.[7]

In view of the importance of collecting accurate, reliable data on respiratory illness morbidity plus our own interest in the impact of childhood respiratory illnesses on the future development of adult obstructive lung disease, we devised a pilot study to compare simultaneously several methods for prospectively collecting historical information on respiratory illness experience of young children. We were interested in developing methods that would fulfill two criteria in particular: (1) Accurate assessment of illnesses not normally reaching attention of medical care providers and (2) adaptability to large sample sizes at minimal cost.

## METHODS

Standardized questionnaires were prepared for use in obtaining background historical data on participating subjects and for use during investigation of individual episodes. A diary for use in the home was also prepared. It was so constructed that a numbered list of symptoms was provided, the appropriate number(s) to be placed in boxes corresponding to the dates on which symptoms occurred. If no symptom occurred, no entries were expected. However, whenever a contact was made with a parent, reminders about the diary were given.

Table 1 summarizes the study protocol. The initial study group consisted of the 183 members of grades 1 to 3 of the Parochial Schools of East Boston, Massachusetts. (East Boston is a geographically definable neighborhood within the city of Boston.) Subjects were randomly assigned to one of two groups. Group A subjects were surveyed by telephone every two weeks and Group B subjects every four weeks. For both groups, parents were instructed on the use of the diary and instructed to call a special phone number for any illnesses in their child. When a parent called to report an illness or when a routine surveillance call was made, a standard, short questionnaire was administered over the telephone. If specified criteria were met, an appointment was arranged to visit the home of the subject for more detailed investigation of the illness. Initial intake interviews were done in a one-week period by three experienced screeners. The surveillance was carried out by only one of these three individuals over a ten-week period from the middle of February through the end of April, 1974.

In addition to the above contacts with study subjects, the school absence lists of the appropriate school were checked each day. If no contact was initiated by a parent within 48 hours after the child first appeared on the list, contact was initiated by the surveillance screener. Since most of the children in East Boston receive their health care through the East Boston Neighborhood Health Center, the logs of the Emergency Room and Triage Desk were surveyed each day for the names of study subjects.

In an attempt to validate our findings, physicians' records were evaluated when available. Also, each of the methods of surveillance used was checked for the consistency of the data obtained.

## RESULTS

Of the 183 subjects selected for study, 162 actually participated, 82 in the group surveyed bi-monthly and 80 in the group surveyed monthly. There were no differences between the two groups with regard to overall participation rates and the distribution of males and females within the groups (Table 2). Only 2 children (both group B) in the study had a chronic respiratory ailment.

There were 149 episodes of illness identified by all the methods utilized, of which 145 were classified as either upper or lower respiratory illness. Table 3 showed the means by which each episode was identified. The only significant difference between the two groups was that fewer

**Table 1**
**Study Population Randomization**

| | Entrance into Study after Randomization | |
|---|---|---|
| | Group A | Group B |
| Diary by parent | + | + |
| Parental call for illness | + | + |
| Monitoring of school absence | + | + |
| Surveillance call q 2 wk | + | – |
| Surveillance call q 4 wk | – | + |
| Home visit for detailed evaluation of respiratory illness | + | + |

group B illnesses were detected by parent initiated contact. For the population as a whole, 41% of episodes (60/145) were detected by the routine surveillance; 35% (51/145) were detected only by evaluation of entries made in the diaries; 12% (17/145) were detected by monitoring of school absence lists. In no instance did monitoring of the Health Center lead to detection of illness episodes that were not detected by either a parent initiated call (4 episodes) or a coincident surveillance call (1 episode). Of the 52 group A illness episodes which could have been investigated in the home by detailed interview, 54 (87%) were actually completed. For group B, 36 of 42 (86%) such episodes were investigated.

Table 4 shows the respiratory illness rates in the two groups. There were no differences between the two groups in the number of illness episodes per week per child (0.094, Gp A; 0.084, Gp B). Group B tended to have fewer children with lower respiratory illnesses (54% vs. 64% for

**Table 2**
**Study Population**

I. Subjects Selected for Study
- 162 (88.5)* participated
- 3 (1.6) refused
- 12 (6.5) not located
- 4 (2.2) no telephone
- 2 (1.1) other

II. Characteristics of Groups A and B.

| | Group A | Group B |
|---|---|---|
| Selected for study | 91 | 92 |
| Participated in study | 82 (90)† | 80 (87) |
| Refused | 1 (1.1) | 2 (2.2) |
| Males | 41 | 40 |
| Median age (range) | 8 (7–12) | 8 (7–12) |
| Females | 41 | 40 |
| Median age (range) | 8 (7–11) | 9 (7–12) |

*Percent
†P > 0.5 for participation Gp A vs. Gp B.

**Table 3**
**Illness Identification**

| Means of Identification | Number of Episodes (%) | | Total |
|---|---|---|---|
| | Group A | Group B | |
| Parent initiated call* | 13 (17) | 4 (6) | 17 (12) |
| Surveillance call | 32 (41) | 28 (42) | 60 (41) |
| Diary | 26 (33) | 25 (37) | 51 (35) |
| School list† | 7 (9) | 10 (15) | 17 (12) |

*Gp A vs. Gp B Standard Normal Deviate = 2.09, P = 0.038.
†Gp A vs. Gp B Standard Normal Deviate = 1.13, P = 0.26.

Gp A; P=0.071) but more children with multiple episodes of all respiratory illness (22.5% vs. 12.2% for Gp A; P=0.084). There was no difference between the groups with regard to number of children with complicated (see Table 4 for definition) lower respiratory tract illness (6.1% Gp A vs. 8.8% Gp B; P=0.52).

The number of illnesses detected by the diary was surprising (Table 4). Of the 162 diaries distributed, 123 (76%) were collected at the end of the 10-week surveillance period. The remaining 39 diaries were lost or the subjects were not at home at the time of the collection. Of the diaries collected, 79 had illness episodes that were also investigated by follow-up questionnaire in the home. Of this group of 79, 59 (75%) diaries contained entries for more than one-half of the illness days reported on questionnaire. Ten (13%) had no entries. In no instance was the categorization of an episode as upper or lower respiratory illness different between the diary and the questionnaire when more than half the illness days were recorded in the diary.

There were 30 episodes of illness attended by a visit to a physician. Records could only be obtained for 17 (57%) of these episodes. In most cases, the descriptions of the illnesses in the physician's records were very sketchy, so that it was difficult to be sure if all the symptoms reported by the patient were recorded. In 9 of the 17 cases (53%) there was agreement between the physicians' findings (symptoms and/or diagnosis) and the findings of the surveillance methods. In 7 cases (41%) there was disagreement. In 5 of these 7 cases, the subjects reported a cough to the surveillance inquiry but not to the physician. In one instance the converse was true. In one case, the criteria for respiratory illness were not fulfilled in the surveillance data; and the physician's record contained only the diagnosis of "URI", without statement of symptoms. Finally, in 2 other of the 17

**Table 4**
**Respiratory Illness Rates over the 10-Week Surveillance Period**

| | Group A | Group B |
|---|---|---|
| Total (illnesses/week/child)* | 0.094 | 0.084 |
| Upper respiratory episodes† | 28 (36%) | 31 (46%) |
| Lower respiratory episodes† | 49 (64%) | 35 (54%) |
| Children with multiple illness‡ | 10 (12.2%) | 18 (22.5%) |
| Children with complicated§ lower respiratory illness** | 5 (6.1%) | 7 (8.8%) |

*Standard Normal Deviate = 0.22, P = 0.83.
†Standard Normal Deviate = 1.80, P = 0.071.
‡Standard Normal Deviate = 1.73, P = 0.084.
§Complicated lower respiratory illness defined as cough with any of the following: wheezing, pleuritic pain, shortness of breath, or confirmed diagnosis of pneumonia.
**Standard Normal Deviate = 0.65, P = 0.52.

cases, insufficient data were available in the physicians' records to make a comparison.

Because of the problem encountered with the physicians' records, we compared our data with the prevalence rates for the Tecumseh Study of Respiratory Illness, which obtained the data from weekly contacts with subjects.[11] For the age group 5 to 9, total respiratory illness rates for the fall and winter months in the Tecumseh Study have been reported to average about 80 to 90 illnesses per week per 1,000 children.[11] This rate is not very different from our rate of 14.5 episodes per week per 162 children (90 episodes per week per 1,000 children), with our data being gathered in later winter and early spring. When we applied the definition of respiratory illnesses used in the Tecumseh Study,[11] we found a higher frequency of lower respiratory illness in this age group than that found in a similar age group over a 190-week period in that study (40.6% vs. 25.6%, $P<0.001$).

## DISCUSSION

The data presented in this study do not have direct clinical relevance. However, population based assessments of the possible morbidity associated with noxious environmental pollutants on the respiratory tract will have to depend on survey/surveillance methods. As such, the principles that can be extrapolated from our data are relevant to measuring the effects of air pollution on human health. This is especially true when the effects being measured are relatively subtle, e.g., small increases in morbidity in certain high risk individuals.

From the standpoint of assessing the incidence of disease occurrence, the data presented indicate that no one method of surveillance is likely to be sufficiently reliable to identify all illnesses; 35% of the identified illness episodes (i.e., those obtained by diary alone) would not have been identified even with surveillance calls as frequent as every two weeks. Use of weekly calls might have improved this but requires added manpower (especially in large population surveys) and still does not guarantee that all episodes are being reported. School absence lists were surprisingly useful (12% of all episodes), especially when the early investigations of illness episodes is necessary. The comparability of our weekly illness rate with that found by telephone or personal contact in the Tecumseh Study[11] suggests that a reasonably accurate estimate of illness rates was obtained in this study.

A major concern for surveillance techniques is the quality of the data collected (accuracy, consistency). We have assessed the accuracy of our data by comparison with other population based data acquired by different methodologies *(vide supra)*. Through the use of a diary and direct investigation of illness episodes, we have a means for continuously monitoring the quality of the data reported. Diagnoses made from diary records were the same as those made from direct questioning (for the distinction between upper and lower respiratory disease) in all instances for which both a direct investigation and a diary were available. The consistency of reporting increases confidence about the diagnosis made, although it does not eliminate the possibility of consistent reporting bias on the part of some parents.

Physicians' records were also used in an attempt to validate our data. Unfortunately, most of the records contained incomplete descriptions of the illnesses reported. Thus, in making a diagnosis of upper or lower respiratory tract infection, it was difficult to apply objective criteria to the reported symptoms. It was also unfortunate that there were no documented episodes of pneumonia, pleuritic rubs, or observed wheezing to compare with the data recorded on the questionnaires. This experience would seem to support the concept that, at least for the more mild forms of respiratory illness, physicians' records may not be as useful as other checks of consistency (e.g., a diary) and that misleading assessments of the accuracy of survey data may result if these records are the only source of data validation. In addition, some skepticism should exist about relying on selected laboratory data such as throat cultures for the definition of upper and uncomplicated lower respiratory illness. The importance of historical data is underscored by the fact that even under optimal conditions etiologic agents cannot be identified in a large proportion of episodes of upper and lower respiratory illnesses in children.[14,15] This distinction has often been made by the adoption of clearly defined combinations of symptom complexes and objective findings on examination.[9,11]

The division of the study population into two groups contacted by telephone either bi-monthly (Gp A) or monthly (Gp B) was completely arbitrary and designed to help us determine the shortest interval that would allow accurate data to be collected at low cost. The only significant difference in the pattern of disease reporting was that those parents contacted bi-monthly were more likely to report illnesses on their own than those in the monthly group (Table 4). This higher rate of reporting is not attributable to an increase in either disease frequency or severity. The types of diseases reported were similar in the two groups, although Gp B parents reported fewer lower respiratory tract illnesses ($P=0.071$) and more children with multiple illnesses ($P=0.84$). The reasonable consistency between the two groups (Table 4) also supports the validity of the approach used.

The data presented represent a pilot study to determine the most efficient method of detecting, over a defined period of time, the frequency and severity of respiratory illnesses in children. As these data are developed over the course of two years — along with data collected from general environmental, meteorological, and home environment monitoring — we will begin to build up a dose-response pattern to environmental exposures. Such information should allow us to sort out the importance of changes in weather, kinds of home heating systems, and general air pollutant effects. The collected data should also enhance our ability to study documented childhood respiratory illness as an additional risk factor in the prospective investigation of the development of chronic respiratory diseases.

## REFERENCES


1. Macklem P: Workshop on Screening Programs for Early Diagnosis of Airway Obstruction. *Am Rev Resp Dis* 109:567, 1974.
2. Douglas J, Waller R: Air Pollution and Respiratory Infection in Children. *Br J Prev Soc Med* 20:1, 1966.
3. Lunn J, Knowelden J, Handyside A: Patterns of Respiratory Illness in Sheffield School Children. *Br J Prev Soc Med* 21:7, 1967.
4. Holland W, Halil T, Bennett A, et al: Factors Influencing the Onset of Chronic Respiratory Disease. *Br Med J* 2:205, 1969.
5. Holland W, Halil T, Bennett A, et al: Indications for Measures to be Taken in Childhood to Prevent Chronic Respiratory Disease. Millbank Memorial Fund Quarterly, *Quarterly*, 47:215, 1969.
6. Shy C, Hasselblad V, Burton K, et al: Air Pollution Effects on Ventilatory Function in US School Children. *Arch Environ Health* 27:124, 1973.
7. French J, Lowrimore G, Nelson W, et al: The Effect of Sulfur Dioxide and Suspended Sulfates on Acute Respiratory Disease. *Arch Environ Health* 27:129, 1973.
8. Leeder S, Woolcock A, Blackburn C: Prevalence and Natural History of Lung Disease in New South Wales School Children. *Intern'tl J Epid* 3:397, 1971.
9. Glezen WP, Loda F, Clyde W, et al: Epidemiologic Patterns of Acute Lower Respiratory Disease of Children in a Pediatric Group Practice. *J Pediatr* 78:397, 1971.
10. Loda F, Glezen WP, Clyde W: Respiratory Disease in a Group Day Care. *J Pediatr* 49:428, 1972.
11. Monto A, Napier J, Metzner H: The Tecumseh Study of Respiratory Illness I. Plan of study and observations on syndromes of acute respiratory disease. *Am J Epidemiol* 94:269, 1971.
12. Fraser R, Paré JAP: *Diagnosis of Diseases of the Chest.* WB Saunders Co, Philadelphia, 1970, p 1051.
13. Monto A: The Tecumseh Study of Respiratory Illness V. Patterns of infection with the parainfluenzeviruses. *Am J Epidemiol* 97:338, 1973.
14. Wannamaker L: Perplexity and Precision in the Diagnosis of Streptococcal Pharyngitis. *Am J Dis Child* 124:352, 1972.
15. Glezen WP, Denny F: Epidemiology of Lower Respiratory Disease in Children. *N Engl J Med* 288:498, 1973.

# 26 Clinical Implications of Air Pollution Research for Illnesses of Children

John H. Knelson, M.D.

What are the clinical applications of air pollution research? I think that it is very difficult to answer this question with our current state of knowledge. It is only within the last five years that a major effort has been mounted in this country, and in the industrialized world in general, to address this problem. It is fair to say that about 95% of what we know now about environmental factors related to clinical medicine has been learned within the last decade and about 80% in the last five years or so. We are still, I think, on the rapidly rising exponential slope of our accumulation of knowledge in this particular area. How, then, can we begin even with the relatively superficial knowledge that we have today to put it into a context suitable for the clinician to apply to his patient population?

There is a basic dichotomy in our ways of thinking. There are those physicians who are faced each day with a waiting room full of patients and those who are faced each day with managing a research program addressing the questions that we have been discussing here. The clinician thinks in terms of the health status of the individuals in his patient population. We in epidemiology think in terms of the public health status of populations. This difference in our way of approaching problems and in our way of thinking deserves to be discussed so that the clinicians can begin to appreciate how the difference in attitude accounts for some of the difficulties in translating what you hear in symposia such as this into clinically applicable information.

With this background statement in mind, I would like to make four basic points and then describe how we can begin to breach this communication gap that I believe exists between those who think in terms of population statistics and those who are concerned every day with the series of individual sick people that they are called upon to take care of.

First I would like to ask the question, "What is the relationship of pulmonary function tests to significant biological change either in populations or in the individual patient?" We have heard papers this afternooon by Chapman and by Kagawa that are pretty convincing. They corroborate many observations by others over the last several years that there are significant changes in some parameters of lung mechanical function associated with changes in ambient air quality. Clinicians could very legitimately ask, "So what?" We know that the usual ways of measuring lung mechanical function are not as sensitive in individual clinical situations as a simple medical history. Changes in spirometry do not regularly precede the onset of symptoms in people with significant respiratory disease. How, then, is it that these statistically significant changes in lung function tests have any clinical significance? I believe that as we relate what was reviewed this afternoon to the important discussions of the session on immunology, we should become convinced of the consistency of the findings, of the reasonableness of the changes in lung mechanical function reported this afternoon with the hypotheses put forth this morning.

In addition, most of us here know that not only do children have changes in lung function tests associated with changing ambient air quality but that experiments in controlled environmental laboratories corroborate the findings of epidemiologists. One must ask the question, "Does measurable change in lung function during the developmental years have any effect on risk factors for lung disease in the adult?" In these controlled environmental laboratory studies the nature of the dose, the duration of the dose, the size of the dose to the individual, and changes in lung function are measurable with exquisite precision compared to the less sophisticated field measurements. The more we do of these controlled environmental laboratory investigations, the more reasonable is the assumption that the findings of our colleagues in the field are not only associated with, but are probably caused by, changes in ambient air quality. We have come relatively close in recent years to fulfilling Koch's postulates with regard to these observations. You need only a certain number of reproducible results and corroborative findings by a variety of investigators in which people put into a controlled environmental laboratory suffer a decrement in lung function, come out of the laboratory and recover lung function, go back in etc., etc., until you begin to believe that there is a cause and effect relationship.

My second point emphasizes the relation between changes in mechanical lung function and the respiratory disease experience of populations. A variety of end points have been discussed in these sessions, but chronic bronchitis, incidence of asthma attacks, and several other parameters of disease experience in both child and adult populations have been recurrent themes. It makes sense that the findings reported by Hammer and by Yoshida relate changes in respiratory disease experience in these populations with changes in mechanical lung function. I would like to pick up the thread that was begun by Dr. Green when he speculated on the mechanisms that make reasonable our observations concerning excess respiratory disease experience and changes in lung function in populations. He discussed the action of oxidants and other atmospheric constituents on cell membranes and how he would expect these to change the integrity of cellular immunity. Dr. Green's contention is beginning to be borne out in some of our carefully controlled experiments on exposure to ozone where we are finding a significant decrement in phagocytic index in human peripheral leucocytes taken from exposed persons. These changes in cellular immune status occur after a very low level, ambient exposure and they persist for 72 hours and then go away — as we would expect, knowing what we do about the life span of these cells.

Third, I would like to refer to Dr. Tager's chapter, which dealt with the problems involved in characterizing the health status of a population. We do need to continue to make progress in our understanding of how to handle large numbers of individuals in our epidemiologic studies, to characterize health status more precisely, to design experiments that are parsimonious so that we can increase the number of subjects to the point where the clinician will begin to have some real confidence in the numbers we are giving him. I am often discouraged when I review the findings of investigators who are attempting to correlate a very complex set of environmental covariates with a complex set of health covariates in populations where the numbers of contrasts and comparisons that are made in the data very rapidly exceed the possibility of dealing with them in a statistically meaningful way. In other words, the size of the population under study is not nearly great enough to allow a sufficient number of degrees of freedom. Consequently, I think the kind of approach proposed by Dr. Tager will give us the information we need in spite of the finite resources we have at our disposal.

The fourth point is in response to a statement by Dr. Green, who expressed his dismay at our lack of precision and our lack of ability to handle the characteriziation of the environmental covariates that we are talking about. The only hope I can give in response to his concern is to report that the Environmental Protection Agency and the investigators working with us throughout the United States have recognized this dilemma and that we have developed a relatively sophisticated air monitoring network devoted to health effects studies. Within the next few years we should be able to characterize the pollutant dose to our populations with increasing precision and give the clinician a much better description of the relationship between the health status of his patients and the physical status of the air and the other environmental factors in which his patients perforce live.

A final point that I would like to make is that, as a pediatrician involved in environmental factors research, I am becoming increasingly impressed, as Dr. Wehrle has already expressed, by the necessity for looking at the influence of many factors on the health of the child because of the influence of these factors on the health of the adult. With the amount of information now available, it becomes quite reasonable to postulate that a variety of respiratory disease experiences in children that can be shown to be associated with changes in environmental quality do influence the ultimate health of these individuals as they go through adult life. In closing I would like to try to leave with you my greatest concern in terms of associating our work with your change in clinical medicine. I believe that if we can change your attitude slightly, change your frame of mind so that you begin to think in terms of public health and begin to consider the very likely possibility that environmental factors in an industrialized society are really importantly associated with the health of your patients, then I will have achieved my mission.

# DISCUSSION: PART FIVE

**Question:** Will someone please tell me what the evidence is that childhood air pollution exposure is the beginning of chronic lung disease in adults.

**Dr. Hammer:** The evidence is not primarily from the air pollution literature and is incomplete. But if you think about the people you see clinically who have chronic respiratory disease as adults, you realize that they don't get that way overnight. It is perfectly reasonable to wonder if histories of repeated lower respiratory disease infections in childhood are indeed related to a higher risk of chronic respiratory conditions in adulthood. To the extent that air pollution also increases respiratory illness, you worry about this risk. We know that children who have repeated respiratory infections also have alterations in airway functions in terms of peak flow rates, and we also know that air pollution adds to this burden. Now, we are not sure that air pollution per se will do this in children, but then we don't even know if it is true for children who have repeated episodes of asthma. I think that there is a lot of evidence in a general sense. Within the next few years we will know for sure.

**Dr. Tager:** I think there is another point here that involves looking at the pollutant burden and the episodic respiratory disease morbidity, once substantial functional consequences have been measurable in the clinical sense, for example $FEV_{1.0.}$ The subsequent progression of the functional changes really bears no relationship to the acute disease morbidity in adults. In children there is a much closer association, at least over the short periods of time that have been measured, between acute disease morbidity and measurable effects. Children who had a history of frequent infections appear to bear a greater alteration in their functional respiratory ability that is not evident in substantially diseased adults. Clearly something has happened earlier on that has influenced subsequent functional status.

**Dr. Goldsmith:** I think one of the most definitive pieces of evidence so far in this country is the series of studies of the military recruits in Chicago by area, because that age is a time of transition between childhood and adulthood. The gradients were quite substantial and were related to whether the people came from the relatively polluted area as opposed to relatively non-polluted areas.

**Dr. Lescoe:** I wonder if Dr. Tager's study with excellent surveillance techniques indicating that there is no increase in incidence in respiratory diseases was correlated at any time with the pollution levels in the area.

**Dr. Tager:** It is going to be done. We wanted to be sure that we at least had a reasonable method of surveillance before we started to study a cohort of about 700 children.

**Dr. Allen:** Doctor Hammer, I presume from your paper that you feel that the present ambient air standards for particulates and $SO_2$ are roughly within the correct range. Or are they too high or too low?

**Dr. Hammer:** I would just say that in addition to the Administrator of the Environmental Protection Agency promulgating standards in the past few years, an august panel of academicians who do not work for the Federal government, or at least not for EPA, made a report to the Office of Management and Budget. Also an august panel of scholars from the American Academy of Sciences had to do the same thing. Both of these groups came to the conclusion that the standards certainly were within the right ball park. Perhaps with new information we shall find that we may not be sufficiently protected.

**Dr. Beard:** At what level of oxidants should there be restriction on athletic activities of children?

**Dr. Wehrle:** I think it is very difficult to respond to this question. It looks as though the newer kinds of information that are now coming along go very much in the same direction as the demonstration that we had in Los Angeles with high school cross-country runners who had discomfort in the chest. I think many youngsters as well as adults have chest discomfort associated with exercise during peak air pollution days but stating specific numbers is rather difficult.

**Dr. Goldsmith:** The evidence that would answer this question is not good. At a concentration of 0.20 ppm of photochemical oxidants, which we used to think was a standard measure and which we now know has got to be adjusted by some calibration factor, the State Health Department in California recommends that school authorities consider restrictions on school athletic activities. This type of action, at a concentration of 0.35 ppm, was recommended first by the Air Pollution Committee of the Los Angeles County Medical Society as a level at which school children should restrict their physical activity. The rationale is that increased physical activity, involving mouth breathing as it does, tends to bring to the alveolar area of the lungs a photochemical oxidant, especially ozone, which is known to be capable at higher concentrations of producing pulmonary edema in experimental circumstances. On this rationale, a cut-off point was recommended originally at 0.35 ppm; it is now the official view of the Air Quality Advisory Committee, on which Professor Beard serves, that this level should be 0.20. This is advisory to the local school administrator, but is not thought at this point to be a level that requires a specific regulation. We would be delighted to find better evidence on what really happens at such oxidant levels to children who are exercising.

**Dr. Knelson:** Let me inject one word of caution. I would like to draw the analogy between this restriction of childhood activity and an old difficulty that we used to have, and I presume still have, of creating cardiac cripples out of children when they in fact have perfectly normal hearts. All the clinicians know what I am talking about. A pediatrician finds a child who has something questionably wrong with his physical examination and history and advises the parents to restrict the child's activity until there is a more thorough medical examination. This advice is amplified way beyond anything intended by the clinician and results in a large number of cardiac cripples among children in our country, absolutely unjustifiably. On the other hand, we do have increasingly convincing evidence that exercise is a very important covariate in the demonstration that shorter and shorter durations of exposure can result in significant clinical decrements when reasonable levels of activity are used as co-stressers. I would inject this note of caution in how you interpret this information on a public scale and balance that with the recognition that the more we learn about oxidant exposure the more we realize that exercise associated with this exposure, as well as others, is a significant covariate.

**Dr. Hayes:** This is going to cause some problems in the Bay area. At the present time we have quite a few oxidant-related health alerts that imply an element of risk for some adults with emphysema and children with asthma. This implication of risk does not apply, at least in our present thinking, to healthy children whether they are exercising or not. I need to know at this time whether I can carry this information to the school district as an official position of the AMA, which means that we are going to have to change our recommendations. In this area for 10 or 12 days out of the year won't we then have to notify specific areas that a smog alert means that they control school physical activity programs and that athletic contests should be curtailed? Second, I want to know if Los Angeles is doing the same thing? I assume that if they use the same number that we do, they will be calling alerts down there 100 days, more or less, every year. Are we going to hear from them as to what they are doing there? It should not be widely different from us. Is this going to be a local or a state-wide recommendation? This is a new number (0.20 ppm/hr) for healthy people and we need some clarification.

**Dr. Wehrle:** You are putting the bee squarely on a number of people in this room. Good for you. I wonder whether Doctor Beard might be a good person to respond to this.

**Dr. Beard:** This question comes from the school health program administrators; looking for an answer, they turn to a logical place, which is the Air Pollution Control District, but we don't have a good basis for providing answers. I think that Dr. Goldsmith has said something that has been helpful — that is, that Los Angeles took the lead in this and, on the basis of informed opinion, arrived at a pollution level that they put into effect, and that has appeared to be satisfactory over a span of some time. I think that this gives us a good basis for giving advice to other places. The problem, of course, is that it is going to be taken as scientific advice and it is really only empirical.

**Dr. Hackney:** I thought that Doctor Beard summed it up well but I would like to amplify two things. Doctor Knelson said that while epidemiologic and empirical information gives really good leads, for further quantitation, you have to appeal to controlled laboratory situations. And so what do we do, Doctor Wehrle, when it would be necessary to give children a deliberate exposure in a controlled laboratory situation in order to get the kind of quantitative information we are asking for?

**Dr. Wehrle:** I think that we are back to the question of ethics and informed consent for the study of children. I see no way with the present guidelines that you can use controlled exposure studies since parents cannot sign consent for their children, and children cannot sign consent for their own actions, and there is no provision for a body of elders or senior scientists to sign off for the rights of a child. I think that we are in a very difficult position at the present time. This is why we need to be as alert as possible to find the natural kinds of experiments that we have been able to take advantage of in the past, to look very carefully for those individuals who happen to move into some of the more polluted areas as well as children whose parents for one reason or another decide to move to an area where the concentration is lower for a particular environmental pollutant. I think it is only in that way that we can work through some of the kinds of considerations that we have.

**Dr. Hackney:** The session here is on children and a point to be made again is that the numbers won't work out the same for normal and for sensitive or reactive children. Given adequate facilities and expertise, I cannot see why the concept of laboratory environmental stress testing in the diagnostic sense cannot be applied to children, or to adults. For example, a specific question for a specific individual can be asked by the individual's physician as to whether health effects at a given level of pollutant indicate a need for activities restriction.

**Dr. Wehrle:** I think Doctor Hackney has made an interesting point. The position of the research committee or the Committee on Human Investigation, as it is properly called in our institution, is that the kind of testing you describe certainly may be done in a child if it is clearly in the interest of that child to have such a procedure done. It is analogous to the case with medication that may not have been tested in children but, on theoretical grounds and on the basis of good evidence, is of benefit to the condition that a child has. Then you can use this therapy even though it may not be approved for children or may not yet have adequate assurance of its safety at that particular age.

# Conference Summary

John R. Goldsmith, M.D.

There are seven things that are scientifically relevant that I have digested from this meeting. One, while lung cancer may not be associated with general urban air pollution, it is true that there is an urban contribution. We are not sure what it is and, moreover, there may be a contribution of point sources to lung cancer epidemiology. Mesothelioma is an example of this and we need to look for other examples.

Two, this problem, like many of the other air pollution health problems, needs occupational health data. We have lived too long in isolation, thinking that air pollution studies could be justified in isolation from occupational health studies.

Three, the role of carbon monoxide in exacerbation of symptoms in arteriosclerotic heart disease and the impairment of performance in normals is proven, as far as I am concerned. There are serious questions about the role of carbon monoxide in the pathogenesis of coronary and other heart disease that have not yet been resolved, and there is a need for further dose-response information for two of the effects that have been shown. The two roles that need further data are the possible role of carbon monoxide exposure in sudden death, which was suggested by the studies of fibrillation, and the possible role in the progression of atherosclerosis, a subject that was hardly raised.

Four, childhood asthma has been increasing in frequency in communities in Japan with pulp and paper mills; when this occurs, it is not different in character from the asthma occurring in children in non-polluted areas. We now have additional evidence that air pollution can trigger an allergic reaction.

Five, air pollution can have an effect on the central nervous system. I am not quite sure I know what to say about this, but I think we are going to have to follow this up with considerable attention.

Six, promising populations for further work in this field, in my judgment, are children and occupational populations, including their families and people who live in the vicinity.

Seven, we have focused much of our attention on individual pollutants alone and that is not sufficient. Most of the realistic effects of pollution occur in combinations.

Now I would suggest that we are really three communities of doctors who have been meeting together — practitioners, those who work for governments, and those who do research. I feel fairly comfortable in speaking in terms of ourselves, collectively, since I have had experience in all of these areas. I think I would not be doing my duty if I did not say that in each of these groups there are achievements and there are criticisms upon which to focus. When the practitioner speaks out on the basis of his concern, he has to be fully informed in order to speak effectively. There are some superb examples among medical societies and environmental health committees in which physicians have done this, but there are too few of them. The practitioner must voice his concerns more often, with better information, and more efficiently.

It is easier to criticize the research scientist. We all know that we research scientists tend to follow whatever interests us and for which money is available, but good research has to follow the public interest. In this respect I must say that as epidemiologists we are always looking at too much and drawing inconclusive results, but the experimentalists are looking at too little and are drawing conclusive results! The real problem is to have convergence; and to an extraordinary extent over the history of meetings such as this, there has evolved a convergence of epidemiologic and experimental information — on carbon monoxide, on lead, somewhat on oxidants, somewhat on sulfur oxide and particulate matter.

Now, I have to raise criticisms of the governmental scientist. More than anything else, I consider myself a governmental scientist. In the first place — and most commendably, if not always unanimously — government scientists have succeeded in providing a rational basis for regulation. Although it has been criticized, this achievement has been re-evaluated by the community of research scientists from the National Academy of Sciences, which has failed to find a serious flaw in it. The major criticism I have, however, has to do with a relatively weak position with respect to centralization of air pollution medical research. I have heard people comment about how little new research is being discussed today. There isn't a very broad base of research data. One of the reasons for this has been that over the last few years the Environmental Protection Agency has planned a great deal of its health effects research in close cooperation with the Coordinating Research Council, which is a body representative of and supported by the motor vehicle and petroleum industry. I think the extent to which this has been done is harmful, and this approach has produced indecisive and inadequate results in some of the areas where good research is needed. Officially, that arrangement has been concluded but its effects are still being felt. I am not opposed to the Coordinating Research Council supporting and carrying out research, but I am opposed to its being the agency that decides which major health problems are to be studied, who gets to study them, who is informed about the results, and when that information is released. Such research planning should also involve cooperation with the general medical community, the physicians in other states, and all open research communities.

Enough of serious matters. A long time ago, bridges were built across the Bay and the people could drive across them in their own cars. In celebration, San Francisco held a World's Fair. About half that long ago, we learned that photochemical pollution from too many people driving around too many California communities could produce some health effects. We celebrated that by establishing some Federal and state research programs that, I think, produced some benefit — better regulation of these pollutants. Today, we now have a way of getting across the Bay without taking the car, and that is called BART (the Bay Area Rapid Transit system). I urge those of you who come from less advanced areas to take a trip on BART before you go back home and find out what can help solve the photochemical air pollution problem. BART won't solve it alone, but it is an important step forward; and the Bay area is celebrating that step by holding the Air Pollution Medical Research Conference here this week in San Francisco.

I think that all of us will agree that pollution harms health and is wasteful and costly. The problem is that we don't know how much it harms health, how much it costs, or how wasteful it is. I hope that the work that has been done and the scientists who reported here have helped to make quantitative statements about that problem; but saving money and health may mean changes in life style. Physicians have a leadership potential. I think that in another decade and a half not only might we accomplish some of the research that we would like to see get done and that we have discussed, but we shall know whether the medical profession has exerted that kind of leadership in helping change life style in areas that can improve health.

# SPECIAL CONTRIBUTIONS

# Earth, Wind, and Fire

Malcom C. Todd, M.D.

There is a currently popular rock music group called "Earth, Wind, and Fire." It seems to me that the name is particularly appropriate for this meeting because of the deep and primal feelings that may be evoked. I am not referring to the passion which such music stirs up among the young, of course. That's been a standard phenomenon for thousands of years now, although today's younger generation — like all of their predecessors — may think they invented both the music and the rapture. And I myself am not a rock music fan, which should tell you either that I've outlived passion, or that I can recognize noise pollution when I hear it. Needless to say, I believe that the latter proposition applies.

Words like *earth, wind, fire, air* and *water* evoke visceral emotions about our environment and our lives. Such words not only define the elements per se, but also the focus of the elemental problems created by environmental waste and pollution. It seems to me that in examining the nature and the causes of those problems — as well as their possible resolution or mitigation — our thinking ought to be equally elemental end direct. Unfortunately, however, this has too rarely been the case.

When you turn to what is being printed and said about who caused what and who ought to do this or that about pollution, it's rather like watching a tennis match. The balls of censure and calls for remedial action bounce merrily back and forth between various organizations both public and private. Because of the obvious health implications of environmental pollution, there are those who charge the medical profession with leading the way in cleaning up the mess. Others claim that businesses and industries, whose by-products sometimes include those which damage the environment, must bear the blame — and the burden of restoration. Still others charge the government with failing to adequately safeguard our environment, so it should be government that bails us out. The difficulty with all this is that there are so many charges and counter-charges, and the players change so fast, that we sometimes despair of winning even a game — much less two sets and the match — in the fight against pollution. Yet when you take that elemental, direct look at the nature and the causes of the problems, and at the possible remedies, the view can be enlightening.

Basically, what has happened is that all of us in America have gotten ourselves into this mess . . . and it's going to take all of us to get ourselves out. It is important to recognize that environmental problems and the social consequences thereof do not stem from spitefulness on the part of any one group or institution. In many cases, in fact, our problems stem from good intentions — and from a desire to live the good life, as Rene Dubos, the eminent microbiologist from Rockfeller University, has made clear. In his 1969 Pulitzer prize-winning book, *So Human an Animal*, Dubos noted that the knowledge of ionizing radiations and of nuclear energy was developed by men who might be regarded as saints in terms of their ideals. Yet much harm has come from such developments. The internal combustion engine was initially a boon to mankind, but today the glut of traffic pollutes the air, disorganizes life, and strangles our cities. Our industries and our businesses have made it possible for us to enjoy many material benefits. But the systematic rape of the earth and its natural resources and the injection of technology into virtually all social activities are beginning to disrupt the dynamic processes which have made earth compatible for mankind.

As Dubos says, "Almost everywhere, the land is being used not as a home, not as an environment for the creation of human culture, but as a source of exploitation and speculation." Even so, he takes pains to emphasize that all Americans have been part of that exploitation. We have all clamored for the second car, the modern appliances, the air-conditioned homes and offices, and the many other material benefits that we enjoy.

Hence the priorities of our institutions — both public and private — also have become skewed economically and biased in favor of the siren song of science and technology. We have been living for today with very little thought for tomorrow — and none at all for the day after.

But within the past few decades, certain intellectual and practical limitations to science and technology have become evident. And we are beginning to understand that our natural resources, too, are limited, and that we can no longer act as if they will last forever. Earlier this year, Russell E. Train, administrator for the U.S. Environmental Protection Agency, underscored that point in an address right here in San Francisco. He said: "We are, it is clear, moving from an age of resource abundance to an age of resource shortages. By choice or by necessity, we are going to have to come to grips with the problems of growth." And as Mr. Train pointed out, this will require more forceful leadership from the government, from business and industry, from medicine, and from the public itself. I might add that we are going to have to adopt saner life styles, not through necessity *or* choice, but as a matter of necessity *and* choice.

A growing number of Americans recognize the necessity for changes in our environmental and social policies and programs, even when the changes are negative insofar as they involve the rationing of our resources. But if those changes are to be truly effective, if they are to be of enduring value, they also will have to be recognized as *positive choices* by a *majority* of Americans. In short, we're going to have to change individual hearts and minds, as well as institutional policies and programs, if we are to maintain the quality of our environment — and hence the quality of our lives.

Unfortunately developments to date have left much to be desired. The energy crisis, for example, has brought new pressures to bear to lower environmental control standards. The Clean Air Act Amendments of 1970 were a major step forward in ecology, but the fuel shortage has resulted in a relaxation of the prohibition against the burning of high-sulfur coal. This has been disappointing for those organizations that *supported* the clean air controls and that *opposed* any slackening of those controls. (And the AMA was among those organizations, by the way.)

Even so, I believe that the setback is only temporary, and that certain elemental forces are at work which will make effective environmental conservation measures mandatory in the long run. To begin with, the economic factors which once stimulated conspicuous consumption are now encouraging conspicuous conservation. The recent exorbitant price increases by the oil-producing nations have heated up the inflationary pressures on the economies of other countries (including the United States) and have stimulated the imposition of energy conservation measures. Our own Congress has imposed a nationwide speed-limit to conserve gasoline, and gas rationing is being seriously discussed in some quarters. Very recently, Senator Percy has proposed a plan which would provide financial rewards to those Americans who purchase smaller cars that are long on mileage and short on energy consumption.

Thus, while the oil shortage is temporarily fostering the use of coal as an energy source, it also has forcefully brought home to us the need for self-restraint in the use of our natural resources.

And environment pollution in general has fostered increased public attention relative to the connection between health and the environment. The serious effects of high ambient oxidant levels on people with respiratory problems, for instance, have increased public consciousness of the relationship between air pollution and health. And beyond the obvious damages to the quality of life, there are economic consequences which reinforce recognition of the necessity to clean up our environment. It is cheaper by far to prevent disease and illness than it is to treat them. Hence the pronounced trend toward trying to prevent chronic and degenerative diseases attributable to environmental hazards such as air and water pollution, accidents, malnutrition, poor living and working conditions, and so forth.

A major problem here, of course, is that there are many physicians who insist that beyond routine immunizations and such, preventive medicine — including a healthier environment — simply is not their responsibility. They say that they have been educated and trained in therapeutic medicine, and that's a full-time job. So someone else will have to worry about preventive medicine. Well, here again there are some hard economic truths which say otherwise. The cost of curative medicine has been rising sharply and will continue to do so, spurred on by inflation and the high costs of science and technology per se.

Professor Dubos, in another book entitled *Man, Medicine and Environment,* has underscored the implications for our profession. As he says, therapeutic medicine is probably now entering a phase of medically diminishing returns because many of our most spectacular and costly achievements are of help to only a few. He concludes that: "The extent of health improvement to be expected from building ultramodern hospitals with highly trained staffs and up-to-date equipment is probably trivial compared with results from the much lower cost of providing infants and children with well-balanced food, sanitary conditions and a stimulating environment."

Dramatic proof of that supposition can be found in the People's Republic of China, which I visited earlier this year along with other members of the AMA delegation. Because of recent improvements in preventive medicine — including environmental aspects such as living and working conditions — old scourges such as VD, tuberculosis, cholera, typhus, and schistosomiasis have been eradicated, or nearly so. You simple don't find any alcoholism, drug abuse, or child abuse; and most Chinese appear to be reasonably well nourished, vigorous, and healthy.

All that is in stark contrast to conditions I found in Micronesia, the South Pacific Islands which have been under U.S. trusteeship since the end of World War II. Many of the health problems found in these islands can be traced to shortages of resources in both therapeutic and preventive medicine. Poor living and working conditions are endemic, resulting in many health-related problems. The VD rate is very high in some areas. And suicide, particularly among young people, is not unusual.

Now let me emphasize that I am not recommending the importation of Chinese methods to cope with problems in Micronesia, or in the United States. We can't equate the Chinese, a very controlled, disciplined and regimented society, with our own. They have a different culture, a different national temperament, and people do as they are told. They do not refuse or question. Nevertheless, it is clear that the time has come for our society to exercise more self-restraint, self-control, and self-discipline in the preservation of our environment and, hence, the quality of our lives. It is equally clear that this will demand the active participation of — and enduring commitments from — medicine, government, business and industry, and the public itself.

Beyond economic and health considerations are the more intangible aspects of environment. The natural "elements" such as "earth, wind, fire, air, and water" are vital not only to our physiological health, but to our emotional health as well, to those deep and timeless instincts and feelings that inhabit our hearts and our minds. It is for those reasons, too, that we are coming to realize that supermarkets, superhighways, and super-lifestyles in general are far too high a price to pay. It is for those reasons that there has been a recent and powerful trend toward reviving the humanities as an educational priority in medical schools and elsewhere. And it is for those reasons that I believe we eventually will prevail in overcoming environmental pollution.

# Air Pollution Is A Social Disease

Congressman George E. Brown, Jr.

Air pollution really is a disease — a disease of the biosphere in which we all live. In a more than symbolic sense we are a part of the biosphere. We help to sustain it, and it helps to nurture us and all other living things. Pollution of that biosphere is as much a disease, in the broadest sense, as are the virulent manifestations of that pollution in the human body, which may include the pollution of emphysema or the pollution of cancer.

Air pollution resembles a social disease — that euphemistic expression for a variety of venereal diseases — because it is generally caused by human beings doing something they think they really enjoy, without considering all of the consequences. That something that causes air pollution is generally excessive consumption, excessive waste, excessive use of high-powered, highly polluting automobiles, and a variety of other excesses in the use of physical products and energy. Just as careless and irresponsible sexual activity can result in venereal disease, careless and irresponsible attitudes toward the use and consumption of nature's material bounty causes air pollution. There is an analogy also between VD and air pollution in our common attitude of refusing to face up to the two problems publicly, to talk about them rationally, to recognize the difficult steps we must take to eradicate them. With VD, shame and embarrassment inhibit us. With air pollution, conventional attitudes about economic growth, the good life, and other valued social goals inhibit our thoughts and actions.

What are some of these difficult steps we must take to eradicate air pollution, and how difficult are they? A few facts will help us understand this problem.

Since the passage of the Clean Air Act Amendments of 1970, the major reduction in pollutants has resulted from the conversion from coal as a fuel to low-sulfur oil and natural gas as a fuel.

Air quality projections for the Los Angeles Air Basin show that even if *all* automobiles met the original Clean Air Act emission standards, air quality would begin to deteriorate again in the early 1980s simply because of the continued growth in automobile use.

In order to protect the public health, research on air pollution has continued to discover new pollutants and new and higher margins of safety levels for pollutants, which in turn demand higher levels of pollution control. The cost, in dollars and in energy, of air pollution control technology increases exponentially as we approach zero emissions.

Energy demand projections by government and non-government sources predict that, without sharp restrictions on energy use or supply, energy consumption will increase by another 50% by 1985 and 150% by the year 2000. Coal use is expected to increase by nearly 100% by 1985, and 300% by the year 2000. Oil consumption is also expected to increase by some 50% by 1985, although the amount of imported oil is expected, hopefully, to be less than that used today. Rainfall patterns near urban centers have changed dramatically, due to the rain-inducing particulates far downwind of cities. Precursors of ozone, the agent most responsible for damage to agricultural products, are so widely dispersed in the lower atmosphere that high ozone levels are now detected in purely rural areas.

While we may hope for breakthroughs in new, clean sources of energy, or in effective and inexpensive pollution control technologies, I do not believe it is wise to count on such technological fixes. The national dilemma is that we are faced with a variety of proposals for continued growth in the use of fossil fuels to produce energy for the next several generations, apparently with the inevitable by-product of hazardous air pollution. As we increase the use of energy, we will most likely further the spread of the disease of air pollution. Thus energy policy is related to environmental policy.

Because air pollution plays a central role in establishing limits to energy use, and because this conference is concerned with air pollution primarily, I think it is appropriate to review some of the background relating to air pollution controls.

While air pollution has been a matter of scientific and social concern for several hundred years, and laws regulating the location and type of fuel burners have been in effect for decades, serious, systematic air pollution control technologies have been applied only very recently. The early air pollution control laws were not based upon any rigorous scientific criteria of emission limitations or ambient air quality standards. Nor were they applied equally throughout the industry or the country. The most stringent controls were, naturally enough, required in the areas that had the most active citizen interest, such as Los Angeles County. In most cases, these areas also had the most severe cases of pollution. Public involvement spread, the areas of control spread, and the basis for those controls became more firmly established in science, law, and administrative practice. Every step of this process involved a long and difficult struggle.

An often unspoken underlying difference of viewpoint in this struggle has been the question of who had the burden of proof in establishing pollution-control standards. Those with an economic stake in on-going practices naturally assume that new regulations should apply only when levels of pollutants are unequivocably demonstrated to have serious harmful effects. Those focusing on the health of the ecosystem believe laws should be drawn with the philosophy that no man-made contaminants should be permitted unless proven in detail to be truly harmless.

Much of the legislation in this field has been written with the relatively narrow view that our technological, market-oriented society would, given the proper stimulus, develop the technology to control pollution without any major change in values. The specific legislative approaches to accomplish this end have varied. The Federal law alone has undergone several major revisions, beginning with the Clean Air Act of 1963, which mandated a Federal role in this field, and climaxing with the primary law in the field of air pollution control, the "Clean Air Act Amendments of 1970" (P.L.91–604; Dec. 31, 1970).

The changes in the Clean Air Act paralleled changes in public attitudes as a whole. For example, there developed increasing recognition that air pollution was not a local problem. This was not only because air pollution crossed state lines, but because the products of our industrialized society were part of a national economy, and pollution was therefore a national product.

Another key development in the 1970 amendments was the resolution of the argument over whether controls should be based on health standards, or on the basis of economic and technological feasibility. The legislative history is instructive here. The report that accompanied the bill when it was sent to the U.S. Senate floor stated:

> In the Committee discussions, considerable concern was expressed regarding the use of the concept of technical feasibility as the basis of ambient air standards. The Committee determined that (1) the health of people is more important than the question of whether the early achievement of ambient air quality standards protective of health is

> technically feasible; and (2) the growth of pollution load in many areas, even with applications of available technology, would still be deleterious to public health. Therefore, the Committee determined that existing sources of pollutants either should meet the standard of the law or be closed down, and in addition that new sources should be controlled to the maximum extent possible to prevent atmospheric emmisions.

This statement of legislative intent provides no room for ambiguity.

We are now at the point in air pollution control in the United States where the laws are strict enough to accomplish the goal, but because enforcement of these laws may cause drastic changes, and conflict with other social values, enforcement agencies are balking at making this effort.

Despite the brave language of the Clean Air Act, we have not eliminated air pollution, nor have we even reduced air pollution to the level required to protect the public health. The final date to attain the ambient air quality standards is 1977, but few, if any, regions of the country appear likely to meet them for all pollutants. The existing law is not perfect, but very few of the attacks on the Clean Air Act have been inspired by those who wish to strengthen it.

The easiest form of attack on the implementation of the Clean Air Act is to criticize the *ambient* air quality standards as being "too stringent." This approach does not attack the existence of "goals" but instead attempts to "correct" those goals. This attack is usually based on the assumption that the burden of scientific proof in enforcing environmental law, especially in the face of economic dislocation, is to show that a given level of a pollutant is without a doubt harmful to health or property. Because much of society agrees that the burden does in fact rest with those who want to regulate pollution, as opposed to those who wish to pollute, this has been a relatively effective line of criticism. The resulting controversy has caused the Senate Public Works Committee, which authored the original act, to commission a National Academy of Sciences study on the Federal ambient air quality standards. Despite the implicit acceptance of the heavy burden of proof, the Executive Summary of that report stated, "In general, the evidence that has accumulated since the promulgation of the Federal ambient air quality standards by the EPA Administrator on April 30, 1971, support those standards. Hence, on balance, the panels found no substantial basis for changing the standards." Presumably this line of attack on the air pollution control strategies mandated by the Clean Air Act will now be weakened.

The second most frequent form of attack is to criticize the *emission* standards of both mobile and stationary sources. These criticisms are also based on scientific arguments, and they have been most vehemently raised by the automobile manufacturers, who believed that the auto emission standards were more restrictive than necessary (especially the NOx standard) in order to achieve the ambient air quality standards. The National Academy of Sciences also looked into this question, and while their findings were somewhat more ambiguous for this study, they still would not recommend changing the auto emission standards for NOx or hydrocarbons. The effort to relax the NOx standard is not over, but it is unlikely that it will be justified on the basis of any sound, scientific arguments. Both of these challenges to the Clean Air Act were more examples of clashing values than of simple disagreements over scientific facts. The determination and interpretation of the standards is, and will continue to be, an area where value judgments enter in. People who have different priorities will give the benefit of the doubt to different sides of the same question. This is a limitation with legislation that is frequently ignored.

This leads us to the most common criticism of the Clean Air Act, and of all environmental controls, for that matter. This is the debate over technological and economic feasibility. These two subjects are usually lumped together, with the implication that if the technology to control a particular pollutant is not economical under a given set of circumstances, then it is not available. However, pollution control technologies can usually be made economical by changes in the tax law, or the pricing mechanisms, provided that the changes are applied to all similar industries. As we approach the need for "zero" emissions in certain industries that use extremely toxic substances, such as radioactive materials, or carcinogens, the claim that the standards may be economically or technologically impossible to obtain, and still maintain production, may be true. In situations such as these, we should look very carefully at the industry in question. The question of technical feasibility then becomes the question of whether we can live without the product, and whether we can cope with the dislocations caused by ending production.

The most frequent recommendation of reformers to make a technology economical is to have a "pollution tax" that would begin the process of "internalizing the externalities." If this is done uniformly across the country, it is argued, the control of pollution would be accomplished naturally within the market system.

The attractiveness of this approach is obvious. The industry that produces a pollutant would have to absorb the costs of controlling it. While I cannot disagree with the intent, or even the logic of this argument, I believe it will, once again, only give us an illusion of a solution. This approach will not solve the root problem, which is the need to accept limits as to what can or should be done with science and technology. Another author described it this way: "The inherent danger of technological solutions to environmental problems is that they give the impression that the problem is being tackled and, in a society geared to growth, this allows the system to continue its headlong rush." The "pollution taxes" will only be truly effective to the extent that they begin to regulate material growth itself.

A very real limit to most environmental legislation, then, is its lack of comprehensiveness, and its inherently narrow approach to the problem. While the claims that pollution controls will cause industry to close down are usually overdrawn, this possibility is not always out of the question, as mentioned above. This is a situation in which values, not scientific facts, are in conflict, and this conflict in values must be recognized.

The environmental movement did embrace the view that growth in production and consumption of material goods is not as important as the quality of life, and that common economic indicators, such as the Gross National Product, did not reflect the quality of life. The Congress, when it passed environmental legislation, voiced some of these same views. However, this was no more than rhetoric until the hard facts of application arose. Now we are faced with the reality that our style of life will really have to change, and that social and economic structures will have to be modified. This prospect has given many former environmental advocates second thoughts. Lewis Mumford, in his monumental book, *The Pentagon of Power,* said:

> Reformers who would treat the campaign against environmental and human degradation solely in terms of improved technological facilities, like the reduction of gasoline exhaust in motor cars, see only a small part of the problem. Nothing less than a profound re-orientation of our vaunted technological "way of life" will save this planet from becoming a lifeless desert . . . For its effective salvation mankind will need to undergo something like a spontaneous religious conversion: one that will replace the mechanical world picture with an organic world picture, and give to the human personality, as the highest known manifestation of life, the precedence it now gives to its machines and computers. This order of change is as hard for most people to conceive as was the change from the classic power complex of Imperial Rome to that of Christianity, or, later, from supernatural medieval Christianity to the machine-modeled ideology of the seventeenth century. But such changes have repeatedly occurred all through history; and under catastrophic pressure they may occur again.

This analysis of the magnitude of change required to shift from the present material growth oriented society to the material steady-state can be depressing. Most of you may doubt that it is necessary, or that if such a change is necessary, doubt that it will be as difficult as Mumford describes. But even if Mumford exaggerates, he does provide us with some interesting areas to explore.

What will a steady-state society that embraces "an organic world picture" be like? First of all, we can be sure that it will be different. We can also be sure that opportunities for non-material growth will be increasingly available. Nevertheless, the implications of the material steady-state are tremendous, and they have not been very thoroughly examined. Given the power of the status quo to enforce its will, it is doubtful that any such revolutionary change will occur without a "catastrophe" to force that change. Even so, some of the policies we are now pursuing do move us in this direction. The pollution standards we have are likely to be followed by pollution taxes. This will probably be followed in short order by a depletion tax (versus the present depletion allowance). These efforts are incremental steps toward the steady-state.

The hard questions associated with these policies are already confronting us. The major question is one of equity. Amory Lovins said of the steady-state:

> Perhaps most difficult will be the need to face the issue of distribution rather than following the ''let them eat growth'' theory; physical stabilization will entail much moral growth, and the recycling of such nearly-extinct societal values as thrift, neighborliness, craftsmanship, and simplicity.

If we do not confront this issue now, in the non-steady-state world, it will only loom larger in the steady-state.

The economy of the material steady-state would differ from that economy which we now have. With a constant or only slowly increasing level of production of physical goods, economic growth could be expected in the non-physical areas of services and leisure. In addition, the physical goods that are produced could be expected to be of a greater quality and have a longer life time of use.

Beyond the physical distribution of goods in the steady-state, there are other questions that we should examine, even in the present society. The steady-state will require a great deal of restraint, which will require more planning for the future than practiced in the growth state, and this will require a great deal of moral growth on the part of government and people. There will be increasing emphasis on opportunity for continued artistic, intellectual, scientific, and spiritual growth. The place of the individual in society could be expected to be elevated in the steady-state. It could be a place where ''creative simplicity'' would flourish.

The concept of the steady-state is not truly new. John Stuart Mill wrote in 1857 that the ''stationary condition of capital and population'' was an inevitable condition that should not be feared because ''there would be as much scope as ever for all kinds of mental culture, and moral and social progress.''

It might also be added that science and technology would play a major role in achieving and maintaining the steady-state. The transition will require the use of the sophisticated tools of society, and our survival will depend on it. However, in the steady-state, technology would remain as a tool that would be carefully used, and controlled.

The material steady-state, then, may be an inevitable condition because of the physical limits to growth, including technological growth, but it does not necessarily have to be a condition to be feared or struggled against.

I have attempted, in these brief remarks, to describe the flaw of the ''technological fix'' to the social disease of air pollution. I have not meant to leave the impression that the ''fixes'' which technology can generate are not worth the effort. Instead, technological controls should be applied along with other controls, and all controls should be improved, as should the research that establishes the need for those controls. But we must realize that more remains to be done.

The limits imposed by pollution and other deleterious consequences of growth are real, and must be confronted. This confrontation has led me to conclude that our current way of life must change to one with greatly diminished emphasis on material growth. This recognition is spreading, and it is gradually being reflected in new laws at the local, state and Federal level.

These laws are creating new conflicts, and new demands. The people at this conference, representing government, universities, and the medical profession, are among the most aware individuals in the country on this problem, and by virtue of your superior knowledge, you have information that is needed by the entire society. You best understand your own findings, and the implications of those findings, as they involve human health; if you spoke out on this issue, and the related issue of growth, you would be listened to. The public has great respect for your professional position, and your motives are generally above reproach. We politicians are not so fortunate.

In my own congressional district there is an excellent example of how members of the medical profession can successfully influence public policy. Representatives of the Riverside County Medical Association have given their time and expertise to public hearings, and to serving on public boards and commissions, directly advising politicians and administrators about air pollution and related issues. This has led them to discuss land use questions, mass public transportation, automotive emission controls, public education and other major public issues, all related to the causes of air pollution. The doctors in my areas have been the most effective speakers on air pollution within the community. Similar actions have occurred at the state and Federal level, but these efforts have been much more diffuse and therefore much less effective.

Your keynote speaker, Dr. Herschel Griffin, participated in one of the major efforts to provide scientific information for use in determining public policy. The Health Effects panel of the National Academy of Sciences study on air pollutants did not extrapolate their findings out to their logical conclusions. I realize that this was not their purpose, but at this time of public confusion, health experts would do the nation a great service if they would defend the existing environmental laws in a convincing manner before public forums. Without such efforts by members of the medical and scientific professions, it is left to politicians such as me to use and explain the findings and the implications of those findings to the public. I, and most other elected public officials, need your help. Between us we can cure the social disease of air pollution. Let's get on with the job.